II International Scientific Practical Conference on Breakthrough Technologies and Communications in Industry and City (BTCI'2019)

IOP Conference Series: Materials Science and Engineering
Volume 828

Volograd, Russia
18 - 23 November 2019

ISBN: 978-1-7138-3163-1
ISSN: 1757-8981

Printed from e-media with permission by:

Curran Associates, Inc.
57 Morehouse Lane
Red Hook, NY 12571

CURRAN ASSOCIATES INC.
proceedings
.com

Some format issues inherent in the e-media version may also appear in this print version.

Institute of Physics
Dirac House, Temple Back
Bristol BS1 6BE UK

Phone: 44 1 17 929 7481
Fax: 44 1 17 920 0979

techtracking@iop.org

Additional copies of this publication are available from:

Curran Associates, Inc.
57 Morehouse Lane
Red Hook, NY 12571 USA
Phone: 845-758-0400
Fax: 845-758-2633
Email: curran@proceedings.com
Web: www.proceedings.com

TABLE OF CONTENTS

IOP Conf. Series: Materials Science and Engineering **828** (2020) 011001 doi:10.1088/1757-899X/828/1/011001

Abstract

II International Scientific Practical Conference "Breakthrough technologies and communications in industry and city" (BTCI'2019), 18-23 November 2019, Volgograd, Russian Federation

PREFACE

The conference was held on 18-23 November 2019 at Volgograd state university (Volgograd, Russian Federation) and provided a forum place for discussion for PhD students, teachers, scientists, researchers from research institutes and industrial enterprises from Volgograd, Moscow, Irkutsk and other cities from Russia and abroad (France, Germany, China).

The conference served as a platform for the researchers to discuss and carry out a research on current production and communication technology in industry and production department. The aim of the conference was to present the innovative approaches to developing convergent solutions in engineering and technology, and also in interdisciplinary scientific researches. As the venue bore interdisciplinary character, the scientists and experts also had an opportunity to cover the applied aspects like training, oral communication, technical translation, interaction and others.

Key topics of the conference covered priority research areas of technical, communication, educational and other related sciences and took into account global trends in social and industrial development. BTCI2019 topics are built on the fundamentals of sustainability, with vigorous attention to innovative technologies for producing and testing permanent joints, modern technologies for producing and processing materials in industry, automation, informatization and management in the enterprise, production communication and ethics, environmental protection, safety and labor protection in enterprises, breakthrough technologies and technology for subsoil development, development of methods to stimulate innovation and investment activity of enterprises.

IOP Conf. Series: Materials Science and Engineering **828** (2020) 011001 doi:10.1088/1757-899X/828/1/011001

PROGRAM COMMITTEE

Program Chair:

Kalinina A. E. - Doctor of Economics, Professor, Vice-Rector of Volgograd State University, Volgograd, Russia

Members:

- Shamne N. L. - Doctor of Philology, Professor, Director of the Institute of Philology and Intercultural Communication, Volgograd State University, Volgograd, Russia
- Cindori S. - Associate Professor, Faculty of Law University of Zagreb, Croatia
- Cernicova-Buca M. - Associate Professor, Politehnica University of Timisoara, Romania
- Larouk O. - Associate Professor, University of Lyon, Lyon, France
- Astafurova T.N. - Doctor of Education, Professor, Volgograd State University, Volgograd, Russia
- Ilinova E.Yu. – Doctor of Philology, Professor, Volgograd State University, Volgograd, Russia
- Kochetova L.A. - Doctor of Philology, Professor, Head of the English philology department, Volgograd State University, Volgograd, Russia
- Minin M. G. - Doctor of Education, Professor, National Research Tomsk Polytechnic University, Tomsk, Russia
- Mityagina V.A. - Doctor of Philology, Professor, Head of the Translation theory and practice department, Volgograd State University, Volgograd, Russia
- Rebrina L.N.– Doctor of Philology, Ass.Professor, Ass. Director of the Institute of Philology and Intercultural Communication, Volgograd State University, Volgograd, Russia
- Zheltukhina M.R. - doctor of philology, member of Russian Academy of Natural Sciences, Professor, Head of the research laboratory "Discourse linguistics", Volgograd socio-Pedagogical University, Volgograd, Russia

Address:

VolSU, 400062, Russia, Volgograd, prospect Universitetsky 100, office 314B, 403B

E-mail: btci_volsu@picconf.ru.net

Website: https://www.picconf.ru.net/volsu-conf-en-tech-scopus

IOP Conf. Series: Materials Science and Engineering **828** (2020) 011001 doi:10.1088/1757-899X/828/1/011001

ORGANIZING COMMITTEE

Organizing Chair:

Malushko E.Yu. - candidate of pedagogical sciences, Deputy Director for IT, Institute of Philology and Intercultural Communication, Volgograd State University, Volgograd, Russia

Members

- Rebrina L.N.– Doctor of Philology, Ass.Professor, Ass. Director of the Institute of Philology and Intercultural Communication, Volgograd State University, Volgograd, Russia
- Eltanskaya E.A. - candidate of pedagogical sciences, Ass.Professor, head of the foreign language communication department, Volgograd State University, Volgograd, Russia
- Maletina O.A. – candidate of philological sciences, Ass.Professor, Volgograd State University, Volgograd, Russia
- Sityna N.A. – candidate of philological sciences, Ass.Professor, Volgograd State University, Volgograd, Russia
- Tsybaneva V.A. – candidate of pedagogical sciences, Ass.Professor, Volgograd State Academy of Postgraduate Education, Volgograd, Russia
- Sviridov A. - candidate of technical sciences, publication manager of company "Atlant", Tomsk, Russia
- Zheltukhina M.R. - doctor of philology, member of Russian Academy of Natural Sciences, Professor, Head of the research laboratory "Discourse linguistics", Volgograd socio-Pedagogical University, Volgograd, Russia

Address:
VolSU, 400062, Russia, Volgograd, prospect Universitetsky 100, office 314B, 403B

E-mail: btci_volsu@picconf.ru.net

Website: https://www.picconf.ru.net/volsu-conf-en-tech-scopus

Peer review statement

All papers published in this volume of *IOP Conference Series: Materials Science and Engineering* have been peer reviewed through processes administered by the proceedings Editors. Reviews were conducted by expert referees to the professional and scientific standards expected of a proceedings journal published by IOP Publishing.

Revisiting assessment of heat and mass transfer in contact apparatus for steam treatment of air

A G Averkin [1], A I Eremkin [1], G I Greysukh [1] and Y A Averkin [1]

[1] Penza State University of Architecture and Construction, 28, Germana Titova str., Penza, 440028, Russia

E-mail: algraw@mail.ru

Abstract. Modified methods are proposed for determining the surface area of heat and mass transfer in the contact apparatus in the humidification, cooling, dehumidification of air with water, as well as dehumidification of air with solid drying agent, based on the number of transfer units. The number of transfer units is determined by the method of graphical integration using I-d-diagram of humid air to evaluate both heat and mass transfer process.

1. Introduction

In air conditioning systems (ACS), contact apparatus is used for steam treatment of air with water when creating a micro-climate of premises of various purposes, etc. Contact apparatus is structurally manufactured in the form of nozzle spray chambers of honeycomb irrigation [1]. In the chambers, the air flow is in direct contact with water by spraying water through mechanical nozzles and the use of irrigated nozzle (sheet-, ring-like, etc.). As a result of heat and mass transfer, the air can be humidified, cooled, or dehumidified. In many cases the use of basic heat and mass transfer equations for the calculation of contact apparatus is complicated due to the impossibility of direct determination of the heat and mass transfer surface value.

Performance efficiency assessment and thermotechnical calculation of nozzle spray chambers and other contact apparatus is done using the efficiency coefficients, based on graphic constructions of the process of steam treatment of air flow at the I-d-diagram of humid air [1, 2]. The efficiency coefficients featuring the ratio between heat transfer in a real process and the maximum possible heat transfer at different orientation of ideal processes have not been obtained as a result of any mathematical conclusion or analysis [2]. The analysis of such methods revealed that they were based on functional dependencies and empirical equations obtained on test benches, simulating a certain type of irrigation nozzle chambers. Typical dependences represent the specific features of the aerodynamic situation in the rain area of the irrigation chamber, being its technical characteristics and are given in the directories and catalogues of such equipment manufacturers. Each of the methods deals with the calculation of a certain type of irrigation nozzle chambers.

2. Research and discussion

E. V. Stefanov proposed a method for evaluating the *performance efficiency* of nozzle chambers when processing the experimental data using the numbers of transfer units of sensible heat $NTU_{я}$ and total

IOP Conf. Series: Materials Science and Engineering **828** (2020) 012001 doi:10.1088/1757-899X/828/1/012001

heat NTU [2, 3]. These indicators are widely used for calculations of processes and apparatus of chemical technology, both abroad [4] and in Russia [5].

It is proposed to use the method of assessing the contact apparatus performance efficiency for steam treatment of air using the digits of transfer units for the purpose of their dimensioning and checking calculation [6].

In Russian publications the letter symbol N or n [5] is adopted to mark the number of transfer units. We shall use the Nt symbol to indicate the number of total heat units transfer, symbol Nt - to indicate the sensible heat units transfer, Nt - to indicate the number of mass units transfer.

It is recommended to determine the number of transfer units by graphic method, in particular, by graphic integration method [5].

This method consists of the following stages:

Stage 1 - construction of the operating and equilibrium lines of the process in the appropriate coordinate system.

It is proposed to use I-d-humid air diagram as a coordinate system. The following shall be the graphic constructions for contact apparatus at the I-d-diagram [6]:

- saturated air curve φ = 100% (equilibrium line);
- a direct process line that characterizes the steam treatment of air (operating line).

Figure 1, a, b represents graphic constructions at I-d-diagram of humid air to determine operating and equilibrium values.

(a) (b)

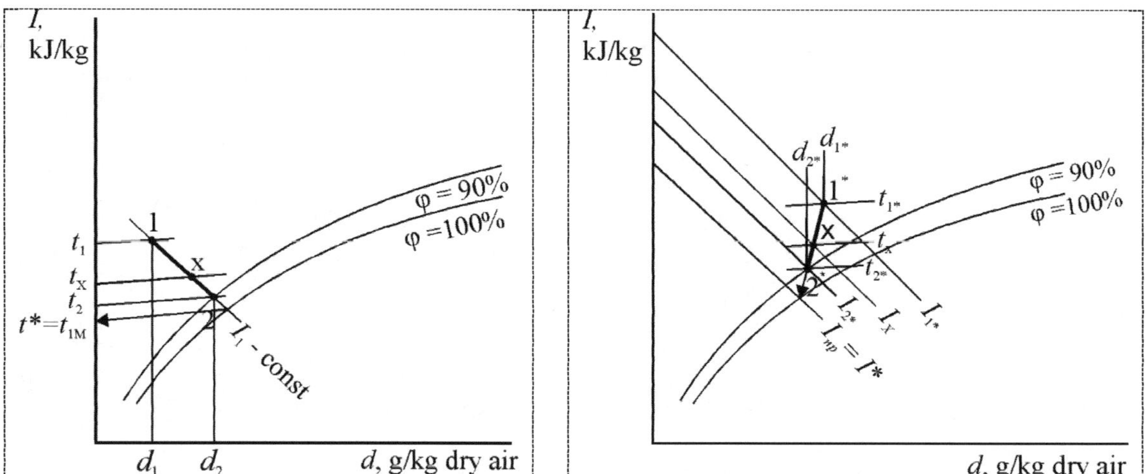

adiabatic humidification and cooling of air (1 - 2)	dehumidification and cooling of air in polytropic conditions (1* - 2*)

Figure 1. *Determination of equilibrium (t*) and operating (t) values of temperature, equilibrium (I*) and operating (I) values of air enthalpy*

As a basic option, Figure 1a presents the construction of the adiabatic (isoenthalpic) process of air cooling and humidification as a straight line 1 - 2. Point 1 features the initial air condition, i.e. at the entrance to the contact apparatus, point 2 corresponds to the final air condition, i.e. at the exit of the apparatus. The determination of equilibrium air temperature values for points 1, 2 and for intermediate X - points is shown generally, in the form of arrows. Determination of the operating and equilibrium values of the air enthalpy is shown in Figure 1b.

Based on this data, graphic dependencies are constructed: $1/(t - t^*) = f(t)$; $1/(I - I^*) = f(I)$.

IOP Conf. Series: Materials Science and Engineering **828** (2020) 012001 doi:10.1088/1757-899X/828/1/012001

Stage 2 - construction of a graphic dependency $\frac{1}{Y - Y^*} = f(Y)$ *based on operating and equilibrium process line* (Figure 2).

Here *Y*, *Y** - respectively, operating and equilibrium parameters, characterizing the air flow condition during the steam treatment of air.

The following are proposed to be determined as operating and equilibrium parameters [6]:

- temperature (to estimate sensible heat transfer);
- enthalpy (to estimate total heat transfer);
- partial steam pressure, humidity content (to assess the mass transfer process)

(a) (b)

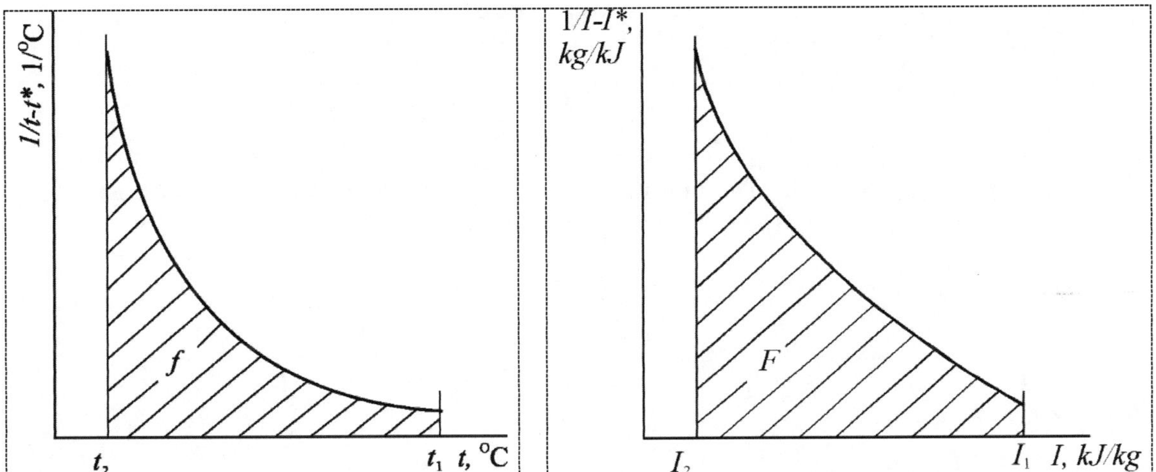

dependency 1/(I-I*) = f(t) | dependency 1/(I-I*) = f(I)

Figure 2. Graphic dependencies for determining the number of transfer units

Determination of equilibrium partial steam pressure values during humidification and dehumidification of air are recommended to be performed based on graphic constructions of processes at the *I-d-* diagram of humid air, shown in Figure 3.

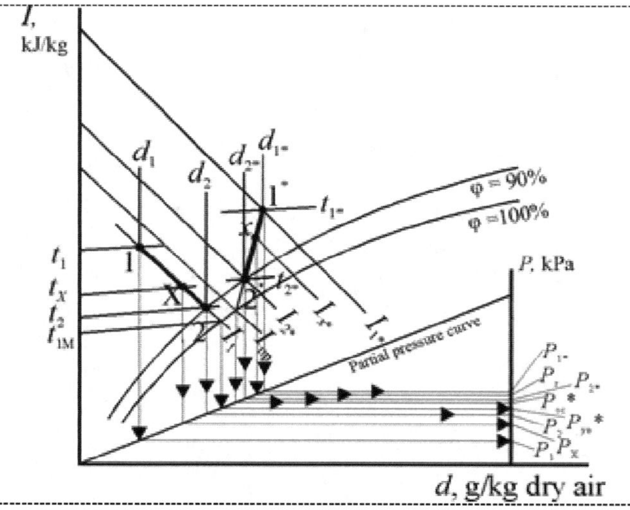

Figure 3. Graphic constructions to determine the operating and the equilibrium partial steam pressure values in the adiabatic humidification mode (line 1-2), polytropic dehumidification of air (line 1*-2*) in the contact apparatus

Stage 3 - determination of the number of transfer units n (or N_t or n_m by equations:

$$n_t = \int_{t_2}^{t_1} \frac{dt}{t - t*} = fm_1 m_2 \; ; \qquad (1)$$

$$N_t = \int_{I_2}^{I_1} \frac{dI}{I - I*} = Fm_3 m_4 \; ; \qquad (2)$$

$$n_{m(p)} = \int_{P_2}^{P_1} \frac{dP}{P - P*} = f_p m_5 m_6 \; , \qquad (3)$$

where t, $t*$ - respectively, operating (current) and equilibrium air temperature, °C (Fig. 1, a); I, $I*$ - respectively, operating (current) and equilibrium air enthalpy, kJ/kg (Fig. 1, b); P, $P*$ - respectively, operating (current) and equilibrium partial steam pressure in the air flow, kPa (Fig 3); f, F, f_p – respectively, area of curved trapezoid within the adopted values (Figure 2); m_1, m_2; m_3, m_4; m_5, m_6 - respectively, the scales of the measured values along X and Y axis (Figure 2).

In works [3, 6] it is shown:

$$n_t = \frac{\alpha F}{Gc} = \frac{t_1 - t_2}{\Delta t_p} \; ; \qquad (4)$$

$$N_t = \frac{\beta F}{G} = \frac{I_1 - I_2}{\Delta I_p} \; ; \qquad (5)$$

$$n_{m(d)} = \frac{\beta_d F}{G} = \frac{d_1 - d_2}{\Delta d_p} \; ; \qquad (6)$$

$$n_{m(d)} = \frac{\beta_p P_{\hat{A}} F}{0.622G} = \frac{P_1 - P_2}{\Delta P_p} \; , \qquad (7)$$

where α - heat exchange coefficient, W/(m^2·°C); β_c, β_d, β_p - mass transfer coefficients, referred, respectively, to the difference in concentrations, humidity content and partial pressures, kg/m^2 (units doubl. force); F- surface heat and mass transfer, m^2; G- air flow, kg/s; c -- heat capacity of humid air, J/(kg·°C); t, I, d, P - respectively, temperature, °C; enthalpy, kJ/kg, air humidity content, g/kg dry. air. and the partial steam pressure, kPa (index 1 corresponds to the initial condition, index 2 - the final condition of air); Δt_p - average temperature difference between air and water (the driving force of the process), °C; ΔI_p - the average enthalpy difference between air and water (the driving force of the process), kJ/kg; Δd_p - the average humidity content difference between air and water (the driving force of the process), g/kg; ΔP_p - the average partial steam pressure difference at the surface of the water phase and in the air, kPa; P_B - barometric pressure, kPa.

Determination of kinetic heat transfer and mass transfer coefficients in the equations (4)-(7) is performed based on the similarity theory using heat and diffusion similarity criteria [5]:

$$\alpha = \mathrm{Nu}\frac{\lambda}{l}; \quad (8)$$

$$\beta = \mathrm{Nu}'\frac{D}{l}, \quad (9)$$

where l - characteristic length, m; λ - air heat conductivity coefficient, W/(m·°C); D - air steam diffusion coefficient, m^2/c, Nu, Nu′- respectively, Nusselt criterion for heat and diffusion (mass transfer) process.

Nusselt criteria are determined by the criterion equations given in [5-7]. To calculate the heat and mass transfer from the drops' surface under forced convection, it is proposed to use the following equations (when Re = 1.. .220) [8]:

$$\mathrm{Nu} = 2 + 1.07\mathrm{Re}^{0.48}\,\mathrm{Pr}^{0.33}\,\mathrm{Gu}^{0.175}; \quad (10)$$

$$\mathrm{Nu}' = 2 + 0.85\mathrm{Re}^{0.52}(\mathrm{Pr}')^{0.33}\mathrm{Gu}^{0.135}, \quad (11)$$

where Re - the Reynolds criterion; Pr, Pr′ respectively, the Prandtl number heat and diffusion; Gu – Gukhman criterion.

Characteristic criteria are calculated by the formulas:

$$\mathrm{Re} = \frac{vl\rho}{\mu} = \frac{vl}{\upsilon}; \quad (12)$$

$$\mathrm{Pr} = \frac{\mu c}{\lambda} = \frac{\upsilon}{a}; \quad (13)$$

$$\mathrm{Gu} = \frac{T_{\tilde{n}} - T_{\grave{\imath}}}{T_{\tilde{n}}}; \quad (14)$$

$$\mathrm{Pr}' = \frac{\mu\rho}{D}, \quad (15)$$

where v - linear speed of air flow, m/s; ρ - air density, kg/m^3; μ - air dynamic viscosity coefficient, PA·c; υ - kinematic air viscosity coefficient, m^2/s; c - specific air heat, J/(kg·°C); a - air thermal conductivity coefficient, m^2/c; T_c - air flow temperature by dry thermometer, K; T_m - air flow temperature by wet thermometer, K.

Stage 4 - defining the heat and mass transfer surface. The required heat transfer surface in the contact apparatus, corresponding to the segment 1-2 (Figure 1), in accordance with equation (4) is:

$$F = \frac{Gcn_t}{\alpha}, \qquad (16)$$

Required mass transfer surface for this process based on equation (7):

$$F = \frac{0.622Gn_m}{\beta_p P_{\hat{A}}}. \qquad (17)$$

For the purpose of complete assessment of heat and mass transfer in ACS devices, E. V. Stefanov introduced the concept of process perfection [2]:

$$\xi = \frac{n_t}{n_m}. \qquad (18)$$

The process of heat and mass transfer in the apparatus is considered perfect if $\xi = 1$. It is emphasized [3] that in practice there are cases when $\xi \neq 1$. Such cases appear when the surface areas of heat and mass transfer are different. The given method additionally provides for the analysis of such situations.

Drying agents based on solid sorbents, i.e. adsorbents, can be used as basic equipment in ACS. Standard apparatus is called adsorbers. In practice, adsorbers with a fixed adsorbent layer are widely used [5, 9].

Silica gel, i.e. hydrated amorphous silica, is the effective adsorbent of steam. Its formula $SiO_2 \cdot nH_20$. It is a granular vitrifaction, obtained by treating liquid glass with mineral acid. To dehumidify the air, KCM silica gel is used with grain sizes of 1...3 mm [9].

The final state of the adsorption process, as well as other mass transfer processes, is the equilibrium state.

Air dehumidification by silica gel is accompanied by the release of adsorption heat and wetting heat, which increases the temperature of the operating bodies, i.e. silica gel and air flow. During heat removal from the adsorber the steam sorption process can be isothermal, in case of no heat removal, the process is isoenthalpic or adiabatic, and the process beam at the I-d- diagram of humid air is depicted by the line I – *const* [9]. Specific features of the sorption process shall be taken into account when applying the method of standard equipment calculation [11].

At the same time, the analysis of the calculation methods of adsorbers with a fixed adsorbent layer reveals that the adsorption isotherm is automatically proposed as the base curve, i.e. the process conditions are not specified [5]. It is obvious that in case of isoenthalpic air dehumidification process isoenthalpy adsorption (adiabatic line) should be used as a base curve.

For graphic construction of the steam isoenthalpy adsorption from the air flow by the silica gel, the authors suggest applying the I-d-diagram of humid air, which has additional isometric lines D^* of equilibrium humidity content the KCM silica gel (Figure 4) [10].

As mentioned above, chemical technology has developed methods for calculating equipment based on the number of transfer units [4, 5]. This methodology is proposed for the conditions described above.

- *benchmark data is set* (adsorber operates without heat removal; dehumidification air flow G, kg/s; initial air parameters, e.g. temperature t_1, °C, relative humidity φ_1, % or other combinations of two air parameters: temperature t_1, °C, humidity content d_1, g/kg; temperature t_1, °C, enthalpy I, kJ/kg, etc.; final air parameter after dehumidification, e.g., temperature t_2, °C or humidity content d_2, g/kg of dry air).
- *the process of air dehumidification is built at the I-d-diagram of humid air,* e.g., line segment 1-2 (Figure 4).

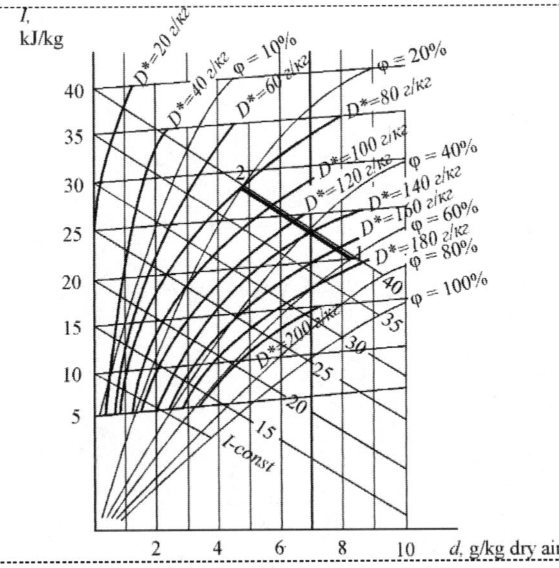

Figure 4. I-d-diagram of humid air with isometric lines of equilibrium humidity content of silica gel (D*)

- adsorber diameter is determined D_a:

$$D_a = \sqrt{\frac{G}{0.785\rho v}}, \qquad (19)$$

where v - air speed in the adsorber section, m/s ($v = 0.25...0.30$ m/s) [5, 7];

- sorbent layer height is determined (silica gel) H:

$$H = nh, \qquad (20)$$

where n - number of transfer units; h - height of a transfer unit, m.

a) *number of transfer units is determined.*

In accordance with the recommendations [7] based on the figure 4, an equilibrium (adsorption isoenthalpy line) is built as well as the process operating line (figure 5).

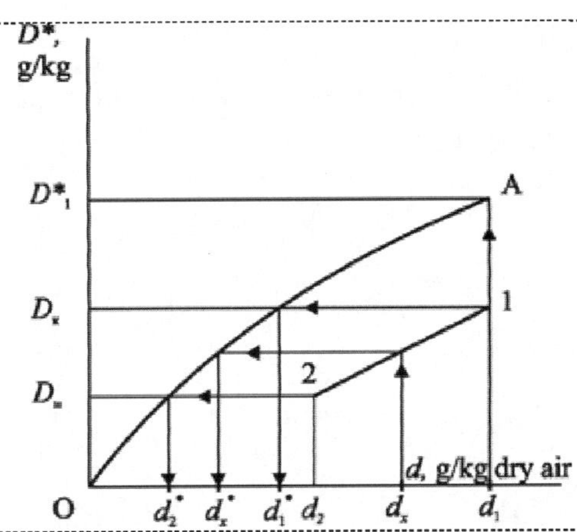

Figure 5. Adsorption isoenthalpy line (OA) and the operating process line (1 - 2): d, D* - air humidity content and corresponding equilibrium humidity contents of silica gel

The calculation of the number of transfer units performed by the formula:

$$n = \int_{d_2}^{d_1} \frac{\ddot{a}(d)}{d - d*}, \tag{21}$$

where *d, d** - current (operating) and equilibrium humidity content of the dehumidified air, g/kg of dry air.

Arrows at Figure 5 show the determination of the equilibrium humidity contents of the air for the corresponding operating humidity content values.

For the numerical determination of the number of transfer units by the formula (21), a graphic dependency is built of air dehumidification with silica gel:

$$\frac{1}{d - d*} = f(d) \tag{22}$$

An overview of the graphic dependency is shown in figure 6.

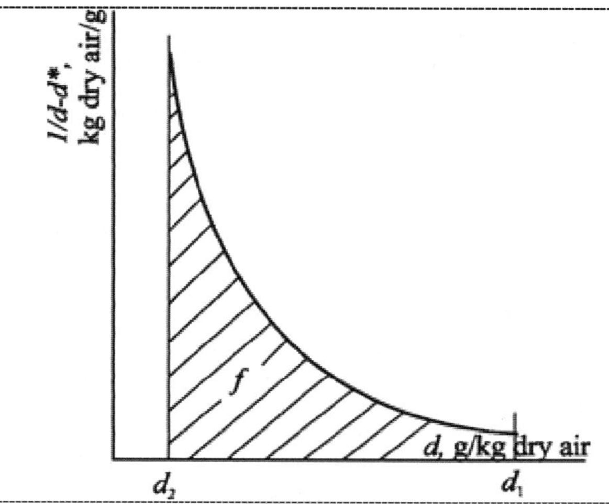

Figure 6. Dependency $1/(d - d*) = f(d)$

The number of transfer units is numerically equal to:

$$n = f \cdot M_1 \cdot M_2, \tag{23}$$

where f - the area of curved trapezoid, M_1, M_2 - respectively, the scale of values along X and Y axis.

b) height of a transfer unit is determined.

The calculation is made by the formula:

$$h = \frac{G}{S_c \beta_y}, \tag{24}$$

where S_c – adsorbent layer section area, m^2; β_y – volumetric mass transfer coefficient in the air, c^{-1}.

$$S_c = 0.785 D_a^2; \tag{25}$$

$$\beta_y = \frac{\mathrm{Nu}' D}{d_{\acute{y}}^2}, \tag{26}$$

where d_e - equivalent silica gel grain diameter, m.

The Nusselt diffusion criterion Nu' is determined depending on the numerical value of the modified Reynolds criterion – Re:

$$\mathrm{Re} = \frac{v d_{\acute{y}} \rho}{\mu \varepsilon_{\acute{i}}}, \tag{27}$$

where ε_n - porosity of the fixed adsorbent layer ($\varepsilon_n = 0.3$ [5]).

$$\text{With Re} < 2,\ \mathrm{Nu}' = 0.51\mathrm{Re}^{0.85}(\mathrm{Pr}')^{0.33}; \tag{28}$$

$$\text{With Re} = 2\ldots30,\ \mathrm{Nu}' = 0.725\mathrm{Re}^{0.47}(\mathrm{Pr}')^{0.33}; \tag{29}$$

$$\text{With Re} > 30,\ \mathrm{Nu}' = 0.395\mathrm{Re}^{0.64}(\mathrm{Pr}')^{0.33}. \tag{30}$$

Other design dimensions of the adsorber (total height, height of the apparatus covers, etc.) are in certain ratios with the calculated basic parameters: diameter and height of the adsorbent layer [5, 7].

3. Conclusions

There were improved and scientifically substantiated methods of calculation of contact apparatus while treatment of air with water, air dehumidifiers using a solid sorbent (silica gel) based on the number of transfer units [12]. Determination of the number of transfer units is performed using the method of graphic integration, the necessary calculations are proposed to be made on the basis of graphic constructions of air treatment processes at the *I-d-* diagram of humid air and its modification, carrying the isometric lines of the equilibrium humidity content of silica gel. The methods allow to calculate separately and perform a comparative assessment of the required surface for the implementation of heat and mass transfer process, to perform calibration and design calculations of contact apparatus for steam treatment of air.

References

[1] Kokorin O Y and Deripasov A M 2002 *Domestic Equipment for Development of Ventilation and Air-Conditioning Systems* (Moscow: IKF "Katalog") p 91
[2] Stefanov E V 2005 *Ventilation and Air Conditioning* (St. Petersburg: "AVOK-Severo-Zapad") p 400
[3] Polushkin V I, Rusak O N, Burtsev S I, Anisimov S M and Vasiliev V F 2002 *Heating, Ventilation and Air Conditioning Part 1 Theory Fundamentals of Creating Micro Climate in the Building* (St. Petersburg: Professiya) p 176
[4] Keys V I and London A L 1967 *Compact Heat Exchangers* (Moscow: Energiya) p 222
[5] Kasatkin A G 2004 *Basic Processes and Apparatus of Chemical Technology* (Moscow: "Alyans" TID LLC) p 753
[6] Averkin A G 2010 *Regional Architecture and Construction* **2** 122-28
[7] Averkin A G and Eremkin A I 2019 *Improvement of air Steam Treatment Devices and Methods of Calculation of Climatic Equipment* (Moscow-Vologda: "Infra-Inzheneria") p 168
[8] Nesterenko A V 1971 *Fundamentals of Thermodynamic Calculations of Ventilation and Air Conditioning* (Moscow: "Vysshaya Shkola") p 460
[9] Sotnikov A G 2005 *Processes, Devices and Systems of Air Conditioning and Ventilation. Theory, Engineering and Design at the Turn of the Century* (St. Petersburg: "AT-RVGSSHNO") p 504
[10] Lizunkov V, Politsinskaya E, Malushko E, Kindaev A, and Minin M 2018 *International journal of energy economics and policy* **8 (3)** 250-257
[11] Averkin A G, Eremkin A I, Ershov E G and Averkin Y A 2017 *Regional Architecture and Construction* **4** 166-72
[12] Ergunova O, Lizunkov V, Blaginin, V, Politsinskaya E, and Shaykina O 2018 *International Journal of Energy Economics and Policy* **8(1)** 122-127

IOP Conf. Series: Materials Science and Engineering **828** (2020) 012002 doi:10.1088/1757-899X/828/1/012002

Environmental monitoring using an unmanned aerial vehicle

O E Bezborodova [1], N A Vinogradova [1], S V Tertychnaya[1], O S Vinogradov [1] and and O B Kachan [1]

[1] «K.G. Razumovsky Moscow State University of technologies and management (the First Cossack University)», Penza 6, Volodarsky st., Penza, Penza Region, Russia, 440005

E-mail: woinova@mail.ru

Abstract. The article discusses routing options for unmanned aerial vehicles during environmental monitoring to save battery power and minimize task execution time. It proposed an algorithm for environmental monitoring using an unmanned aerial vehicle. For its implementation, the authors developed and justified a network model of work. The article presents the results of network modelling for environmental monitoring in reconnaissance mode during the emergency case in the enterprise.

1. Introduction

Experts put forward more and more new ideas for the use of unmanned aerial vehicles (UAV) in various fields, including environmental monitoring (EM). Environmental experts widely use UAV equipped with various electro-optical sensors to monitor the condition and use of natural objects: poaching, illegal logging, detection of fires. There are projects for the use of specially equipped UAV for measuring environmental parameters: environmental control, patrolling at industrial enterprises in normal operation (mode 1) and reconnaissance, detection of sources of accidental emissions in emergency zones (mode 2) [1]. In these situations, UAV provides data collection from objects of different distances and dangers with minimal risk for laboratory staff.

For the science-based planning of the use of UAV during environmental monitoring (EM) in emergency zones, it is advisable to use network models (NM) [2, 3].

This planning method consists of a graphical representation of the total amount of work on EM and gives the following advantages:

- identify the necessary measures for the EM;
- identify the start and end time of the EM according to the standards for certain types of work and the results of calculations, draw up a schedule;
- identify a specific contractor for each EM event.

Analytical laboratories located on the AUV improve the level and quality of EM. Expanding the functional capabilities of UAV by equipping them with systems of radiation, chemical and biological control significantly increase their value as a means of remote analytical control. In addition to instrumentation, such UAVs have means of communication with a stationary computer for processing and storing the received information. Thus, the UAV sometimes has a significant instrumental load which imposes additional requirements on the battery of the UAV [4], its charge will be spent not only on the movement of the UAV itself but also on the operation of devices. Therefore, saving battery power by optimizing the route and runtime of the UAV task is relevant.

IOP Conf. Series: Materials Science and Engineering **828** (2020) 012002 doi:10.1088/1757-899X/828/1/012002

2. Theoretical part

To build the optimal route, it is necessary to ensure the movement of the UAV in a way that [5, 6]:

- the movement time was sufficient for the triggering of control sensors, but also minimal for increasing efficiency;
- without re-examination of the terrain and missed sites;
- the data collected was necessary and sufficient for making a managerial decision based on the results of the EM;
- eliminate the failure of the UAV due to the discharge of the battery.

Using modern management tools for the UAV group [7–9] allows achieving high efficiency of their application.

UAV in mode 1, patrol mode, has two possible route options (Fig. 1) depending on the location of environmental impact sources. In this case, the movement is a closed-circuit starting and ending at the starting point (SP) determined by the ground control point (GCP).

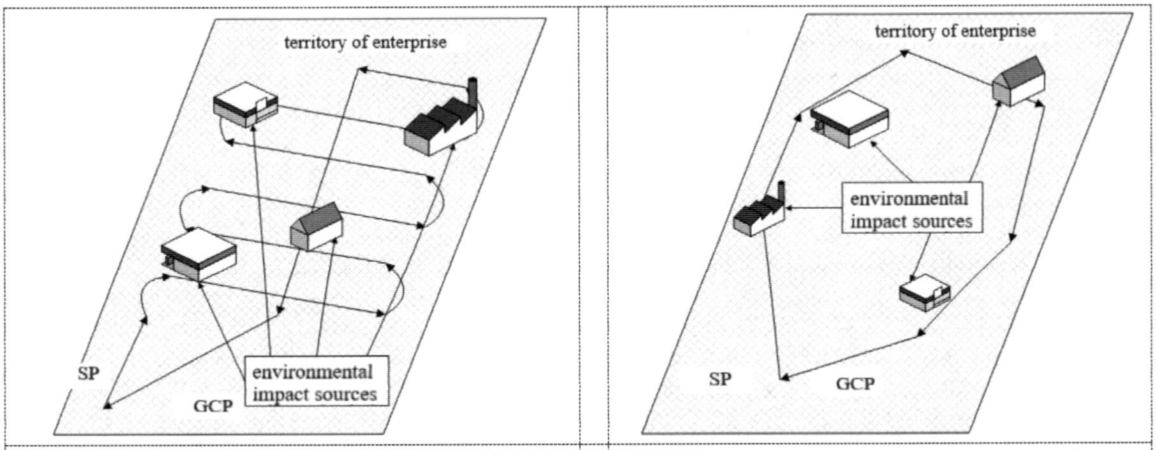

a) zigzag route b) circular route.

Figure 1. Route of the unmanned aerial vehicle in mode 1.

The zigzag route (Fig. 1a) is most effective when the territory of the enterprise has a dense uniform development. In this case, the UAV moves in a zigzag so that the straight sections of motion (l) pass as far as possible through the maximum number of sources of environmental impact, and the UAV has a turning radius (r) over the free sections.

In this case, we determined the length of the UAV movement route (d) and the area of the examined territory (S) by the formulas:

$$d = 8r + 2l(n + 1) + 2\pi r(2n + 1), \quad (1)$$

$$S = 16r^2 + 4lr(n + 1) + 4\pi r^2(2n + 1), \quad (2)$$

where r is the turning radius of the UAV, m; l is the length of the straight section of the route, m; n is the number of curves of the route.

If sources of environmental impact are located along the perimeter of the enterprise territory (Fig. 1b), then a circular route is preferable, the parameters of which (d and S) are determined by the formulas:

$$d = 2\pi r(n^2 + n + 1), \quad (3)$$

$$S = 4\pi r^2(n^2 + n + 1). \quad (4)$$

The UAV operating in mode 2, reconnaissance in the emergency zone, has also two possible routes of movement (Fig. 2) that depend on the configuration of the source of the damaging factor (DF).

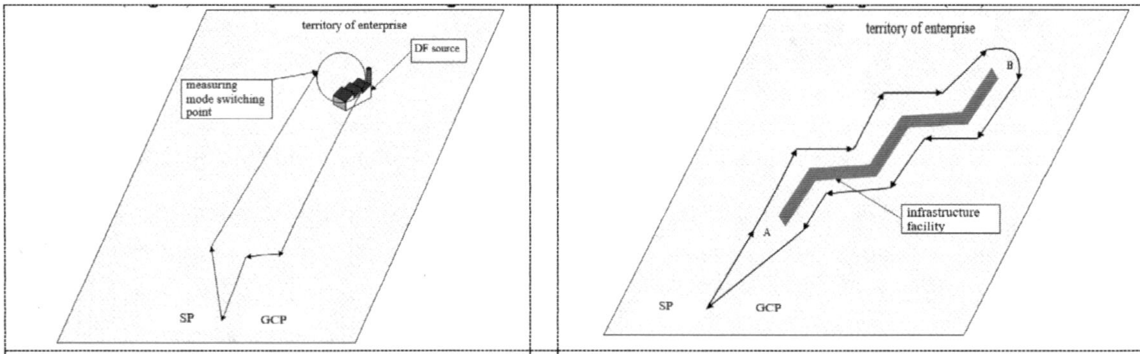

a) point source b) linear source

Figure 2. The Route of an unmanned aerial vehicle in mode 2.

The flight around a given point object is required during the environmental monitoring of specific objects (Fig. 2a). It is used in cases with known coordinates of the object that require clarification of its state. And the length of the optimal route consists of the path to the source and back (linear sections) and the survey route of a point object (circle):

$$d = 2\pi r + 2 \cdot l. \quad (5)$$

A fly-around of a linear object is necessary to control extended objects in conditions of their unambiguous position, for example, the state of pipelines (Fig. 2b).

When implementing this method, all UAV flight route from the starting point A to the ending point B separates into sections where the boundaries are coordinate points at the turning points of the linear object. Based on the task, at the B endpoint of flight, the UAV goes to the SP and the total route length is determined as

$$\mathbf{d} = 2 \cdot \sum_{i=1}^{k} l_i, \quad (6)$$

where k is the number of individual sections of a linear object; li is the length of individual sections of a linear object.

Considering the manual introduction of the coordinates of the route points and the parameters of their overflights into the UAV information system, it is necessary to construct the route of the UAV movement in a way as to conserve battery power most economically (mode 1) or complete the task in the shortest time (mode 2).

The network planning and management system are based on an NM which is an informational dynamic model, of the relationships and results of all the work necessary to achieve the goal - to draw up a plan for the implementation of EM measures considering multivariate developments.

NM allows most rationally to build a plan of the EM, to establish the sequence and priority in the implementation of all actions. NM makes possible to determine the time limits of each event with sufficient accuracy and, therefore, the time to achieve the result, the final event; optimize the use of resources.

For EM, the authors propose to use specially equipped helicopter-type bi-directional aircraft with vertical take-off and landing, which will perform the following types of work:

IOP Conf. Series: Materials Science and Engineering **828** (2020) 012002 doi:10.1088/1757-899X/828/1/012002

- movement on a set route with a set speed;
- transmission of the detected information to the GCP;
- execution of the operator's commands while moving.

NM are compiled at the initial stage of planning. Initially, the process planning requires separating into several works, drawing up a list of works and events, thinking over their logical connections and the sequence of execution, assigning work to responsible executors. It allows for evaluating the duration of each work. Then it is possible to compile the NM. After that, specialists calculate the parameters of events and activities, determine the time reserves and the critical path.

Then they analyse and optimize of the NM, and, if necessary, re-draws with a recount of the parameters of events and works.

The event on the NM is represented by circles and activities by arrows demonstrating the relationship between activities. Experts use natural numbers to indicate events and the codes of the initial and final events for activities.

To analyze and monitor the implementation of both individual NM works and the entire set of measures of the EM, it is necessary to determine the NM parameters: temporal characteristics of paths, works and events.

The NM parameters include the average duration, the early and late dates of events beginning, the reserves of time of events and works, road and work tension coefficient.

The average duration of work is determined by the formula [2]:

$$t_{ij} = \frac{2t_{min} + 2t_{max}}{5}, \tag{7}$$

where t_{max} and t_{min} are the maximum and minimum estimates of the work duration.

Each NM event has two dates for completion: early -t_{pi} and late -t_{ni}.

The early completion date of an event is the time required to complete all the work preceding the given event since this event will be after the completion of all work for which it is final and is determined by the formula:

$$t_{pj} = max\begin{Bmatrix} t_{pi} + t_{ij} \\ t_{ps} + t_{sj} \\ t_{pm} + t_{mj} \end{Bmatrix} \tag{8}$$

The late date for the completion of an event is such a term exceeding which will delay the beginning of the final NM event. Therefore, the late dates of events are calculated by the formula:

$$t_{nj} = min\begin{Bmatrix} t_{ni} - t_{ij} \\ t_{nk} - t_{ik} \\ t_{nf} - t_{if} \end{Bmatrix} \tag{9}$$

The time reserve of an event R_i is the period for a possible delay in the execution of this event without violating the critical path:

$$R_i = t_{ni} - t_{pi} \tag{10}$$

The full reserve of work time R_{nij} is the maximum period for increasing the duration of this work without changing the critical path:

$$R_{nij} = t_{nj} - t_{pi} - t_{ij} \tag{11}$$

An important property of this reserve is the possibility of its distribution between works lying on the next path. It is the reserve of the entire subsequent path.

Free reserve of work time R_{cij} is a period for the possible extension of the end of this work without changing the early dates for the start of subsequent work

$$R_{cij} = t_{pj} - t_{pi} - t_{ij} \tag{12}$$

Reserves of work time allow manoeuvring the start and end dates of the work, setting the most favourable timelines for the work in terms of rational loading of resources allocated to achieve the goal. Work reserves also provide an opportunity to identify the critical path. Presenting a chain connection of works, it goes through works that do not have reserves.

One of the most important operations in the analysis of the calculated parameters of the NM is to determine the coefficients of work tension and the probability of the final event completion in a set period.

The tension coefficients of paths K_L^{H} and works K_{ij}^{H} allow identifying the critical zone of NM, along with time reserves.

The path tension coefficient K_L^{H} is determined by the ratio of the durations of mismatched sections of the path L and the critical path L_{cr}:

$$K_L^{\text{H}} = \frac{t(L) - t'_{cr}(L)}{t(L_{cr}) - t'_{cr}(L)} = 1 - \frac{R_L}{T_{cr} - t'_{cr}(L)} \tag{13}$$

where is the duration of critical work on the path L, Tcr is the duration of the critical path.

The work tension coefficient K_{ij}^{H} is equal to the path tension coefficient of maximum duration passing through work (i, j):

$$K_{ij}^{\text{H}} = L_{Lm}^{\text{H}} = \frac{t(L_m) - t'_{cr}(L_m)}{t(L_{cr}) - t'_{cr}(L_m)} = 1 - \frac{r_{ij}^{n}}{T_{cr} - t'_{cr}(L_m)} \tag{14}$$

The magnitude of the tension coefficients lies within $0 < K^{\text{H}} \leq 1$, reaching one score only in critical works. The closer the tension coefficient is to 1, the more difficult it is to complete this work in a set period. The closer the tension coefficient is to 0, the greater the relative reserve the maximum path has through this work.

The calculated tension factors allow an additional classification of the work. Depending on the size, K^H there are three categories of work: critical ($K^H > 0{,}8$); subcritical ($0{,}6 \leq K^{\text{H}} \leq 0{,}8$); reserve ($K^H < 0{,}6$).

If the total time reserves of the two jobs are equal, the one with the highest tension coefficient will have the greatest chance of becoming critical. To optimize the NM, first, it is necessary to use the reserves of work that have minimal tension factors.

3. Practical implementation

We examined the construction of NMs using the example of planning EM measures for emergencies at the local level with the destruction of buildings and structures and the spill of an emergency chemically hazardous substance (ECHS).

Our study considers local emergencies when the situation in a certain territory arose as a result of an accident in which DF did not go beyond the facility's boundaries and can be eliminated by forces of emergency rescue teams, when no more than four people died, no more than 10 people were injured, living conditions of no more than 100 people were violated, direct material damage is not more than 1000 minimum wages on the day of emergencies [10].

For EM in the emergency zone, we identified four groups of activities in which it is advisable to use UAV (Table. 1).

The time for the EM event with UAV depends on the trajectory, the speed of the UAV and the response time of the instruments and equipment installed on it.

Table 1. Groups of activities to use UAV

Work type	Work content	Time to complete the work, min
	Preparatory activities	
1 - 2	An emergency message arriving at the Unified Dispatch Duty Service	5
2 - 3	Entering UAV flight parameters by GCP operator and launching UAV with special equipment	5
3 - 4	The arrival of UAV to the border of the emergency zone	5
	Initial data collection activities in various parts of the emergency zone	
4 - 5	Flight around objects in the emergency zone in patrol mode to collect information on warning emergency services about the state of objects: visible damage, ECHS spills (mode 1)	$t = {}^{d}/_{v}$
5 - 6	Exchange of information with the operator GCP	1
5 - 7	Return to base	5
	Search activities	
6 - 8	Transmission of coordinates of damaged UAV objects	1
8 - 9	UAV flight around the territory to search for DF sources at objects (mode 1)	$t = {}^{d}/_{v}$
9 - 10	Detection of DF sources at objects by UAV (mode 1)	k
10 - 6	Exchange of information with the operator GCP	1
10 - 7	Return to base	5
	Preparatory activities	
6 - 11	Transmission of coordinates of DF sources by UAV	1
11 - 12	Monitoring of DF sources by UAV (mode 2)	$t = {}^{d}/_{v}$
12 - 6	Exchange of information with the operator GCP	1
12 - 7	Return to base	5

Note: k is the number of DF sources that allows UAV using.

Not for all EM events, it is possible to calculate works time according to the methodology given in [2, 3]. First, it depends on the scale of the emergency, the affected area, the severity of the emergency consequences, the extent of the damage, the levels of DF, the response time of the devices installed on the UAV.

For the EM, it is preferable to use a cargo UAV, for example, the Versadrones Heavy Lift Octocopter brand which has the following tactical and technical characteristics: overall dimensions (in millimeters) - 1200 x 1200 x 400; UAV weight - 4000 grams; total lifting weight - 12000 grams; the maximum speed is 70 kilometers per hour [11].

The duration of the EM measures with UAV is calculated and optimized according to formulas (7) - (9) using data from [12].

Works 1, 2, 3 are the Preparatory activities and their duration is determined by the standard time of arrival of the first unit in the emergency zone and the operator's ability to prepare and launch the UAV. At speeds of up to 70 kilometres per hour, the UAV will arrive in the work area in 5 to 10 minutes.

Works 4 and 5 are the Initial Information Collection Activities in various parts of the emergency zone. Given that UAV in video mode requires an optimal speed of 20 kilometres per hour, flying around an emergency zone along a zigzag or circular route will depend on its length and speed of the UAV considering the speed of the devices work and will take time. $t = {}^{d}/_{v}$

The time required for exchanging information with the operator of the GCP depends on the conditions of reception and transmission in the emergency zone (work 5 - 6, 6 - 8, 10 - 6, 6 - 11 and 12 - 6).

Fig. 3 shows the network model built based on these calculations.

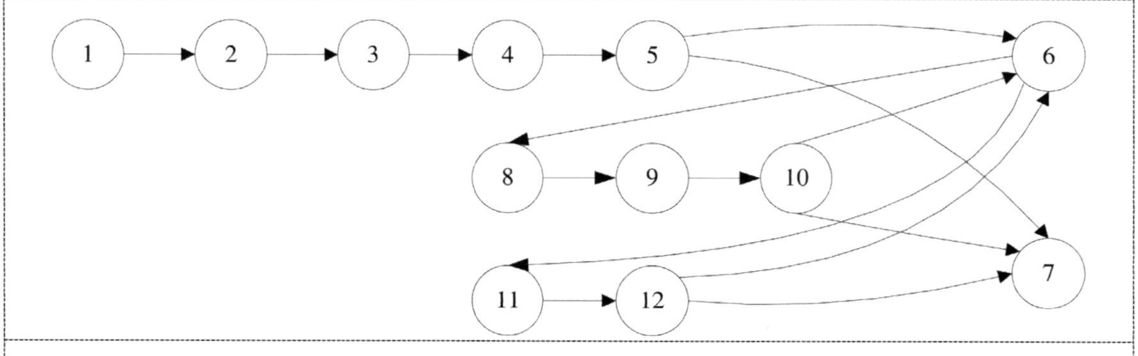

Figure 3. Network model that implements the algorithm for search and rescue operations in the emergency zone using UAV.

The algorithm for using UAV in EM begins with preparatory works 1, 2, 3 that consists in entering flight parameters and UAV starting with special equipment.

Works 4 and 5 are the collection of information necessary for the further operation of the UAV, clarification of the situation in the emergency zone and decision-making on the need and scope of the search activities related to the detection of objects with ECHS leak.

Based on this information, the UAV controls the concentrations of ECHS at the locations of depressurized objects. This information is the source for the rescuers to work to minimize the levels of DF at a given object of the emergency zone.

NM shows that UAVs perform some groups of measures sequentially, and some in parallel. Thus, the total time for search and rescue operations is reduced even further.

4. Conclusion

Thus, depending on the operating mode of the UAV, it is possible to set various driving routes for it which minimize the distance travelled, the time and resources spent on the UAV battery for the task.

For reconnaissance of the territory to find out the general situation in the emergency zone, transmitting a video signal, the EM for searching for DF sources, a zigzag flight path with rectilinear inserts (Fig. 1a) and a circular route is optimal for BVS (Fig. 1b).

The circular route is optimal for performing work on minimizing DF levels with known coordinates of the DF source. Moreover, depending on the parameters of the DF source (point or linear), it can consist of either straight-line segments and a semicircle (Figure 2a), or only linear sections (Figure 2b).

References

[1] Vinogradov O S, Vinogradova N A, Bezborodova O E and Sherstnev V V 2018 *Science of Krasnoyarsk* **4(7)** 7–24 DOI: 10.12731/2070-7568-2018-4-7-26

[2] Tairova E V 2007 Network *Planning Methods in Organizing Work Packages: Textbook* (Irkutsk: Irkutsk State University of Railway Engineering) p 95

[3] Novitsky N I 2014 *Network Planning and Production Management: Educational-Practical Guide* (M .: New knowledge) p 159

[4] Sherstnev V V, Bodin O N, Bezborodova O E, Polosin V G and Kharitonov S A 2019 *Methods, Means and Technologies for Obtaining and Processing Measuring Information: Materials Intern. Scientific and Technical Conf. "Schlandin Readings - 2019"* (Penza: Publishing house of PSU) pp 161–165

[5] Melnikov A V, Gaidai V A and Rogozin E A 2017 *Computer science, computer engineering and management, Bulletin of the Voronezh Institute of the Ministry of Internal Affairs of Russia* **1** 52–62

[6] Arzamastsev A A 2015 *Bulletin of TSU* **2(20)** 465-8

[7] Diveev A I 2017 *Reliability and Quality of Complex Systems* **4(20)** 64–72 DOI: 10.21685/2307-4205-2017-4-9
[8] Semenets V O 2018 *Reliability and Quality of Complex Systems* **3(23)** 65–70 DOI: 10.21685/2307-4205-2018-3-9
[9] Kanashchenkov A I 2018 *Reliability and Quality of Complex Systems* **3(23)** 71–84 DOI: 10.21685/2307-4205-2018-3-10.
[10] *Decree of the Government of the Russian Federation of May 21, 2007 No. 304 On the classification of emergency situations of natural and man-caused nature* URL: https://cis-legislation.com/document.fwx?rgn=16830
[11] The Versadrones Website *Cargo Octocopter* URL: http://versadrones.com/products/heavy-lift-octocopter
[12] Akhtyamov R G, Elizariev A N, Vdovina I V, Planida Yu M and Khaertdinova E S 2012 *Scientific and educational problems of civil protection* **2** 29–34

IOP Conf. Series: Materials Science and Engineering **828** (2020) 012003 doi:10.1088/1757-899X/828/1/012003

Digital environment, information systems and robotics: an absolute benefit or a new economic and political threat?

A K Botasheva [1], S V Anufrienko [1], I B Andish [1], O V Yefimova [1] and M A Dyakonova [2]

[1] Pyatigorsk State University, Kalinin av., 9, Pyatigorsk, 357532, Russia

[2] The State University of management, Ryazansky prospect, 9, Moscow, 109542, Russia

E-mail: ssidorova@list.ru

Abstract. The introduction of new technologies into all spheres of human activity is rightly perceived by the scientific community as a qualitative leap in the development of human civilization. The effect of the introduction of information achievements, breakthrough technologies and communications, the spread of electronic data exchange, digitalization of the environment, etc. contributes not only to the growth of interdependence of the world economy, but also to the rapid improvement of the life quality. Therefore, the accepted set of values defines the central role of new technologies, which are often considered an absolute good. But the authors wonder whether this is an absolute good, or whether behind the facade of the actively developing process of exploiting the world and expanding science and technology into all types of human activity there is a fundamental change in the means of production and ways of human exploitation. The authors also conclude that the digitalization of the urban environment, appearance of autopilots and navigation systems, development of modern IT and nano-technologies, improvement of communication and production technologies open the questions not only of a military, but of an economic and political nature as well. The pursuit of technological efficiency does not take into account the "human dimension", and the humanistic and sociocultural meaning of many technological achievements gets lost. **Keywords:** new information technologies, information systems, robotic technology, artificial intelligence, breakthrough technologies and communication, "smart cities", "smart house"

1. Introduction

N. Bostrom, a lecturer at the University of Oxford, was the first to systematically explore the cultural and philosophical premises of a person's desire to acquire new abilities. In 2005, in the work "The History of Transhumanian Thought", the scientist revealed the desire of a man to go beyond the natural borders within the framework of positivism [1]. N. Bostrom's research is extensive and deals with a class of new threats associated with breakthrough technology development. According to the researcher, there is a number of potential ethical and social problems.

Considering the phenomenon, K. Schwab and R. Samans operate with the term "Fourth Industrial Revolution" and predict the forthcoming increase in the social inequality because of the mass unemployment with the advent of robots [2]. The researchers fully agree with N. Bostrom that governments must prepare for the possible negative consequences of automated work spread in the society.

The same approach to the social problem of digitalization, robotization and development of information systems is shared by M. Ford who assumes that the introduction of new technologies brings benefits to the oligarchic top of countries, and the society will suffer from growing unemployment. The scientist proposes to solve the problem by ensuring a universal basic income [3].

E. Brynjolfsson and T. Mitchell have made a significant contribution to the study of the interdependance between economic and political risks from the introduction of new technologies into everyday reality. After an in-depth analysis, the researchers concluded that the negative consequences of automated work spread in the society were extremely undervalued [4].

The researchers, S. Russell, S. Hauert, R. Altman and M. Veloso, focus on the invention and implementation of deadly autonomous weapons systems. The scientists express fear that the development of new technologies of that character is directed against the civilian population [5]. The assumption for discussion is put forward by R. Arkin, P. Ulam and A. Wagner, implying the benefit of replacing alive soldiers on the battlefield with robots, as robots programmed not to violate the war laws (legal requirements for warfare) will be more useful [6]. The term "techno-Leviaphan", first introduced in B. Scott's paper on the problems associated with the introduction of crypto [7], is, in our opinion, able to describe not only increasingly introduced deadly autonomous systems, but for the whole process of new technologies application, which resembles a situation when children pick up a dangerous toy to play with. Iyad Rahwan is deeply confident that artificial intelligence algorithms affecting our lives should be "transparent, fair and accountable". He offers a conceptual framework for regulating the impact of artificial intelligence on the society. The scientist proposes "a pact between various human stakeholders mediated by machines" [8].

An in-depth analysis of the ethical perspectives of robotics and artificial intelligence development is presented in J. Torresen's extensive scientific paper "A Review of Future and Ethical Perspectives of Robotics and AI". The researcher concludes that misuse of new technologies carries more dangers than the technology itself, getting out of control [9]. Considering the benefits and potential dangers of introduction of engineering innovations in different spheres of human life, the robotics lecturer at the University of Bristol, S. Hauert, insists on having a "balanced discussion" on a social media platform, because "denying robotics is not productive" [10].

Some researchers take the opposite view. For example, D. Acemoglu and P. Restrepo claim that in parallel with the development of technology, the ways will be found to support the welfare of the unemployed and to cope with the risks of "smart cities" [11]. Russian researchers also positively assess the forthcoming total automation of production. Thus, the works of M.I. Velyada, V.I. Fedorova, I.S. Bondarenko [12], S.A. Tolkachev and A.D. Kulakov [13] present an analysis of the international market of robotics in terms of influence on labor resources. The researchers bring the evidence that the introduction of robotics can only have a positive impact, the automation has led to an overall increase in labour demand. Assessing the socio-economic consequences of the robotics industry rise in the USA, S.A. Tolkachev and A.D. Kulakov assume that there is no direct link between the introduction of automation and the growth of unemployment and write about "futuristic fears" [13].

Despite the fact that a significant number of scientific works has been devoted to the most important problems of the consequences of the digital environment development, information systems and robotics, the scientific explanation of the phenomenon remains insufficient, and a number of key positions are in opposition to each other. As the researchers of Edinburgh University of Napier rightly point out, most of the scientific research that deals with this problem are technological in nature. So, they emphasize that scientific research lacks social intelligence, cultural artifacts, and environmental attributes [14].

2. Research Methods

The aim of our study is to analyze the negative consequences of actively undergoing world development and the expansion of science and technology to all types of human activity. In connection with which the research objectives include:

IOP Conf. Series: Materials Science and Engineering **828** (2020) 012003 doi:10.1088/1757-899X/828/1/012003

- interpretation of the digital environment essence;
- analysis of the trends associated with the digitalization of the human environment ("smart city", "Internet things", etc.);
- study of the negative consequences of the use of unmanned aerial vehicles for the safety of civil airlines;
- study of the negative consequences of the autonomous weapons systems and unmanned aerial vehicles use by terrorist organizations;
- study of the processes of robotization and the use of artificial intelligence as factors enhancing the processes of economic and political imbalance in the future;
- identification of a new class of threats that carry new technologies related to the activities of different countries in outer space.

In general, the research questions are as follow: How much does humanity benefit from the growth of technology? Does the process of robotization, automation, computerization, digitalization, etc., have long-term consequences in the form of economic and political negative factors? This is the problem area for analysis of the risks of new technologies development that needs more attention from the scientific community.

In accordance with the purpose and objectives of the study, we have applied a set of general scientific methodological principles and concepts.

Thus, a number of key positions were developed based on a systematic analysis of the development of new technologies in modern conditions.

The choice of functional comparative analysis as the main methodology in this study helps to focus attention on the common features and tendencies of modern technologies development in the field of digital media, information systems and robotics, to reveal the functional entity of the further explosive growth of the processes of automation, computerization, digitization, etc. The principle of empirical verifiability and objectivity, as the benchmark for which particular types of technical innovations serve, has allowed uncovering a number of possible consequences of their implementation in the form of economic and/or political risks of different levels and scales.

3. Results

Let's start with the tempting vision of the future: the so-called "smart city", which has many technologies, such as free wi-fi provided by the street lamps, controlled from a single center of traffic lights, lighting, parking, automated garbage collection, smart meters for utilities and public transport, information platforms, etc. All of them are connected and working online. We agree that social integration and equality are part of a reasonable and sustainable city [15], but are forced to note that the question "How hacker safe is our city?" remains open. Because of the openness of the "digital cities" to hackers, R. Kitchin and M. Dodge criticize the current concept of smart cities, as technologies contain critical vulnerabilities and are exposed to the risk of hacking exploits (computer programs designed to identify vulnerabilities in the software and used for cyberattacks) [16].

Exploits can be used at a distance, and the attack can be masked, which reduces the probability of detecting and catching criminals. In particular, in 2016 the hackers blocked the Municipal Transportation Agency in San Francisco, extorting free travel for all passengers [17]. In 2017, 35% of hospital trusts in the UK suffered from ransomware WannaCry focused on the UK health care system [18]. In 2017, in Washington, D.C., hackers attacked police surveillance cameras on the eve of the inauguration of the 45th President of the USA, D. Trump [19]. 2016 was marked by the hacker attack on the urban transport system of Moscow. Denis Legezo, an employee of Kaspersky Lab's, easily proved the vulnerability of the system to control traffic in the capital and manipulated the transport sensors [20]. The hacker confessed that it was easy because the manufacturer's instructions for the sensors were available on the Internet.

Cesar Cerrudo, a CTO of the research security company IOActive Labs, decided to affect remotely the operation systems of the traffic control to identify the vulnerability in the software of "smart cities" [21]. Having selected as a test platform a number of high-tech cities worldwide, the researcher found that by using unmanned aerial vehicles (UAVs) it is possible to control the switching of the traffic lights in every capital of the world. The general opinion of the specialists is that the greatest damage that can be caused is the capture of the city energy system (when nuclear reactors, airports, metro, mobile communication stop functioning, it also affects hospitals, malfunctions occur with the water supply, etc.) [22]. For example, in 2003, in the United States, several people died after a power outage in several cities of Ohio because of a failure in the software. The financial damage then amounted to $ 6 billion. The leaders of the grouping "Brigades of Abu Hafsa" that belong to the al-Qaeda terrorist organization, said that it was the "operation quick lightning" carried out by the order of Osama bin Laden, "The Soldiers of Allah plunged the lives of Americans into darkness, just as the American fugitives plunged into darkness the lives of the Muslim peoples in Iraq, Afghanistan and Palestine. The Americans lived a black day that will never be erased from their memory, a day of fear, horror and anxiety. Now they know what anarchy and chaos are" [23].

Scientists rightfully express concerns that city authorities and governments that purchase software do not test the security systems. Moreover, there is no proper level of data encryption and a general lack of strict security measures. Given that by 2020 about 50 million devices will have been connected to a global network of the "Internet of things". It is easy to imagine what a global opportunity attackers have, for they are able to manipulate the intelligent system of water supply or electricity, ATMs, wastewater facilities, traffic control, etc. In the case of a large-scale cyber attack the digital system's failure threatens to create chaos.

Another acute socio-political discourse was started after the events in London's Gatwick airport (the second busiest airport after Heathrow) in December 2018. It had been closed for 36 hours because of the unidentified UAV, unmanned aerial vehicles (drones). According to the operating director of the airport, Chris Woodroofe, only the measures taken by the military men, allowed "a small part of the passengers" to take their flights [24]. The remaining 120 thousand passengers stayed in the terminal for more than a day, but despite the chaos and collapse that prevented most passengers to get home for the Christmas holidays, passengers noted that they preferred to obey the security decisions [25]. The UK government offered a reward of $ 63,000 for information about the criminals who used a UAV, but they were never found. The airport was subsequently forced to spend $ 6.3 million dollars to install the UAV detection systems. Experts say that seemingly innocuous drones can be very dangerous. A small in its size RC quadcopter can be a cause of an airplane crash.

In Canada, in 2007, a drone collided with an aircraft, whereby the aircraft's wing was damaged. Although the aircraft was able to land successfully, the head of the company QuantumAviation working in the field of aviation safety, says that at first glance a fragile quadrocopter contains large batteries, and getting into an engine or a fuselage, drones become many times more dangerous than birds [26]. Robot researchers at Imperial College London claim that "a drone weighing more than two kilograms can damage a cockpit glass" [27]. The USA Federal Office of Civil Aviation, together with the experts from the NATO Alliance, dealing with security of unmanned aerial vehicles, came to the conclusion that a quadcopter on the way of an aircraft is a high risk, since the drone stuffing of lithium-ion batteries is not destroyed, but becomes the cause of fire in the body of the aircraft. The world community is worried about the fact that only in 2017 there were 92 cases of drone and aircraft dangerous approaches recorded [27].

A great public resonance was caused by the actions of the ISIS militants who use the UAV for intelligence, strikes and propaganda [16]. To be fair it should be noted that the pioneer in the use of controlled drones that carry ammunition was Israel, still having armed drone "Harpy" (carrying munitions drone of IAI type). How military experts write, such drones patrol in the air for more than 7 hours and can be operated at the distance up to 500 km. The operator on a control station finds a target and sends a "Harpy" to it and explodes a warhead weighing up to 32 kg. ISIS fighters had the opportunity to use the gadget for transferring of TNT weighing up to 1 kg with fuse percussion, and manually via a standard control channel Lightbridge to strike the Syrian provinces of Damascus, Idlib, Homs. UAVs of the "aircraft" type are able to stay in the air for more than an hour, and the radius of their destruction is 100 km.

The use of "helicopter" type quadcopters (Phanton II and III class) by the ISIS fighters prove to be more successful since they have a positioning system in the absence of a GPS signal (the "aircraft" type loses its position in the failure of a GPS signal). Phanton II and Phanton III carry up to 1300 grams of load, which indicates a relatively high carrying capacity, and with two explosive devices can easily maneuver in any urban area up to 25 minutes. As a result, with the help of drones, the militants were able to destroy a "soft" fuel tank in Mosul from New Jersey, USA (manufactured by Aero Tec Laboratories, Inc), several units of tankers and lightly armored vehicles. In January, 2018, with the help of a drone bomber, an explosive charge was delivered to the open hatch of the Iraqi Hammer, which was transporting ammunition for the Iraqi army. It should be noted that to teach to use drones ISIS organized "schools for media training and development" for all comers, including teenagers.

Many previous achievements in the development of drones can be surpassed by a unique tiny FlyCroTug drone developed at Stanford University. FlyCroTug can penetrate into any room through the ventilation hatches, then put a loop on the door handle of the locked room, fix itself on the wall and pull the cable effectively to open the door. The engineers and scientists who developed the drone claim that it is intended to help people during accidents and blockages: move the first-aid kit, deliver water, means of communication, tools, etc. since it is capable to move objects 40 times heavier than it is [28]. But it is so easy to imagine how, where, by whom and for what purposes these drones will be used after their launch into the commercial production.

In turn, the firms involved in the development and sale of drones do not offer the ways to neutralize the software and hardware of drones in the zones of limited use, since they fear of a decrease of the demand for the products. The above factors indicate that commercial production and sale of drones refers to focusing events, that is, those events that should not remain without an appropriate political reaction, that should direct public attention to various aspects of the problem. Another aspect of this problem is the lack of extensive UAV detection systems, which are currently deployed only in the territory of superdeveloped states.

To the factors enhancing the processes of economic and political imbalance in the future, we relate the process of robotization and the use of artificial intelligence in industrial countries. Such a statement, at first glance, seems fantastic. However, the Director of the IMF (International Monetary Fund) Christine Lagarde, delivering her speech in the high-level Panel of the UN on economic empowerment of women in the New York Council on International Affairs in September 2017, noted that the introduction of modern technologies will negatively affect the economic sphere of several countries, including some highly developed ones. In this case the indicator of the negative phenomena would be unemployment. The IMF Director attributed the jobs occupied by women to a particular risk group, "the processes of robotization and the use of artificial intelligence in the coming years will lead to the disappearance of some jobs, and women will be affected stronger than men", said Lagarde. According to the head of the IMF, "11% of jobs that are now occupied by women worldwide, will be affected in varying degrees, or even disappear", and the processes of robotization will impact, in varying degrees, 7% of jobs occupied by men. "In other words, in 30 industrialized countries, the processes of robotization will affect 28 million women, and if we extrapolate these estimates to the whole world, in varying degrees, they will affect 280 million women, unless some action is being taken to retrain them for other professions... and to solve this social problem a political solution should be worked out", said the head of the IMF [29].

But we can rightly say that under the risk not only women are, as there are some recent developments in the field of male labour replacement. The Ministry of business, energy and industrial strategy of the United Kingdom has invested about $ 9 million [30]. The research project is titled "micro-Robots". On a competitive basis, this research is delegated to four leading universities in the USA, and the first achievements in this field will be used for servicing underground communication networks of the UK. The solution to this problem will cause a real revolution in the repair of city infrastructure around the world, as the scientific research in this area has been conducted for a long time, and different versions of robots-repairers intended for autonomous check of underground pipes exist in different countries. But such works as digging in urban areas, demolishing of asphalt pavements and repair of damaged pipes, until the recent time, remained a daunting task for modern robots. In the long term the project involves the creation of "inspection bots" of a size less than a centimeter. The army of bots will have bigger supporters to repair the found damages. Robot inspectors carry diagnostic equipment and are able not only to move quickly on any surface, but also to swim and fly. In contrast, the "working" robots will have the energy and the working substance to clean and to install patches on the pipe using the cable. The economic effect is expected to be huge, because only in the United Kingdom annually they have 1.5 million roads excavations, leading from idle transport time to billions of dollars in damages (up to 6.3 billion a year). This leaves a question: Does this multi-billion-dollar economy with the use of robots mean good for the workers who were previously employed in these industries?

With the rapid development of robotics all areas of physical labor application in the world will gradually get extinct. For example, today there is a full version of a robot bricklayer "Hadrian X", which is able to build a one-bedroom house literally from a scratch. The Australian company completed Fast Brick robot construction in 2015. And the company can be proud of it because the laser alignment system of a robot bricklayer involves laying blocks and bricks of different sizes and combinations with an accuracy of 0.5 mm. "Hadrian X" is equipped with a 3D model software, allowing to calculate the position of each block or brick in a building under construction. It is noteworthy that the concrete slab construction of the house is inspected with the laser scanner, which critically evaluates the construction according to all the requirements and standards. The robot Builder is accompanied by a 28-metre articulated telescopic boom crane that loads and stacks bricks together with the help of special glue. Therefore, we can assume that the construction companies that are interested in quality construction in the near future will purchase these robots, which will lead to significant male unemployment around the world.

Risks of a larger scale are represented with new technologies associated with the activities of different countries in the outer space. At the end of February, 2019, the chief of the U.S. Air Force Staff said that the war in the space would be possible in a few years. Unfortunately, the general is absolutely right: intelligence and communication capabilities are in the outer space, as well as on land and at sea. National interests allowed US to use the opportunity: in August, 2019, the President of the United States, Donald Trump, announced about the establishment of the U.S. Space Forces. The comment of the President was brief, "The enemies of the United States with the help of new technologies threaten American satellites" [31]. Analyzing this position, we note that the militarization of space and the competition for superiority in the space represent a threat to everyone's safety. And given the plans of the American company, Space Tango, to build a space robotic factory, "ST-42", in the earth orbit in the next few years, we understand that the time has come for new ways of controlling the environment. On the other hand, humanity hopes that the new technology, including the organization of production in the space based on new possibilities of the environment, will truly bring salvation to people from certain incurable diseases, will bring additional opportunities and comfort into the lives of people.

We have discussed some of the technogenic factors, which could negatively affect the future of the mankind. Along with the digitalization of the urban environment, "smart cities" and "smart houses", technically advanced household appliances and cars, autopilots and navigation systems, there are still vital and open topics for discussion, namely: How will the rights and freedoms of a man in the age of information transparency be ensured (provided a chip implantation and tracking tools, managed via a satellite, will be implemented)?; What will the limits of the authorities control be?; Will the communities become more integrated and interact outside the boundaries of the countries?; What will be the consequences of new technologies application and growing capabilities of criminal groups?, etc. Here is also an important question: How is it possible to use most of the technologies of a "smart city" as some people have no computer skills, or possibilities to buy expensive devices and gadgets? And how will the disappearance of a number of professions affect the growth of mass unemployment because of the epochal leap of robotics? In our opinion, the social stratification of the population will become deeper in the future and the social gaps will only increase.

But, according to the researchers of the Napier University in Edinburgh, a large portion of scientific researches, that are related to this problem, are purely technological by their nature. Unfortunately, they emphasize, that there is a lack of social intelligence, cultural artifacts, and environmental attributes. Therefore, a look through the prism of chaos of the globalization era at the very complex and controversial socio-economic and political processes of our time leads to the understanding: the scientific progress and the positive aspects and advantages of globalization in the form of a single information field and a growing exchange of technological innovations lose their attractiveness due to the growing inequality in the development of countries, the increasing gap between the rich and the poor, as well as the unpredictability of political events in separate countries and regions.

4. Conclusion

Given the foregoing discussion, we have come to the following conclusions:

1) we are witnessing an actively developing process of exploiting the world and the expansion of science and technology to all types of human activity;

2) the means of production and methods of exploitation of a man by a man are radically changing, moving into an artificial environment;

3) in the pursuit of technological efficiency and / or commercial profit, manufacturers do not take into account the "human dimension", the humanistic and sociocultural meaning of technological achievements gets lost;

4) the cycles of technogenic processes are often many times higher than the rate of restoration of natural resources and landscapes;

5) the issue of female and male unemployment is acute around the world;

6) the cities become more vulnerable to hacker attacks by terrorists;

7) the outer space is becoming an arena of people's activity, focused on the domination over it and the development of the military-industrial complex.

In this regard, we should consider the following:

- firstly, political decisions, regarding the growth of mass unemployment around the world and the critical attitude towards manufacturers, promoting technical devices on the mass market, are potentially dangerous when used by dishonest people;
- secondly, governments when purchasing software solutions for "smart cities" should rely not only on whizz-bang functions, but also strive to provide real security for cities through testing for hacker penetration;
- thirdly, in each city there should be an operational team for responding to emergency situations related to the city software.

All of the above is associated with social transformations in various fields in the age of new technologies and makes political decisions relevant, the purpose of which, paradoxically as it sounds, is to protect people from their own achievements.

References

[1] Bostrom N 2014 *Superintelligence: Paths, Dangers, Strategies* (Oxford: Oxford University Press) p 411
[2] Schwab K and Samans R 2016 *World Economic Forum The Future of Jobs* p 7
[3] Ford M 2015 *Rise of the Robots: Technology and the Threat of a Jobless Future* (New York: Basic Books) p 7
[4] Brynjolfsson E and Mitchell T 2017 *Science* **358(6370)** 1530-34
[5] Russell S, Hauert S, Altman R and Veloso M 2015 *Nature* **521** 415
[6] Arkin R C, Ulam P and Wagner A R 2012 *Proceedings of the IEEE* **3** 571-589
[7] Scott B 2014 *E-International Relations* **Jun 1**
[8] Rahwan I 2018 *Ethics and Information Technology* **20(1)** 5-14
[9] Torresen J 2018 *Front. Robot. AI* **4(75)**

[10] Hauert S 2015 *Nature* **521(7553)** 416-17
[11] Acemoglu D and Restrepo P 2018 *American Economic Review* **108(6)** 1488-1542
[12] Velada M, Fedorova V and Bondarenko I 2018 *Bulletin of the Moscow State University* **4(27)** 91-98
[13] Tolkachev S and Kulakov A 2016 *The World of the new Economy* **2** 79-87
[14] Mora L, Bolici R and Deakin M 2017 *J. of Urban Technology* **24(1)** 3-27
[15] Albino V, Berardi U and Dangelico R 2015 *J. of Urban Technology* **22** 13-21
[16] Kitchin R and Dodge M 2017 *J. of Urban Technology* **26(2)** 47-65
[17] Taylor H 2016 *CNBC* **Nov 28**
[18] Ghafur S, Kristensen S, Honeyford K, Martin G et al 2019 *NPG Digital Medicine* **2(98)**
[19] Hermann P 2018 *The Washington Post* **July 28**
[20] Leyden J 2016 *The Register* **22 Apr**
[21] Cerrudo C 2015 *IOActive* p 19
[22] Pardo T and Nam T 2016 *Smarter As the New Urban Agenda: A Comprehensive View of the 21st Century City* (Berlin: Springer) pp 23-47
[23] Torres-Soriano M R 2015 *Media War and Conflict* **8(2)** 1-18
[24] Calder S 2019 *Independent* **Jan 22**
[25] Singh K. 2018 *Reuters* **Dec 24**
[26] Hern A and Topham 2018 *The Guardian* **Dec 20**
[27] Hsu J 2017 *Scientific American* **Nov 8**
[28] Kubota T 2018 *Stanford.edu* **Oct 24**
[29] Lagarde C 2017 *International Monetary Fund* **Sept 19**
[30] Houser K 2018 *Futurism* **Dec 31**
[31] Botasheva A, Miller I and Rumachik I 2019 *Advances in Intelligent Systems and Computing* **726** 585-592

Problems of increasing the innovation activity of industrial enterprises in the context of global competition

H-Ch Brauweiler[1], V V Kurchenkov[2], O V Fetisova[2], L V Ponomareva[2] and E A Kurchenkova[2]

[1]Western Saxon University of Applied Sciences, Kornmarkt 1, D-08056, Zwickau, Germany
[2]Volgograd State University, Universitetskiy av., 100, 400062, Volgograd, Russia

E-mail: kurchenkov@volsu.ru

Abstract. The article deals with the increasing innovation activity of Russian enterprises in the context of global competition. It describes the rise in intensity of competition as a result of the decrease in the main market players and the increase in the scale of their economic activity. The authors analyze the tendency of growth of the incubation cycle duration of new developments in industrial enterprises of high-tech industries which makes it commensurate with the life cycle of new technology or a new product. It is proved that innovation activity of the industrial enterprise, on the one hand, can, under certain conditions, bring extra income due to the growth of labor productivity and reduction of internal costs. On the other hand, extra income is brought by obtaining a new market segment through the implementation of the innovation product. The article also describes the influence of innovation activity of industrial enterprises on the structure of investments in fixed assets and the duration of the investment cycle.

1. Introduction

In modern conditions, innovation is becoming the major factor in the competitiveness among large industrial enterprises. In the face of growing global competition, major transnational corporations with significant market shares, high concentration of capital, and substantial scientific and innovative potential often become major players in the world markets. Indeed, industrial enterprises in the leading countries of the world show the high level of innovation activity. In particular, in Germany more than 70% of enterprises are actively innovating [1], in France they are more than 65%, and in the EU as a whole they constitute 60%. Unfortunately, in Russia today this indicator is not high enough and does not exceed 30-40% [2].

The main reasons for low innovation activity of enterprises and organizations in Russia include lack of investment resources for innovation, lack of motivation of the major stakeholders for production process updating, lack of the innovation infrastructure development, and the legal vulnerability of innovation-active enterprises.

Also, along with the above-mentioned reasons for the low innovation activity of Russian industrial enterprises, we should highlight underdevelopment of full-fledged competitive relations and high level of monopolization within the country, especially among large manufacturers. In this connection, a somewhat paradoxical situation arises. Namely, large enterprises are able to concentrate significant financial and material resources that can be used for innovation development, but do not have an

IOP Conf. Series: Materials Science and Engineering **828** (2020) 012004 doi:10.1088/1757-899X/828/1/012004

external competitive factor due to the monopolistic position. While small and medium enterprises exist in a competitive environment, but do not have enough resources, primarily investment ones, for innovation development.

This situation significantly inhibits the growth in innovation activity of economic entities in various scales of economic activity in the modern Russian economy [3].

However, it should be assumed that global competition can become an external factor of increasing the innovation activity of large domestic industrial enterprises. At the same time, under certain conditions related to the extension of the innovation cycle, on the one hand, and shortening of the life cycle of the new product on the market, on the other hand, this competition can restrain the growth of innovation activity, compensating for the obtaining additional income by the growth of aggregate costs of advanced innovative development.

2. Results and discussion

As it was noted before, in the global economy, large industrial corporations prove to be the most innovative ones [4]. This is due, on the one hand, to the high concentration of financial capital in them, which is used for innovation development. On the other hand, the presence of the external stimulating factor associated with competitive relations. The aggravation of global competition forces or stimulates large companies to use their super-profits for new developments. As J. Schumpeter wrote: "Frequently, if not in most cases, a going concern does not simply face the question whether or not to adopt a definite new method of production that is the best thing out and, in the form immediately available, can be expected to retain that position for some length of time. A new type of machine is in general but a link in a chain of improvements and may presently become obsolete" [5].

Indeed, today the intensity of R & D in large industrial companies is increasing significantly. At the same time, this trend persists steadily over the last time, starting from the second half of the last century. In particular, already in the late 1970s, the share of the 100 largest corporations in developed countries accounted from 55% to 90% of industry expenditures on research and development [6], and eight powerful concerns accounted for more than a third of the appropriations for these purposes. Companies with less than 500 employees in the Federal Republic of Germany accounted for 3.3%, from 500 to 2000 people - 9.3%, and for firms with the number of employees over 2000 people - 87.4% of R & D expenses. At the end of the last century, 600 US corporations accounted for 99% of R & D funding. The share of 8 of them is 1/3 [7].

It should be noted that large companies also lead to an aggravation of competitive relations. Since a small quantity of large players in global markets does not reduce the level of competition, but increases its intensity. Basically, all global world markets (automotive, aircraft, shipbuilding, instrument engineering, mining) today have an oligopolistic market structure, which implies a limited number of participants, and this circumstance leads to increase in the intensity of competition, unlike other competitive market structures. As some authors point out: "The intensity of competition can be measured by the value of 1/n, where n is the number of market participants. The more market participants, the less competition intensity is, and vice versa, the smaller the number of market participants, the higher and sharper the competition is. With oligopolistic market structure, where the number of economic units is 4, after dropping out of one participant in competitive struggle, for each of the remaining ones market share will increase by 25%" [8].

However, it should be mentioned here that the intensity of competition in oligopolies can be much greater if we consider that not one participant, but just three participants can drop out of the market as a result of competition because most oligopolists lose their positions in the market in favor of one leader. In conditions of perfect market competition, only a few players are eliminated as a result of competitive rivalry, and their market share is insignificant to take it into account for the assessment of the level of competition intensity.

In addition, it should be assumed that the intensity, in this case, is not only the expression of the quantitative ratio of market volume and price sufficient for the analysis of static equilibrium, but also the expression of the qualitatively innovative competitive mutual adaptation of large industrial

IOP Conf. Series: Materials Science and Engineering **828** (2020) 012004 doi:10.1088/1757-899X/828/1/012004

enterprises in the perspective of dynamic development. The intensity of competition of the limited number of large producers is determined by the interdependence, a high degree of uncertainty, which is associated with technological scale changes and is essentially dynamic. And since leadership in oligopoly is associated with the tendency towards constant innovation in technology and organization, the intensity of competition ultimately reflects the intensity of immanent, advanced innovative development, as a condition for competitive mutual adaptation in oligopoly.

In accordance with this, a large industrial corporation, introducing innovations, receives a double benefit. On the one hand, due to the introduction of new technologies (technological innovations), production costs and, accordingly, the prime cost of the product is reduced [9]. On the other hand, due to the introduction of a new product, competitors are being forced out of the market. In conditions of oligopolistic competition, the effect of these external and internal factors of enterprise development are interdependent, when savings on domestic reserves and production scale make it possible to receive savings due to strategic advantages, by obtaining additional market share (figure 1).

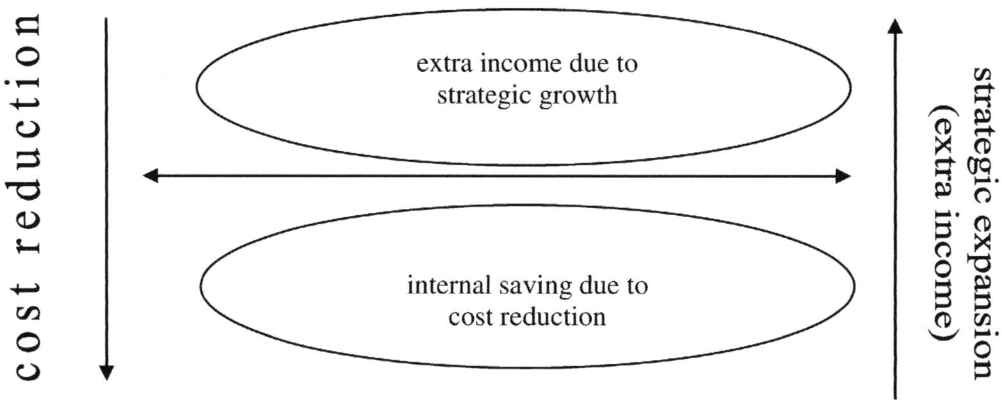

Figure 1. Interconnection between savings on internal reserves and the strategy of industrial enterprises

However, it should be noted that these opportunities for obtaining benefits from external competitive advantages are limited by the trends related to the scientific and technical progress, increasing of complexity of industrial products and the growth of investment costs in the framework of innovative activities of the corporation. In particular, in accordance with the leading nature of the innovative development of industrial corporations, the intensity of competition between them in oligopolistic conditions is determined, on the one hand, by reduction in the life cycle of technology (product), on the other hand, by increase in the incubation cycle for developing a new technology.

Table 1. Duration of the phases of incubation development of some basic technologies of the new technological order [10]

Production	Phase duration (years)
Mainframes	15-20
NC machinery	10-15
Industrial robotics	10-15
Microprocessors	5-10
Personal Computers	2-3

As shown in Table 1, the incubation cycle for developing a new technology or a new product in the leading industries is, on average, 10-15 years. Simultaneously, these products have a relatively short life cycle, which, under accelerated conditions of scientific and technical progress, has a tendency to further shortening, and today also averages no more than 15 years. Accordingly, the trend of technological development in modern conditions is that the incubation cycle for developing a new technology becomes commensurate in duration with its life cycle in the market (figure 2).

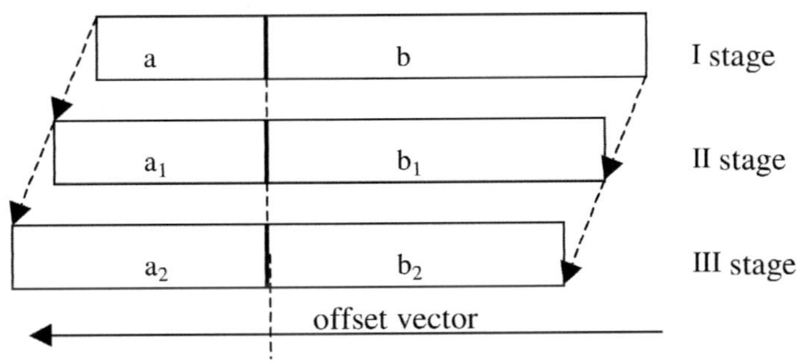

Figure 2. The trend of changes in the duration of development cycles and the use of new technology in global competition: a - incubation cycle for developing a new technology; b - the life cycle of this technology

Also, along with the equality in time periods, there is a tendency to equalize the amount of development costs and the amount of monopoly superprofits, provided by new technology during its life cycle (figure 3).

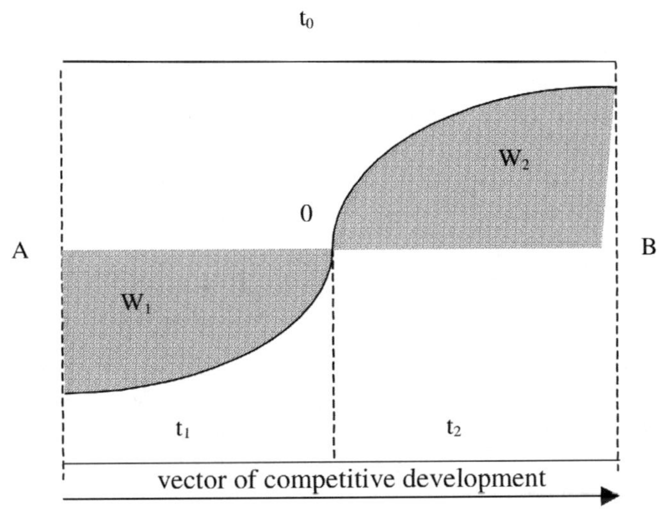

Figure 3. The cycle of innovation development of the industrial enterprise in the context of global competition (oligopoly): t_0 - individual cycle of monopolistic competition; t_1 - new technology development cycle; t_2 - monopolistic cycle of this technology; W_1 - the cost of developing a new technology; W_2 - the value of monopoly accumulation (monopoly superprofits)

In this case, within the framework of the single technological structure, the profits made by large industrial companies in conditions of oligopolistic competition should be invested in innovative developments, reducing the overall efficiency of the company in the short term. The exceptions can be situations of changing the technological structure, or the so-called "technological discontinuities" [11], which arise in some industries.

It should also be noted that the reduction of the life cycle of a new technology as a result of aggravation of global competition naturally leads to the fact that the obsolescence of equipment in industrial enterprises is far ahead of physical wear and tear of equipment [12].

Consequently, a large industrial corporation, in order not to be an outsider in the conditions of oligopolistic competition, is forced to switch to a new technology before the expiration of the normal depreciation period of old equipment. In this regard, it will inevitably have to incur double costs: the cost of developing and implementing a new technology, including neutralizing the risks of its market adaptation, and the cost of old equipment, which expenses are not refunded in full (in the form of depreciation) in a given period of time. In this situation, the source of undistributed superprofit, obtained as a result of monopoly limitation in the previous period, becomes a necessary investment link for the further development of this corporation.

Such a sum of expenses can be realized only based on substantial savings that a normal profit cannot provide since it was not originally designed for accelerated large-scale renovation of industrial production. The possibility of an accelerated transition to a new technology, a new production environment, for a large corporation is made up of the real value of monopoly accumulation over the previous period. In the case of technological shift, monopoly accumulation is necessary due to the incompatibility of the proportions of natural, normal accumulation and compensation.

If a large industrial company in the conditions of oligopoly does not switch to a new technology before the period of reimbursement of the cost of the old technology, which is dictated by the conditions of technical development, it will be done by other participants. As a result, this company will not only loose additional profit but will also suffer losses due to compulsory reduction in the share of the sales market. Because the principle of dependence in oligopolistic competition is associated with the principle of substitution.

3. Conclusion

Thus, the level of innovation activity is a crucial factor in the growth of competitiveness, as for a separate industrial enterprise, so for the national economy as a whole. Russian economy still lags behind in terms of innovation activity from the leading countries of the world. The main reasons for the lack of innovation activity of Russian industrial enterprises today are: lack of investment resources, the suboptimal structure of industrial production with the high share of the materials sector, disinterest of new owners in increasing of innovation activity of enterprises, underdevelopment of innovation infrastructure, etc. [13, 14].

Along with these reasons, it is important, on the one hand, to improve competitive relations within the country, which act as an external stimulating factor for innovation activity. On the other hand, facilitating the inclusion of large Russian industrial enterprises in the system of global competition, primarily in the markets of high-tech products [15]. To a large extent, limiting the growth of innovation activity is associated with the change in the incubation cycle for developing a new technology, on the one hand, and the life cycle of this technology or product innovation, on the other. This causes equalization of additional costs for R & D and monopoly superprofits during the period of market realization of new product or technology.

Increase in the intensity of global competition as an external factor for stimulating innovation activity is associated with the oligopolistic structure of the global industrial markets. Under these conditions, as noted, the conjunct market conditions are the basis of “competitive mutual adjustment” of a small number of its participants. A favorable conjuncture is created by the largest industrial company, as a result of its successful innovation development, and the conjuncture conditions for oligopolists present at the same time competitive conditions for development. In this regard, in the conditions of modern oligopoly, the possibility of obtaining additional income by reducing internal costs, as a result of innovation development, is organically linked to the possibility of getting additional income, by obtaining strategic advantages, and increasing market share under oligopoly conditions.

IOP Conf. Series: Materials Science and Engineering **828** (2020) 012004 doi:10.1088/1757-899X/828/1/012004

References

[1] Brauweiler H-Ch 2002 *Innovationen im peripheren Raum. Eine Analyse am Beispiel der Grenzräume Süd- und Ostsachsens.* (Wiesbaden: Dt. Univ.-Verlag) p. 141
[2] Kurchenkov V V, Brauweiler H-Ch and Ponomareva L V 2018 *Advances in Economics, Business and Management Research (AEBMR)* (Amsterdam – Paris: Atlantic Press) **39** 22–25
[3] Kurchenkov V V, Fetisova O V, Orlova A A and Gladkaya E A 2017 *Revista Galega de Economia* **26 (3)** 33-44
[4] Rubino F E 2016 *International Journal of Learning and Intellectual Capital* **13 (2-3)** 119-34
[5] Schumpeter J 1994 *Capitalism, Socialism and Democracy* (Routledge: London, New York) p. 98
[6] Demin A A, Raskov N V, Lavrov S B et al 1983 *Modern Concern: Political and Economic aspect* (Leningrad: Leningrad State University) p. 101
[7] Gorokhova K G and Semenko G V 1986 *State-monopoly Regulation of the Economies of the Main Capitalist Countries* (Moskow: Znanie) p. 68
[8] Galkin A A, Kotov V N, Krasin Yu A, Menshikov S M et al 1989 *Capitalism Today: Development Paradoxes* (Moskow: Progress) p. 110
[9] Zedtwitz M 2016 *J. of Prod. Innovat. Manag.* **32 (1)** 12-28
[10] Glazyev S Yu 1990 *Economic Theory of Technical Development* (Moskow: Nauka) p. 42
[11] Foster R 1988 *Innovation: The Attacker's Advantage* (Mandaluyong:Summit Books)
[12] Grabowska M 2014 *App. Mech. and Mater* **718** 50-55 DOI: https://doi.org/10.4028/www.scientific.net/AMM.718.50
[13] Mochalova K M and Petergova A V 2015 *News of the Moscow State Technical University MAMI* **5 4 (26)** 237-41
[14] Duplenko N G and Smirnova O S 2016 *Bulletin of the Baltic Federal University by I. Kant. Hum.and Soc. Sci.* **2** 85-95
[15] Pavlycheva A V 2016 *Novainfo.ru* **1(46)** 67-72

IOP Conf. Series: Materials Science and Engineering **828** (2020) 012005 doi:10.1088/1757-899X/828/1/012005

Synthesis of nanofibers organosiloxane structures under hydrothermal conditions

N I Cherkashina[1], Z V Pavlenko[1], E S Samoilova[1] and N I Novosadov[1]

[1]Belgorod State Technological University named after V.G. Shoukhov, Kostyukov str., 46, Belgorod, 308012, Russia

E-mail: Natalipv13@mail.ru

Abstract. The paper presents the possibility of producing polyhedral oligomeric silsesquioxanes (POSS) under hydrothermal conditions in a high-pressure reactor. POSS was prepared by hydrolytic polycondensation of cyclohexyl trichlorosilane under conditions of elevated temperature and pressure. Morphology and structure of the final products were characterized by X-ray diffraction (XRD), thermal analysis (TGA and DTA) in an atmosphere of argon, Fourier transform infrared spectrum (FT-IR) and scanning electron micrograph (SEM). The resulting compound is a nanofibers of a polyhedral oligomeric silsesquioxane. The diameter of the obtained nanofibers is 50 ... 150 nm, and the length is from 200 nm to 30 μm.

1. Introduction

Organosiloxane materials are widely used as a filler to create various functional polymer composites [1,2]. Composites with additives of organosiloxane structures were widely used in the space industry [3-6]. Due to their chemical nature, organosiloxanes are highly resistant to the flow of atomic oxygen in outer space [7–9]. The Si atom, which is part of organosiloxanes, plays the main role in creating stable protective structures in the form of non-volatile oxides or glass-like substances on the surface when such materials are in an aggressive environment. The improved resistance of these materials to the action of strong oxidizing agents (including atomic oxygen in outer space) is based on the surface conversion mechanism, i.e. the formation of a new dense oxide protective layer.

One of the most promising organosiloxane fillers for space composites is polyhedral oligomeric silsesquioxanes (POSS), the chemical structure of which can be represented by the general formula $(RSiO_{1.5})_n$ with n = 6, 8, 10, ..., where R is an organic radical [10-11]. The term "silsesquioxanes" indicates the ratio between the number of oxygen and silicon atoms: *sesqui* = 1.5. The most accessible and widely used are octahedral compounds (n = 8).

POSS synthesis is the subject of many works [12-14]. One of the main methods for producing POSS is the hydrolytic polycondensation of trifunctional monomers of the $XSiY_3$ type at room temperature [15–16]. However, the reaction time is several days, and the product yield reaches 70%. To increase the reaction rate, it is possible to use catalysts or increase the temperature of the process. In [17], to accelerate the course of the reaction for obtaining POSS-N_3, the process temperature was increased to 80 ° C.

IOP Conf. Series: Materials Science and Engineering **828** (2020) 012005 doi:10.1088/1757-899X/828/1/012005

This paper presents the possibility of obtaining POSS structures by hydrolytic polycondensation of cyclohexyl trichlorosilane under conditions of elevated temperature and pressure.

2. Materials and methods

2.1. Materials

Cyclohexyl trichlorosilane $C_6H_{11}SiCl_3$ (manufacturer PJSC Khimprom, Novocheboksarsk, Russia) was used as a reagent. Cyclohexyl trichlorosilane is a colorless liquid with a boiling point of 206 ° C and a density of 1.23 g/cm^3 (99.2% purity).

Silicon tetrachloride with the chemical formula $SiCl_4$ was produced by KHORST LLC, Moscow, Russia. $SiCl_4$ is a colorless liquid with a boiling point of 57 ° C and a density of 1.48 g/cm^3 (99.9995% purity).

POSS synthesis was carried out in acetone solution. The chemical formula of acetone is C_3H_6O, boiling point 56 ° C, density 0.79 g/cm^3. Acetone was produced by Component-Reagent LLC, Moscow, Russia (99.75% purity).

Distilled water was used for the reaction according to GOST 6709-72 "Distilled water - Technical specifications" (pH = 6.6).

2.2. Synthesis POSS

To obtain silsesquioxanes of a fully condensed structure (POSS), at the first stage, the synthesis of polyhedral oligosilsesquioxanes of a partially condensed structure was carried out. Scheme (1) shows the reaction of the formation of polyhedral oligosilesesquioxanes of a partially condensed structure by hydrolytic condensation of cyclohexyltrichlorosilane at room temperature (20 ° C) for 24 hours:

$$nC_6H_{11}SiCl_3+1.5nH2O \rightarrow (C_6H_{11}SiO_{1,5})n+3nHCl \quad (1)$$

Figure 1 shows the structural formulas of partially condensed oligosilsesesquioxanes, which can be obtained as a result of the reaction according to scheme (1).

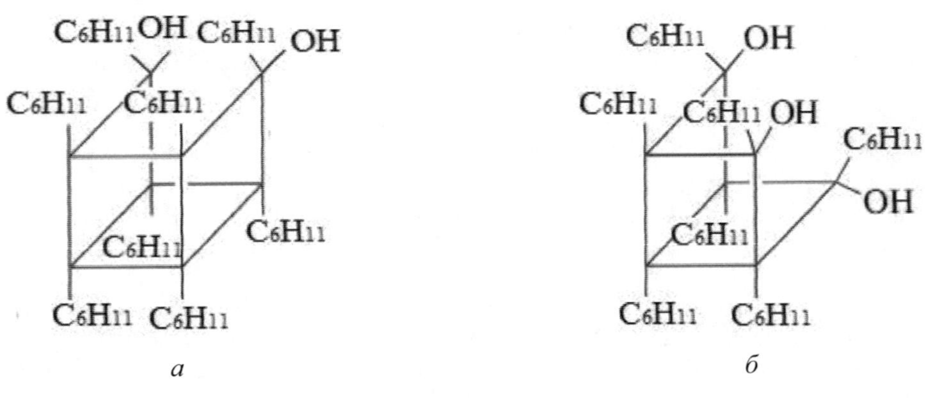

Figure 1. Structural formulas of partially condensed oligosilsesesquioxanes

The resulting partially condensed oligosylsesquioxoxanes are a white precipitate. The resulting precipitate was washed from acetone. For this, centrifugation was used at high speeds of 6000 rpm to separate the solid phase from the starting reaction products, followed by washing. The washing and centrifugation operation was repeated several times.

The synthesis of monofunctional polyhedral condensed oligosilesesquioxoxanes was carried out according to the sol-gel technology according to scheme (Fig. 2) by the interaction of partially condensed oligosilesesquioxoxanes with silicon tetrachloride in a GSA-0.3 high-pressure reactor for 4 hours.

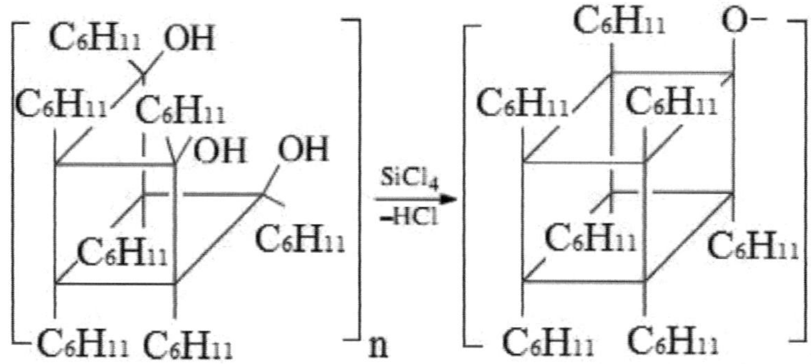

Figure 2. The synthesis of monofunctional polyhedral condensed oligosilesesquioxoxanes

2.3. Research Methods

The Fourier transform IR spectrum was recorded using a IR spectrometer VERTEX 70. The specimen was pressed using a spectroscopically pure KBr matrix.

X-ray phase analysis was performed using the DRON-3 diffract meter (CuKα radiation, Ni filter) according to the standard method. The tube voltage amounted to 20 kV, the anode current was 20 mA, and the detector rotation rate was 2.4 degrees per minute with 1° angular pitch. The measurement rate limit reached 1000–4000 pulses per second. Data analysis and preliminary processing was performed with PDWin software (DrWin, Qual) using the PDF JCPDS database (version 2.02 1999).

Microphotography was performed using electronic microscope Tescan Mira 3 LMU.A thermal analyzer, STA 449 F1 Jupiter (NETZSCH), was used for thermal analysis. The materials were investigated in the temperature range from 20 to 1000° C. The heating rate was 5 K/min.

3. Result and discussion

The FT-IR spectra of the hydrothermally prepared compound are presented in Fig. 3. The very strong and predominant absorbance peak at 1080 cm^{-1} can be assigned to Si–O–Si asymmetric stretching vibrations. The IR band at 450-470 cm^{-1} is due to O–Si–O bending vibrations. The IR band at 780-800 cm^{-1} is assigned to Si–OH group. The IR band at 1440–1480 cm^{-1} corresponds to the C–H bond in the cyclohexane radical ($–C_6H_{11}$). The IR band at 2950–2970 cm^{-1} corresponds to the $–CH_2–$ structural fragment in the structure of the cyclohexane radical. It should be noted that on the IR band of the obtained substance there are no bands at 3450-3500 cm^{-1}, which indicates the absence of physically adsorbed water (H_2O). Also, there is no IR band at 1650–1680 cm^{-1} in the IR band responsible for the –OH bond. Thus, it can be argued that the chemical bonds in the IR band of the obtained substance correspond to the bonds that must be present in the IR band of the polyhedral oligomeric silsesquioxane. The absence of physically adsorbed water and the –OH bond indicates the production of a fully condensed polyhedral oligomeric silsesquiocane.

IOP Conf. Series: Materials Science and Engineering **828** (2020) 012005 doi:10.1088/1757-899X/828/1/012005

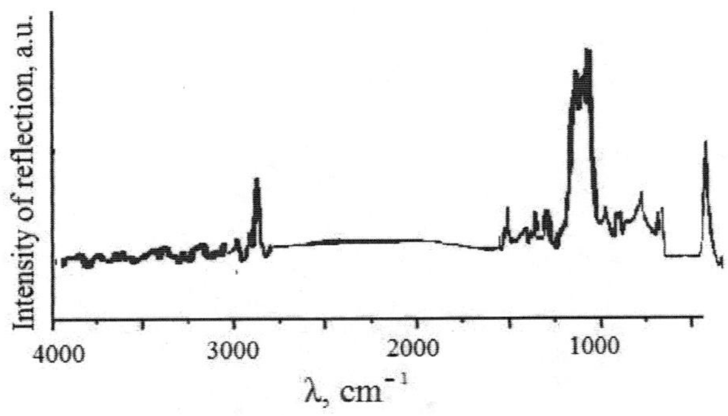

Figure 3. FT-IR spectra of a substance synthesized under hydrothermal conditions

The X-ray powder diffractogram of the obtained substance is shown in Figure 4. Analysis of the diffraction pattern showed that the synthesized substance has a pronounced crystalline structure. There is no information (card) in the PDF2 database about polyhedral oligomeric silsexivoxanes. However, the obtained diffraction pattern results are consistent with the known literature data indicated in [17]. The characteristic crystal features of synthesized compound show four main reflections at $2\theta=9°$, 12°, 12.5°, 19° corresponding to the lattice spacing of 10.4 Å, 7.9 Å, 6.9 Å, and 4.3 Å, respectively.

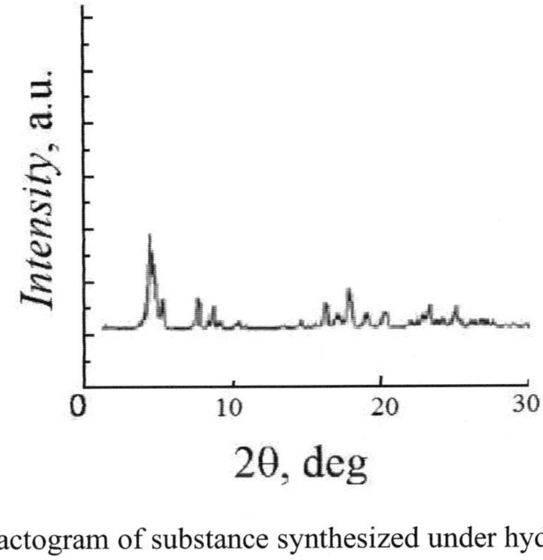

Figure 4. X-ray diffractogram of substance synthesized under hydrothermal conditions.

Figure 5 shows a SEM image of a substance synthesized under hydrothermal conditions. An analysis of the data in Figure 5 showed that the resulting substance is a nanofibers. The diameter of nanofibers is from 50 to 150 nm. The length of the nanofibers varies from 200 nm and can reach 30 μm.

IOP Conf. Series: Materials Science and Engineering **828** (2020) 012005 doi:10.1088/1757-899X/828/1/012005

Figure 5. SEM image of substance synthesized under hydrothermal conditions.

Thermal analysis of the synthesized substance was carried out in the temperature range from 20 to 800 ° C in an argon atmosphere. The Al_2O_3 hitch served as a reference. Figure 6a shows the obtained TGA curve of the synthesized substance. Figure 6a shows that the synthesized substance is thermostable to a temperature of 300 ° C without loss of mass on the TGA curve. The temperature at which mass loss ends is 580 ° C. After a temperature of 580 ° C, the residual mass of the sample did not change and amounted to 40%.

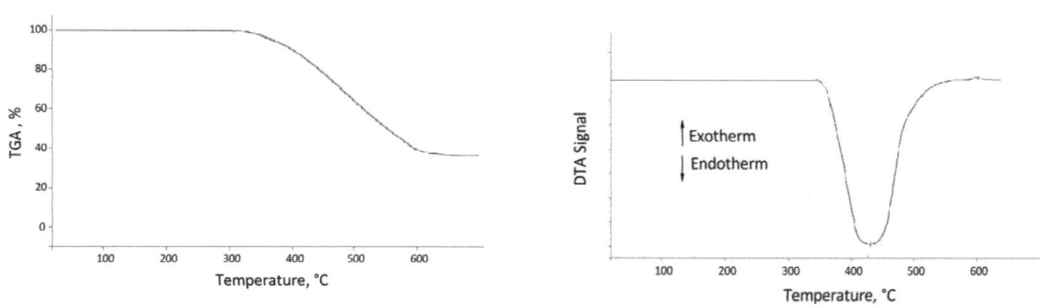

a b

Figure 6. TGA curve (a) and DTA curve (b) of substance synthesized under hydrothermal conditions.

Figure 6b shows the DTA curve of the synthesized substance. The DTA curve in Figure 6b is characterized by a long exothermic effect with a pronounced maximum at 420 ° C. The peak at a given temperature corresponds to the maximum weight loss of the test substance. Thus, the resulting substance has high thermal stability. The obtained results on the thermal analysis of the obtained substance are consistent with published data on the thermal stability of the polyhedral oligomeric silsesquioxane in [18].

4. Conclusion

The hydrolytic polycondensation of cyclohexyl trichlorosilane was synthesized under hydrothermal conditions. It was found that the compound obtained is a nanofibre of a polyhedral oligomeric silsesquioxane. The diameter of the obtained nanofibers is 50 ... 150 nm, and the length is from 200 nm to 30 μm.

It was established that the obtained POSS nanofibers have a pronounced crystalline structure. The characteristic crystal features of synthesized compound show four main reflections at 2θ=9°, 12°, 12.5°, 19° corresponding to the lattice spacing of 10.4 Å, 7.9 Å, 6.9 Å, and 4.3 Å, respectively.

The resulting nanofibers are thermostable to a temperature of 300 ° C. The temperature at which mass loss ends is 580 ° C. After a temperature of 580 ° C, the residual mass of POSS does not change and remains 40%.

Acknowledgments
The work was supported by a project of the Russian Science Foundation (19-79-10064).

References

[1] Dulay M T, Zaman N, Jaramillo D, Mody A C and Zare R N 2018 *J. Carbon Res.* **4** 29
[2] Tatarinova E, Vasilenko N and Muzafarov A 2017 *Molecules* **22** 1768
[3] Pavlenko V I and Cherkashina N I 2018 IOP Conf. Series: Mat. Sci. Eng. **450** 022017
[4] Wu B, Zhang Y, Yang D, Yang Y, Yu Q, Che L and Liu J 2019 *Polymers* **11** 1013
[5] Duo S, Chang Y C, Liu T and Zhang H 2013 *Phys. Procedia* **50** 337-342
[6] Song M M, Duo S W and Liu T Z 2011 *Adv. Mater. Res.* **239-242** 1368-1371
[7] Cherkashina N I, Pavlenko V I, Sidelnikov R V and Besedin P V 2018 *Bulletin of the Belgorod State Technological University named after V.G. Shukhov* **10** 110-118
[8] Minton T K, Wright M E, Tomczak S J, Marquez S A, Shen L, Brunsvold A L, Cooper R, Zhang J, Vij V, Guenthner A J and Petteys B J 2012 *ACS Appl. Mater. Interfaces.* **4** 492-502
[9] Pavlenko V I, Novikov L S, Bondarenko G G, Chernik V N, Gaidar A I, Edamenko O D and Cherkashina N I 2013 *Inorg. Mater. Appl. Res.* **4** 169–173
[10] Wu J and Mather P T 2009 POSS *Polym. Rev.* **49** 25-63
[11] Zhou H,Ye Q and Xu J 2017 *Mater. Chem. Front.* **1** 212-230
[12] Sierke J K and Ellis A V 2015 *Inorg. Chem. Commun.* **60** 41-43
[13] Rahimifard M, Mohammadi Ziarani G, Badiei A and Yazdian F 2017 *J. Inorg. Organomet. Polym.* **27** 1037–1044
[14] Gunkara O T 2019 *Phosphorus, Sulfur, and Silicon and the Related Elements* **194** 909-916
[15] Brown J F and Vogt L H 1965 *J. Am. Chem. Soc.* **87** 4313
[16] Feher J F, Newman D A and Walzer J F 1989 *J. Am. Chem. Soc.* **111** 1741
[17] Bach L G, Islam Md R, Nga T T, Binh M T, Hong S-S, Gal Y S and Lim K T 2013 *J. Nanosci. Nanotechno.* **13** 1970-1973
[18] Wei S, Meng L, Liu W, Guo S and Zhang X 2019 *Mat. Res.* **22** e20180278

Prediction model of the efficacy and the implementation time of transportation intelligent systems

A. L. Chupin[1], Zh. S. Chupina[1], N. N. Morozova[1], T. M. Vorotyntseva [1] and E. V. Levinskay[2]

[1]RUDN University, 117198, 6 Miklukho-Maklaya street, Moscow, Russia
[2]RANEPA, 119571, 82 Vernadsky Avenue, Moscow, Russia

E-mail: vSe.1@mail.ru, chupin_al@pfur.ru

Abstract. The article deals with a formalized model of the efficiency forecast based on the analysis of different models of the efficiency forecast and the implementation time of intelligent systems. This model is associated with mathematical counting of the future value of the process. This model is based on the representation of renewal and modernization in the form of a stochastic process. The mathematically-oriented understanding of this process is based on the sum of random variables, the result of which reflects the occurrence of events that lead to the failure of the implementation time of intelligent systems and the failures during modernization.
Keywords: IS, equipment upgrade, transport, timing, forecast, system.

1. Introduction

Nowadays, the Russian Federation is moving towards the entire development like all the states-members of the Council of Europe, but without the powerful development of information technology, the movement is problematic. However, currently, according to the Federal State Statistics Service, the dependence of technologies is more than 90%. This fact is associated with that the 21st century will mark the age of economic pharmacy, the introduction of information technologies into all spheres of life. The percentage of enterprises that develop and implement technological innovations does not exceed 10%. So, at present and in the foreseeable future, the most important task of the Russian economy is the development of information technologies, specifically in the transport infrastructure. The initial step to develop technologies is to modernise relevant industrial enterprises. In market conditions, the development is aimed at ensuring the ability to respond quickly to changes in the circumstances of the product markets. These changes are directly supported by equipment upgrades and technological process improvements. The changes are associated with the span time, the materials and the conditions which are typical for industrial enterprises of the labour resources deficit. So, an important element to justify the changes is the indicator of the efficiency of implementation and determination of the end of the process which will lead the enterprise to release of updating and modernization. The unpredictability of this process will be a characteristic feature. It will be manifested in disruptions in the delivery time of equipment, the modernization failures and the resulting changes in both time and cost parameters of the studied processes. The indicator of events, their number, leading to failures of terms, of course, is not known. In this regard, the formation and development of management decisions in the field of technology implementation of the industrial equipment, the determination of its cost and time parameters will have to be based on the prediction.

IOP Conf. Series: Materials Science and Engineering **828** (2020) 012006 doi:10.1088/1757-899X/828/1/012006

The effective tool to predict is mathematical models. In the types of mathematical models, the possibility to observe this forecast is provided by a stochastic model, and the result is the sum of random variables, the result of which reflects the occurrence of events that lead to the failure of the implementation of intelligent transport systems and failures of modernization of production processes. The model and its processes are based on theoretical approaches, which are substantiated by the authors B.V. Gnedenko and A.N. Kolmogorov in their work of "Limit distributions for the sums of independent random variables". Also, in this work and others, the models assuming the presence of limit distributions for the sums of independent random variables are considered. An important part of an accurate forecast to manage the modernization of an industrial enterprise proves the need to take into account the limitation of the random number of these destructive events. The study is aimed to develop a model that provides a forecast of the efficiency and timing of the development of intelligent systems.

2. The formalized prediction model

In general, the problem of the efficiency and timing prediction of intelligent transport systems development comprises follows. Regarding an enterprise whose interest is the development of information technology, the development of equipment and process modernization is the main goal of any production. During the process, some destructive events may occur, which lead to delays and failures of the modernization process. The result of each event is increasing the time to complete the whole upgrade process and the efficiency upgrading by a value of G and by a value of H. From this formalization model, the values of X and Y are reasonable to consider as random variables. The task of this forecast is to determine the probability that the process of information technologies implementing (intelligent systems) and the production process modernization of the enterprise will be completed within the scheduled term (scheduled time T) and the probability that the effectiveness of the implementation will not exceed the set value S. The condition will be considered as T and S exceed the ideal time conditions T0 (T zero) and the effectiveness of the implementationS_0 necessary for the modernization of the enterprise. Thus, under the conditions of the problem, it is required to determine the probability

$$F_1(\alpha) = P(\alpha \le T - T_0) \quad (1)$$

$$F_2(S) = P(\beta \le S - S_0), \quad (2)$$

where α is a random variable equal to the sum of n random variables G;

β is a random variable equal to the sum of n random variables H.

Determining the probabilities (1) and (2) includes evaluating the distribution functions of random variables n, G, H. Usually, some enterprises have a limited amount of statistical information. Therefore, only the mathematical expectation μ of the time interval between the destructive events leading to an increase in the completion time of the modernization process of an industrial enterprise, as well as the mathematical expectations $\hat{G}$, $\hat{H}$ of the values G and H caused by these events of production functions. In such an information situation, based on the principle of maximum entropy (how much we do not know), the time interval between successive events and the values of G, H should be assumed to be distributed over exponential distributions (exponential). The number of n destructive events is distributed according to Poisson's law

$$P(r{=}n) = \frac{v^n}{n!}e^{-v}, \quad (3)$$

where P(r=n) is the probability of that the number r of destructive events that took place during the modernization of the enterprise process is equal to n;

$v = \frac{1}{\mu}$ is the parameter of the Poisson distribution.

The density function of the distribution of a random variable G has the form:

$$f_1(g) = \lambda_1 e^{-\lambda_1 g} \quad , g \in \{0, 1\} \quad (4)$$

where $\lambda_1 = 1\frac{1}{\hat{G}}$ is a parameter of exponential distribution of a random variable of G increase of term of the introduction of intelligent systems owing to the approach of the next destructive event.

The density function of the distribution of a random variable H has the form

$$f_2(h) = ƛ_2 e^{-\lambda_1 h} \quad , h \in\{0, 1\} \tag{5}$$

where $ƛ_2 = 1\frac{1}{\hat{H}}$ is the exponential distribution parameter of the random variable Y.

The efficiency of the process of modernization development of the enterprise due to the occurrence of the next destructive event.

Taking into account the relations, (3), (5), the random variables α and β are the sums of the Poisson number of exponentially distributed random variables G and H. The apparatus of characteristic functions comprises the following peculiarities.

For a random variable α the characteristic function has the form:

$$\varphi_\alpha(t) = \int_{-\infty}^{\infty} e^{it\alpha} dF_1(\alpha) \tag{6}$$

For a random variable β the characteristic function has the form:

$$\varphi_\beta(t) = \int_{-\infty}^{\infty} e^{it\beta} dF_2(\beta) \tag{7}$$

In the relations (6), (7) t is a real variable, i – an imaginary unit.

Since the very purpose of using characteristic functions is predetermined by the fact that they have the properties of uniqueness and multiplicativity. The multiplicativity property is that the characteristic function of the sum of independent random variables is equal to the product of their characteristic functions. The uniqueness property determines that the characteristic function$\varphi_z(t)$ некоторой случайной величины Z is absolutely intertwined (integrable) on the entire real axis, and the function F(z) has a continuous derivativef(z) (distribution density), then there are relations (inversion formula):

$$f(z) = \frac{1}{2\pi}\int_{-\infty}^{\infty} e^{itz}\varphi_z(t)dt . \tag{8}$$

The relation (8) indicates a one-to-one correspondence between the characteristic function of a random variable Z and the density function of its distribution.

The characteristic function of the exponential distribution (4) has the form:

$$\varphi_1(t) = (1 - it\lambda_1^{-1})^{-1}. \tag{9}$$

Given (3), (6), the characteristic distribution function$\varphi_\alpha(t)$ (1) is:

$$\varphi_\alpha(t) = \sum_{n=0}^{\infty} \frac{v^n e^{-v}}{n!} (1 - it\lambda_1^{-1})^{-n} \quad (10).$$

Using the inversion formula, we determine the density function of the random variable distribution α:

$$f(\alpha) == \frac{1}{2\pi}\int_{-\infty}^{\infty} e^{itz}\varphi_\alpha(t)dt = \frac{1}{2\pi}\int_{-\infty}^{\infty} e^{itz} \sum_{n=0}^{\infty}\frac{v^n e^{-v}}{n!} (1 - it\lambda_1^{-1})^{-n}dt \quad (11).$$

After replacing the summation and integration order in (11), we can obtain:

$$f(\alpha) = \frac{1}{2\pi}\sum_{n=0}^{\infty}\frac{v^n e^{-v}}{n!}\int_{-\infty}^{\infty} e^{itz}(1 - it\lambda_1^{-1})^{-n}dt \quad (12).$$

From (12), using the integral from the table of integrals, sums, series and studies of authors I.S. Gradstein, I.M. Ryzhik, we obtain:

$$f(\alpha) = \sum_{n=0}^{\infty}\frac{v^n e^{-v}}{n!}\frac{(\lambda_1 a)^{n-1}e^{-\lambda_1 a}}{\lambda_1\Gamma(n)} \quad (13),$$

where $\Gamma(n)$ is the gamma function.

From (13) it follows that the probability (1) that the process of implementation and development of intelligent systems will be completed in the scheduled time and will be equal to:

$$f_1(\alpha) = P(\alpha \le T - T_0) = \sum_{n=0}^{\infty}\frac{v^n e^{-v}}{n!}\int_0^{T-T_0}\frac{(\lambda_1 a)^{n-1}e^{-\lambda_1 a}}{\lambda_1\Gamma(n)}dt \quad (14).$$

The characteristic function of the exponential distribution (5) has the form:

$$\varphi_2(t) = (1 - it\lambda_2^{-1})^{-n} \quad (15).$$

Using the inversion formula, we can determine the density function of the distribution of a random variable β:

$$f(\beta) == \frac{1}{2\pi}\int_{-\infty}^{\infty} e^{it\beta}\varphi_\beta(t)dt = \frac{1}{2\pi}\int_{-\infty}^{\infty} e^{it\beta}\sum_{n=0}^{\infty}\frac{v^n e^{-v}}{n!}(1 - it\lambda_2^{-1})^{-n}dt \quad (16).$$

IOP Conf. Series: Materials Science and Engineering **828** (2020) 012006 doi:10.1088/1757-899X/828/1/012006

Further, by performing transformations similar to (12), (13), we obtain the density function of the distribution of the random variable β in the form:

$$f(\beta) = \sum_{n=0}^{\infty} \frac{v^n e^{-v}}{n!} \frac{(\lambda_1 \beta)^{n-1} e^{-\lambda_1 \beta}}{\lambda_1 \Gamma(n)} \quad (17).$$

It follows from (18) that the probability (2) that the efficiency of the implementation of intelligent systems will not exceed the set value S and it is equal to:

$$f_2(\beta) = P(\beta \leq S - S_0) = \sum_{n=0}^{\infty} \frac{v^n e^{-v}}{n!} \int_0^{S - S_0} \frac{(\lambda_2 a)^{n-1} e^{-\lambda_2 \beta}}{\lambda_2 \Gamma(n)} d\alpha \quad (18).$$

3. Conclusions

The main elements of managerial decisions in the model development which provides the prediction of the effectiveness and timing of the intelligent systems development are the assessment of the cost of implementing and developing intelligent systems [10], [11]. At the early stages to formulate these decisions, the determination of the cost and time parameters of the update and implementation of the modernization process should be based on forecasting. This prediction is carried out in conditions of limited statistical information. Well-known methods do not fully take into account the specifics of the studied processes, so they can lead to enormous errors in forecasting efficiency. The relations (14) and (18) obtained in the present work allow estimating more adequately the introduction and development of intelligent systems.

References

[1] Kremer N S 2004 *The Theory of Probability and Mathematical Statistics: The teaching guide.* (M.) pp. 37-39

[2] Kochetkov E S 2001 *The Theory of Probability and Mathematical Statistics: The teaching guide.* (M.) pp. 78-79

[3] Petrochenko A A 2019 *Journal of Management Studies* **5(2)** 57-65

[4] Mizintseva M F, Sardarian A R and Petrochenko A A 2018 *2nd International Conference on Social, Economic and Academic Leadership (ICSEAL 2018)* **217** 231-237. DOI: 10.2991/icseal-18.2018.33

[5] Chupin A L, Medvedev Yu V and Mizintseva M F 2017 *Journal of Economy and Entrepreneurship* **11** 12-1

[6] Chupin A L 2017 *Technical Sciences: Problems and Solutions: the Materials of the VI International Scientific and Practical Conference "Technical Sciences: Problems and Solutions"* **6(5)** (M., Ed. "Internauka") pp. 64-69

[7] Chupin A L, Medvedev Yu V and Lemesheva Zh S 2019 *Intelligent Transport Systems for Sustainable Mobility: Monograph* 154 p.

[8] Emirov N D and Emirova A E 2017 *Vestnik. SPSUTD. Ser.3* **1** 37-43

[9] Anisimov E G, Anisimov V G, Bogoeva E M, Lemesheva Zh S and Saurenko T N 2014 *Theory and Practice of Risk-Based Approach in the Performance of Control (Supervisory) Functions of State Bodies: the Materials of the Interdepartmental Scientific and Practical Conference of the Russian Customs Academy* (Federal Customs Service, Russian Customs Academy) pp. 25-30.

[10] Ergunova O, Lizunkov V, Blaginin, V, Politsinskaya E, and Shaykina O 2018 *International Journal of Energy Economics and Policy* **8(1)** 122-127

[11] Ergunova O T, Lizunkov V G, Malushko E Yu, Marchuk V I, and Ignatenko A Yu 2017 *OP Conference Series: Materials Science and Engineering* **177(1)** 012046. DOI: 10.1088/1757-899X/177/1/012046

IOP Conf. Series: Materials Science and Engineering **828** (2020) 012007 doi:10.1088/1757-899X/828/1/012007

Stocks of organic substance in post-agrogenic lands in Leningrad region

D A Danilov [1,2], S V Navalikhin [1] and A A Vayman [2]

[1] Saint-Petersburg State Forest Technical University named after S.M. Kirov, 5 U, Institutskiy per., Saint-Petersburg, 194021, Russia

[2] Leningrad research institute for agricultural science «Belogorka», 1, Institutskiy st., Belogorka, 188338, Russia

E-mail: stown200@mail.ru

Abstract. We have analysed factors affecting the formation of organic matter stocks in sod-podzolic soils and post-agrogenic lands in the south-western part of the Leningrad region. There are many factors influencing the process of humus formation, which does not allow to reliably simulate the process of transformation of organic residues and the accumulation of humus in areas previously used for agricultural production. There is a connection in the differences in the specific density of the soil and the unequal content of humus in the genetic horizons of fallow soil compared to the arable layer of the field, where these horizons are mixed into a homogeneous mass. The data obtained show a high correlation between the humus content in the former arable horizon of soil with a clay fraction and pHKCI. At different successional stages of accumulation of organic matter in post-agrogenic lands is not the same, as the formation of different species composition of vegetation makes adjustments to this process. When determining humus stock in post-agrogenic lands, an error can reach to 30%. The main reason is the differences in the specific density of the soil and the unequal content of humus in the genetic horizons of fallow soil compared to the arable layer of the field, where these horizons are mixed into a homogeneous mass. The conclusion was made about different methods of soil sampling from arable lands and from post-agrogenic lands.

1. Introduction

Land use has a major impact on soil carbon and nitrogen dynamics. Dominant paradigms argue that arable land cultivation results in long - term and large losses of soil carbon compared to fallow fields and forests. It is assumed that tillage, mainly the conversion of pastures to arable land, leads to significant losses of organic carbon. Agricultural land after the cessation of their active cultivation is resumed by natural vegetation, and the soil again accumulates organic carbon. [1, p. 31].

This accumulation process reverses the effects responsible for the loss of organic carbon in the soil from the moment it is deprived of perennial vegetation. There are significant differences in the time and rate of carbon accumulation in the soil, due to the productivity of regenerating vegetation, physical and biological conditions in the soil and the past history of soil enrichment with organic matter and physical effects on it. The maximum accumulation of organic matter during the initial stage of growth perennial vegetation, although significant, is usually much less than in the previous history of organic carbon entering the soil [1]. Average accumulation rates are the same for forest lands or former pastures and amount to 33.8 g C m-2 y -1 and 33.2 g C m2 y -1 respectively. These rates of organic carbon accumulation in the soil, combined with a small amount of occupied area, are insufficient to account for a significant proportion of the missing carbon in the global carbon cycle that accumulates in the soils of former agricultural lands [3, 4, 5].

Previous studies of the carbon cycle have recognized the importance of land-use changes, logging and natural regeneration of tree species, as well as associated variations in the carbon content of above-ground vegetation biomass. Radiocarbon dating of the upper humus horizons of 10 cm thickness of the zonal series of Eurasian soils from the red soil of the humid subtropics to the Arctic tundra gave characteristic periods of organic matter formation in the range of 60-350 years. Thus, the dominant part of the organic matter of the upper fertile soil layers is deposited there for a rather short time. Only removed from the surface or buried organic substrates can be preserved for centuries and even millennia, as evidenced by radiocarbon dating of individual fractions of soil humus. Natural soils quickly lose organic carbon when they are involved in agricultural circulation. Revealing the role and fate of the organic matter of the soil was a critical issue for scientists. There is much debate about definitions of soil organic matter among researchers. It is believed that the amount of organic carbon contained in a particular soil is a function of the balance between the rate of deposition of plant residues in the soil and the rate of mineralization of residual carbon of soil biota. However, the organic matter in the soil is always in a very dynamic condition, where biological and plant remains are constantly being transformed. The mechanisms of biological stabilization of organic carbon in the soil depend on the decomposition of the mineral phase of the soil and the chemical structure of organic residues added to the soil. A fairly accurate integral indicator of organic matter quality, on which the intensity of its decomposition depends, is the ratio of carbon to nitrogen. Plant residues with a wide ratio of carbon to nitrogen do not provide sufficient nitrogen for the metabolism of microorganisms at their high activity. Numerous studies have shown differences in the change of organic matter at different stages of soil restoration to the natural state after the cessation of active agricultural use.

The regional climate has the greatest impact on carbon storage. It is noted that primary productivity, biomass, and soil carbon content are associated with warm, humid climates and are limited in warm, dry climates, while warm temperatures are also associated with higher rates of decomposition and reduced soil carbon storage [6]. Studies conducted in Sweden showed that former arable soils had higher pH, higher nitrate concentrations and higher soil density compared to natural soils under the forest. However, there was lower organic matter and lower concentrations of ammonium in the topsoil (0-5 cm) and these differences decreased with soil depth. Previous land use is a major factor contributing to changes in Corg in the soil after reforestation. On the former arable lands, the change of Corg. differed between the soil horizons and was significant up to 20 % in the soil layer 0-10 cm. Afforestation of former meadows had a small negative effect, as indicated by a slight change in Corg. after land use change in the region. Studies conducted in the Czech Republic on land with restored woody vegetation showed that the content of organic matter in the upper layer of the soil mineral horizon, compared with arable soils, increased significantly. Soil development has resulted in an increase in total carbon content from 1.6-1.8%to 15%, with a range of 4.2 to 15.2% [7]. The results of fallow lands study in Latvia confirm the theory that after afforestation in the long term, the soil accumulates organic carbon (Corg.). During the first 30 years, the content of Corg. in the upper soil layer did not increase and ranged from 42 to 43 t/ha -1 [8]. A more rapid increase in the content of Corg was observed where the age of forest plantations exceeded 30 years and the average content of Corg in plantations aged 31-60 years was 51.3 t/ha -1.

A study conducted in Lithuania showed that the restoration of pine on nutrient-poor sandy soils could lead to an increase in the stocks of Corg. in the soil for a short period compared to the deposits of arable land without natural renewal [9]. The content of organic carbon on fallow arable land (3.81 kg cm-2) was 1.7 times less than in pine plantations (6.52 kg cm-2). It was concluded that the positive effect of reforestation on soil Corg. stocks is more pronounced in soils of arable land than in pastures or natural pastures. It is assumed that broad-leaved trees have a greater ability to restore Corg. in the soil than conifers. Restoration of woody vegetation in the boreal climatic zone leads to small losses of

IOP Conf. Series: Materials Science and Engineering **828** (2020) 012007 doi:10.1088/1757-899X/828/1/012007

organic soil carbon compared to other climatic zones, probably because trees grow slower in these conditions, although this does not exclude gains over time after conversion.

In the boreal zone of Russia the most significant transformation of soil properties in the conditions of changing vegetation was revealed for poor sandy soils of the southern taiga [10]. The degree of change in the content and reserves of organic carbon, enrichment of humus with nitrogen and acidity in the soil layer from 0 to 20 cm during postagrogenic evolution decreases from North to South.

Bringing the arable lands out of agricultural use is caused, as a rule, by economic reasons, and it is explained by a lower level of expected income over expenses on growing crops. Such a situation emerges due to a decrease in soil fertility as a result of its long-term exploitation ("soil fatigue"), caused by a number of reasons: permanent agriculture (monoculture), insufficient efficiency of crop rotation and green manuring applied, limited introduction of organic fertilizers, that is, everything that ensures the supply of organic residues in the soil, the decomposition of which adds to the stock of humus [11]. Studies conducted show that arable lands are undergoing the most significant deterioration of natural soil fertility. In the first period after the cessation of land use, it goes through the stage of eluvial-gley process. It consists in the destruction of the humus horizon and compaction of the remaining soil. Erosion processes occur, in the form of flushing the upper soil horizon to the bottom of the hills and ridges where swamping processes begin. In addition, from top to bottom, the thickness of the remaining part of humus decreases due to cramming and indenting, soil density and humidity grow.

In the North-West of Russia, after 6-7 years after the withdrawal of the site from economic use, the restoration of the morphological appearance of disturbed soils can begin: low-power sod (0.5 – 1 cm) appears in the watershed areas and up to 4 cm in the lower part of the slope. In various landscape positions, the restoration of the humus horizon begins: in the upland areas, it passes through the sod process; in the lowered areas, through waterlogging. Full restoration of physical and water-physical properties of the soil in all elements of the relief takes place in 17 years. In landscape conditions of hilly-ridge relief the provision of sod-podzolic soils of light mechanical composition with humus and elements of mineral nutrition reaches the natural level not earlier than 16-17 years [12].

2. Methods & objects

The objective of this study was to determine the amount of accumulation of humus in the soil while maintaining the natural conditions of the soil-forming process without any human intervention. The objects of the study were unused agricultural land and adjacent sparse deciduous stands, vacant lots, glades on which soil samples were taken. To perform this task, samples of the most common sod-podzolic soils of the South-West of the Leningrad region in Gatchina, Luzhsky and Tosnensky administrative districts were collected and analyzed (59.5407,30.8777; 58.7470,29.8665; 59.5652, 30.1282).

The mixed soil sample was composed of at least 15 individual samples. The sample weight of 250-300 grams. In the selected soil samples, the content of organic matter, acid-base indicators were determined, the granulometric composition was specified. The humus content was determined by the Tyurin method in the Nikitin modification. Determination of granulometric composition is carried out with a Pipette of N. A. Kachinsky. Soil pH was measured by potentiometric method.

3. Results & Discussion

The results of the study showed that the main factor in the evolutionary development of the soil-forming process is the vegetation cover. In addition, the physical conditions of soil formation affect the accumulation of humus, the relief, soil-forming rock, water-air properties of the soil, heat and humidity supply.

The processes of accumulation, movement and decomposition of humus are the basis of soil formation. Each plant formation corresponds to its own type of soil humus, characterized by certain limits of its quantitative accumulation and qualitative composition. The quantitative accumulation of humus is determined by the size of the annual intake into the soil of both ground litter and the underground part (roots) of the vegetation cover. The nature of biological and biochemical transformations of these residues depends on the composition of microorganisms, the direction of microbiological processes, and, finally, on the intensity of decomposition of the humus itself [3]. Multifactorial dependence of humus formation does not allow to reliably simulate the process of transformation of organic residues and accumulation of humus in areas previously used for agricultural production. For this reason, only a simplified version can be presented: the relationship of humus accumulation with the most stable features of parent rock – clay fraction content in parent rock and soil acidity [3]. For drained soils of the meadow-forest sod-podzolic zone, this dependence can be expressed by the formula:

$$C_{hum} = \lg (\% \text{ Cl.})*pH_{(KCI)}* 5.8$$

where:

C_{hum} – mass of carbon (humus) of organic matter in the 0-20 cm layer of soil in tons ha^{-1};

lg (% Cl.) – the logarithm of percentage (%) of the content of physical clay (particles d <0.01 mm) in the soil forming rock pH (KCI) – exchange (salt) acidity 5.8 – conversion factor of soil organic matter weight into soil humus carbon (Chum) expressed in tons ha^{-1}.

Humus expressed by means of carbon weight (C_{hum}) is a commonly accepted format used in calculating the accumulation, transformation and movement of organic matter carbon in the biosphere. The results of C_{hum} determination in the 0-20 cm layer of soil, depending on the clay fraction content and soil acidity, are given in table 1. The most widely used indicator of soil fertility is the percentage of humus in the 0-20 cm layer of soil. This indicator can be calculated by weight of C_{hum} in the 0-20 cm layer of soil expressed in tons/ha (table 1):

$$\text{humus } \% = \frac{C_{hum} \text{ (in the 0-20 cm layer)}}{20*0.8*\text{spec. density}}$$

where

C_{hum} – soil carbon mass of humus in the 0-20 cm layer of soil in tons/ha (table 1)

20 – 0-20 cm layer of soil

0.58 – conversion factor of C_{hum} in the humus content in the soil expressed in % (carbon (C) content in humus is 58 %)spec. density. – average specific density of soil (g/cm^3) for the whole 0-20 cm layer:

1.2 – of sandy loam soil

1.15 – light loamy soil

1.1 – loamy soil.

The results of the calculation of % humus in a layer of 0-20cm soil are given in table 2.

For comparative characteristics, indicators of humus stocks in the upper 20 cm layer of soil given in table 3 are used.

Table 1. Stock of humus carbon C_{hum}, (tons ha^{-1}) in the 20 cm soil layer depending on the clay fraction content (%) and acidity (pH_{KCI}) (%)

Clay, %	pH_{KCI}							
	3.6	3.8	4.0	4.5	5.0	5.5	6.0	6.5
15	24.6	26.0	28.2	30.8	34.2	37.6	41	44.5
20	27.1	28.6	30.2	33.9	37.9	41.4	45.2	49.0
25	29.2	30.9	32.5	36.6	40.6	44.6	48.7	52.8
30	30.9	32.6	34.3	38.6	42.9	47.2	51.5	55.8
35	32.2	33.9	35.8	40.2	44.6	49.1	53.6	58.1
40	33.4	35.3	37.1	41.8	46.4	51.0	55.7	60.3

Table 2. Humus content in the 20 cm soil layer depending on the clay fraction content (Cl,%) and acidity (pH_{KCl}), %

Clay, %	Values of pH_{KCl}							
	3.6	3.8	4.0	4.5	5.0	5.5	6.0	6.5
15	1.8	1.9	2.0	2.2	2.5	2.7	2.9	3.2
20	1.9	2.1	2.2	2.4	2.7	3.0	3.2	3.5
25	2.2	2.3	2.4	2.7	3.0	3.3	3.7	4.0
30	2.3	2.4	2.6	2.9	3.2	3.5	3.9	4.2
35	2.5	2.7	2.8	3.1	3.5	3.8	4.2	4.6
40	2.6	2.8	2.9	3.3	3.6	4.0	4.4	4.7

Table 3. Stock of humus in the 20 cm soil layer depending on the clay fraction content (Cl, %) and acidity (pH_{KCl}), tons ha^{-1}

Clay, %	Values of pH_{KCl}							
	3.6	3.8	4.0	4.5	5.0	5.5	6.0	6.5
15	42	45	47	53	59	65	70	77
20	46	50	52	58	65	71	78	85
25	50	54	56	63	70	77	83	91
30	54	56	59	66	74	82	89	96
35	56	58	62	71	77	85	93	100
40	58	61	64	72	80	88	96	104

The data obtained show a high correlation between the content of humus in the former arable soil horizon with clay fraction (R=0.87-0.93) and pH_{KCl} (R=0.85-0.90) in the former surface horizon. These humus reserves in the soil belong to non-forest lands with a formed growth cover and a well-defined morphology of the soil profile. In other words, the studied lands have an ecosystem close to the meadow phytocenosis, which will be approached by the developing and changing biocenosis of arable lands that have come out of agricultural use. Errors in determining the stock of humus do not exceed 10-15%, if samples are taken from the arable layer of a functioning field. They are explained by various reasons. In field work it can be soil sampling from fields of different soil moisture regime, or different granulometric composition. In laboratory studies – these are the conditions for the preparation of soil samples (drying, grinding, sifting storage), failure to meet deadlines for using the prepared reagents Before starting soil sampling on fallow lands, it is necessary to conduct a survey and make measurements of the power of genetic horizons in each soil difference identified during the survey. On the lands lying in the fallow for a long time (over 20 years), soil sampling should be done according to genetic horizons if they are well expressed, and with their indeterminacy – by the 10 cm layers. If this condition is not met, the calculated stocks may significantly differ from the actual accumulation. Thus, for example, the upper 20 cm layer of arable soil has a density of 1.1 and humus stock of 65 tons ha^{-1}. The soil of post-agrogenic lands in fallow over 20 years old, having no clearly expressed genetic horizons, has the same level of fertility. In the upper 10 cm layer of the fallow soil, humus stock is 50 tons ha^{-1}, while in the 10-20 cm layer – 15 tons ha^{-1}; specific density of the upper layer – 0.8, and of the 10-20 cm layer – 1.2. The total stock of humus in the 0-20 cm layer will be 65 tons ha^{-1}. However, if we limit ourselves to the sampling from only one layer, and make recalculation for the whole 20 cm layer, then data scattering on the stock of humus in the 0-20cm layer will be from 30 to 100 tons ha^{-1}, which distorts the true content.

IOP Conf. Series: Materials Science and Engineering **828** (2020) 012007 doi:10.1088/1757-899X/828/1/012007

4. Conclusion

The accumulation of organic matter on postagrogenic lands occurs differently at different successional stages since the formation of different vegetation cover makes adjustments to this process.

Errors in determining the stock of humus on postagrogenic lands can exceed 30%. The main reason is the differences in the specific density of the soil and the unequal content of humus in the genetic horizons of fallow soil compared to the arable layer of the field, where these horizons are mixed into a homogeneous mass. From this it follows that the methodology for the selection of soil samples from arable land and from post-agrogenic lands should be different.

References

[1] Baeva Yu I, Kurganova I N, Lopez de Gerenyu V O and Telesnina V M 2017 *PEMME* **XXVIII (2)** 29-39 DOI: 10.21513/0207-2564-2017-2-27-39

[2] Nyawira S S, Nabel J E M S, Don A, Brovkin V and Pongratz J 2016 *Biogeosciences* **13** 5661-75

[3] Seppelt R and Voinov A 2003 *Ecological modelling* **168(6)** 217-31

[4] Schumacher J and Gensior A 2011 *Glob. Change Biol.* **17** 2415–27

[5] Fernandez-Martinez M, Vicca S, Janssens I A, Sardans J, Luyssaert S, Campioli M, Chapin F S III, Ciais P, Malhi Y, Obersteiner M, Papale D, Piao S L, Reichstein M, Rodà F and Peñuelas J 2014 *Nature Climate Change* **4(6)** 471–6

[6] Guo L B and Gifford R M 2002 *Global Change Biology* **4(8)** 345–60

[7] Holubík O, Podrázský V, Vopravil J, Khel T and Remeš J 2014 *Soil & Water Res* **9(4)**

[8] Ruskule A, Nikodemus O, Kasparinskis R et al 2016 *New Forests* **47** 587

[9] Armolaitis K, Aleinikovienë J, Baniûnienë A, Lubytë J and Bëkaitë V 2007 *Baltic Forestry* **13(2)** 169178

[10] Ryzhova I M, Erohova A A, Podvezennaja M A 2014 *Soil science* **12** 1426-35

[11] Golubeva L V 2015 *Forest-Ecological Transformation of Post-Agrogenic Lands on Carbonate Sediments in the Sub-Zone of the Middle Taiga of the Arkhangelsk Region: the Dissertation of the Candidate of Agricultural Sciences* (Arkhangelsk) p 160

[12] Litvinovich A V, Pavlova O Yu and Chernov D V 2002 *Report of the Russian Academy of Agricultural Sciences* **6** 26-28

[13] Bankin M P, Bankina T A and Korobeynikova L P 2005 *Physical and Chemical Methods in Agrochemistry and Soil Biology* (St. Petersburg: Publishing house of St. Petersburg State)

Modern effective technologies applied for constructing external walls

O A Gnezdilova [1], V A Kudryavtseva [1] and N Y Terenteva [1]

[1] Irkutsk State Transport University, Chernyshevskogo str., Irkutsk, 664074, Russia

E-mail: kudravera@mail.ru

Abstract. The development of the construction industry should be based on resource and energy-saving principles. The energy efficiency issue is being discussed at various levels, including the government one. The issue of energy saving technologies and structures is crucial for the construction industry, as they can reduce the amount of energy consumed and save money. One of the main components of the energy efficiency of buildings is thermal protection. In the construction industry, one of the simplest and most rational ways to save energy is to reduce heat losses through the building envelopes. One of the options for increasing the energy efficiency of building envelopes is effective heaters. It is difficult to choose materials and structural designs of supporting structures since they determine all other structural elements of the building. For Russia whose regions are located outside the climate control zone, energy efficiency issues are crucial. Currently, energy conservation is one of the government priorities. This is due to the shortage of basic energy resources, increasing energy production costs and global environmental problems.

1. Introduction

The construction industry is one of the most important sectors of the national economy ensuring the constitutional rights of citizens and integrating many related economic sectors.

The current level of scientific and technological development requires resource-saving technologies when performing construction works. The construction industry is one of the energy- and resource-intensive industries. However, it is necessary to reduce the cost of construction material production [1].

In the construction and housing sectors, the principles of sustainable development were declared by the International Construction Council - CIB (Conceal International du Batiment) created in 1953 under the auspices of the UN. The concept is based on the limited consumption of natural resources, the efficient use of the potential of related industries, and energy-efficient solutions. This approach is the main one for the national economic programs of a large number of countries, including Russia.

Russia with its harsh climate is one of the largest energy producers and consumers. An important task of all the national long-term development programs focused on energy saving is to ensure the most effective thermal insulation of building envelopes.

2. Materials and methods

Development of the construction industry requires new materials, technologies and approaches. In the construction industry, global trends are resource efficiency and a reduced construction period. However, in Russia, due to its climatic features (temperature extremes, permafrost, seismicity and wind), it is also necessary to build very warm wall structures.

In addition to the thermal insulation rules developed in the Soviet era, much attention is paid to sustainable development.

The solution that can reduce materials consumption and improve energy intensity is Poroplast CF02 as an insulant for multilayer building envelopes [2, 3]. Liquid urea-formaldehyde foam has been applied in foreign countries for a long time [4, 5]. In Russia, it is produced using reliable production technologies [3].

Poroplast CF02 is produced from available inexpensive materials on-location. The main physical and mechanical and operational properties of the composite are presented in Table 1.

Table 1. The main physical, mechanical and operational properties of the composite "Poroplast CF02".

Name of the parameter	Measurement unit	Standard
Density, within	kg/m3	8-30
Compressive strength at 10% linear strain, not less than	MPa	0,01
Estimated Mass Moisture	%	15
Sorption moisture by mass, not more than	%	18
Water absorption by volume when completely immersed in water for 24 hours, no more than	%	16
Maximum permissible increment of moisture, by weight	%	50
Thermal conductivity	Wt/(m·oC)	0,029 - 0,035
Vapor permeability, no more than	mg/(m·h·Pa)	0,27
Combustibility group		G1, incombustible
Durability, not less than	year	50

The thermal insulation is carried out by pouring the foam into the cavity of the building envelope without joints and gaps. The advantage of the composite over plate insulants is its ability to restore insulating properties of the building envelope by re-pumping the insulation material into the voids through the technological holes.

When designing and operating the multi-layer walls, the most important problem is moisture condensation inside the structure. It can lead to excessive wetting of the insulation layer and the gradual loss of its thermal properties.

3. Results

To test properties of the building envelope consisting of Poroplast CF02, we determined moisture content under unsteady heating and mass transfer using a MODEN 2.0.

The calculation was performed for the external wall which consists of four layers: A - cement-sand plaster, B - clay brick, C - Poroplast CF02, D - ceramic brick. The diagram is shown in Figure 1 (where:
d_i – thickness of the i-th layer, m; γ_i – density of the i-th layer, kg/m^3; c_i – heat capacity of the i-th layer, kJ / (kg^{0}C);λ_i – thermal conductivity of the i-th layer, W/(m·^{o}C); μ_i – vapor permeability of the i-th layer, mg/(m·h·Pa); β_i – moisture conductivity coefficient, kg / (m·s); T_i – temperature on the axis of the i-th layer, ^{o}C; T_{int}, T_{ext} – temperature of internal and external air, respectively, ^{o}C; T_0, T_{10} – temperature on the inner and outer surfaces, respectively, ^{o}C; e_i – partial pressure of water vapor on the i-th layer, Pa; e_{int} , e_{ext} – partial pressure of water vapor of internal and external air, Pa; E_i – saturated water vapor pressure on the axis of the i-th layer, Pa; E_{int}, E_{ext} – saturated water vapor pressure corresponding to the temperature of internal and external air, Pa; I – the boundary between the room air and the inner surface of the 1st layer; II, III .... IX - the boundaries of the joints of the layers, X - the boundary between the outer surface of the 9th layer and the external air).

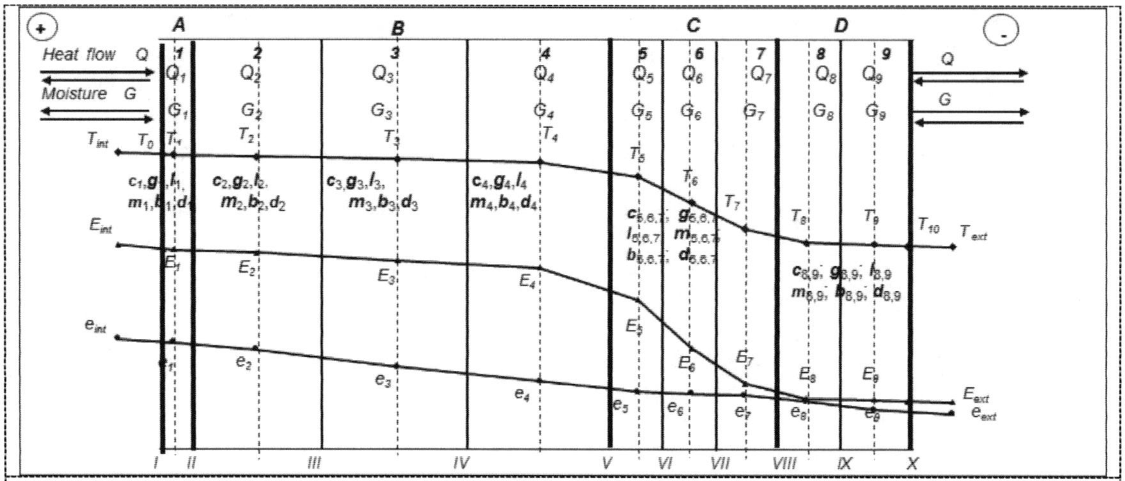

Figure 1. The design scheme.

The thermal characteristics of the materials are summarized in Table 2.

Table 2. Thermotechnical characteristics of materials.

Name of the material	Density γ, kg/m^3	Thickness d, m	Design characteristics of materials under operating conditions A		
			Thermal conductivity coefficient λ, W/(m·°C)	Vapor permeability coefficient μ, mg/(m·h·Pa)	Specific heat capacity in a dry state, kJ / (kg·°C)
Cement-sand mortar	1800	0,02	0,076	0,091	0,84
Clay brick	1800	0,25	0,7	0,11	0,88
Poroplast CF02	15	0,14	0,030	0,27	1,63
Ceramic brick	1600	0,12		0,14	0,88

Figure 2 shows moisture distribution over the building envelope during the conditional annual cycle (heat insulation density: 15 kg/m^3 and relative moisture of external air: a - 55%; b - 60%; I - moisture at the beginning of the wetting period (November); II - moisture at the end of the wetting period; III - moisture during the period of maximum moisture (February); IV - maximum sorption moisture). Temperature and moisture conditions comply with GOST 30494-2011. In the Siberian region, the average moisture content in Poroplast is within the sorption moisture value. Thermophysical properties of the envelope materials do not change. A moisture content increases only in the layer adjacent to the outer milestone of the façade brick. It can increase thermal conductivity of Poroplast by 9%.

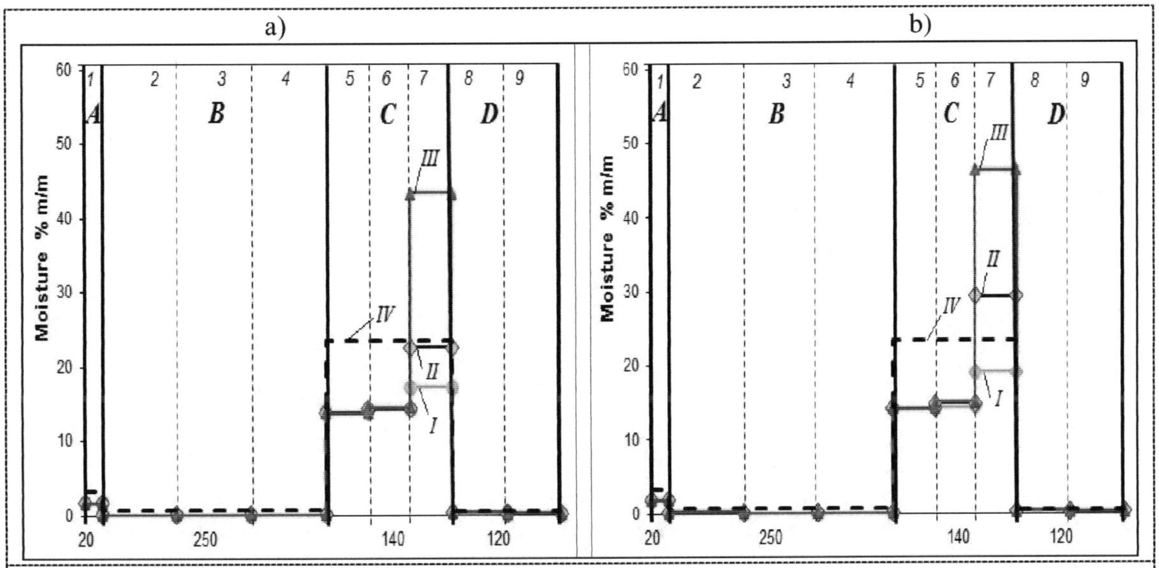

Figure 2. The diagram of moisture in the external wall.

Therefore, the multilayer building envelope consisting of Poroplast CF02 can retain its energy-saving properties for a long period

4. Discussion

The role of effective thermal insulation in constructing walls is crucial. Based on the world experience, one can conclude that 1 m^3 of effective thermal insulation saves 1.4-1.6 tons of standard fuel per year. The thermal layer increases the resistance to heat transfer, saves energy costs for heating buildings and decreases the consumption of materials and laboriousness of construction works.

Density of foam and porous plastics is 6-10 times lower than the density of mineral wool, 4-8 times – than mineral wool slabs, 25-50 times – than wood, 30-60 times – than brick and 100-200 times – than concrete and reinforced concrete. Foam and porous plastics can increase thermal resistance and reduce weight of the building envelope.

Table 3 shows the data on the heat transfer resistance of external building envelopes (walls and roofs) of residential buildings in various countries in accordance with the construction area.

Table 3. Comparison of the values of required resistance to heat transfer.

Country	Required resistance m^2C/W	
	Wall	Roof
Germany	2,0-2,5	3,0-3,6
Denmark	3,3-5,0	5,0-7,0
Canada	3,0-4,1	4,5-6,15
Norway	4,0	4,35
Russia	2,1-5,6	2,8-7,3
USA	1,08-2,5	3,57-7,14
Finland	2,9-3,57	2,2-4,55
Sweden	3,3-4,0	5,0-5,9

In order to ensure the required heat transfer rates, the wall thickness should be 1.5-2.8 m. Accordingly, such a thickness will increase the consumption of materials and labor and construction costs. Therefore, it is necessary to construct multilayer structures made from structural materials and effective insulants [6,7,8,9,10,11,12].

Figure 3 shows the comparative cost characteristics of 100 m^2 of the facade insulation system.

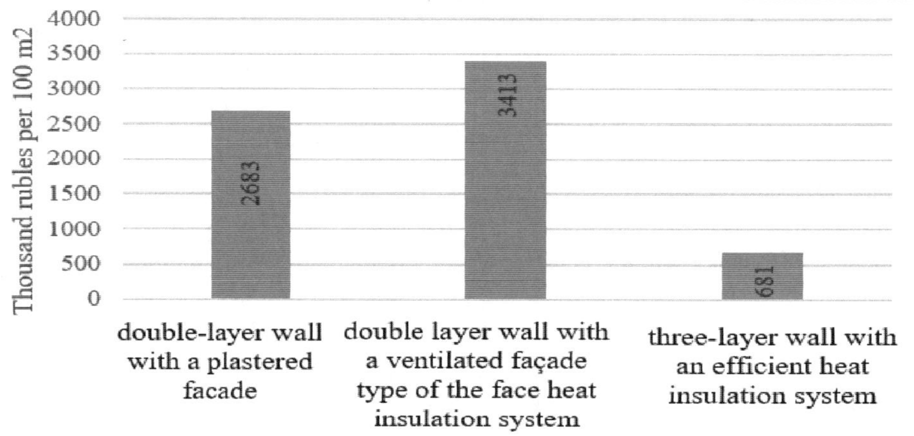

Figure 3. The cost of facade insulation systems per 100 m2, thousand rubles [13].

Thus, the three-layer structure can decrease construction costs

5. Conclusion

In Russia, the issues of energy efficiency and energy saving are crucial due to harsh climatic conditions and regular natural disasters. It is advisable to implement program-targeted mechanisms and scientific achievements.

However, despite the technological progress, it is difficult to ensure the durability of materials used in building envelopes and the economic unprofitability of the latter in the long term.

One of the attempts to solve the problem is the use of Poroplast CF02 as an insulant. Field surveys of the building envelope consisting of Poroplast CF02 showed that the moisture content is 17-24% which allows us to consider Poroplast CF02 as a reliable and stable insulant for multilayer building envelopes.

References

[1] Medvedeva G and Sekarina O 2017 *Science studies* **5(9)** 1–11
[2] Moskvitin V 2005 *Construction Technologies* **7(4)** 110–5
[3] Moskvitin V 2006 *Housing construction* **3** 24–6
[4] Barret P 1984 *Structural Survey* **2 (4)** 369–73
[5] Rossiter W, Mathey R, Burch D and Pierce T 1977 *Urea-Formaldehyde Based Foam Insulations: an Assessment of Their Properties and Performance* (Washington: National Bureau of standards department of commerce) p 92
[6] Snyder R and Melo-Abreu P 2005 *Frost Protection: Fundamentals, Practice and Economics* (Italy: Publishing Management Service Information Division) p 12
[7] OHegarty R, McCormack S and Kinnane O 2016 *Solar Energy* **135** 408–22
[8] Planas C, Cuerva E and Alavedra P 2018 *Ain shams engineering journal* **9 (4)** 3325–34
[9] Baetens R, Jelle B and Gustavsen A 2011 *Energy and Buildings* **43** 761–69
[10] Barbero S, Marco D, Ferrua C and Pereno A 2014 *Energy and Buildings* **77** 40–7
[11] Cuce E, Cuce P, Wood C and Riffat S 2014 *Renew Sustain Energy Reviews* **34** 273–99

[12] Loss C and Davison B 2017 *Engineering Structures* **132** 695-713
[13] Gnezdilova O and Kudryavtseva V 2018 *Transport infrastructure of the Siberian region* **10** 589-94

Comparative evaluation of air-water mixers operating with flotation plants

B M Grishin[1], M V Bikunova[1], A I Shein[1], NN Laskov[1] and I I Sokolov[2]

[1] Penza State University of Architecture and Construction, 28 Germana Titova, Penza, 440028, Russia
[2] Volgograd State Technical University, 28 V I Lenin prospect, Volgograd, 400005, Russia

E-mail: bgrishin@ rambler.ru

Abstract. The article deals with theoretical and experimental studies of oily industrial wastewater flotation treatment using a saturator and vortex mixing devices (VMD) for preparing a water-air mixture. The use of VMD allows, in comparison with the saturator, to improve the efficiency of oil product removal on the flotator by 5-7% due to the greater gas saturation of the flotation volume. The energy consumption for preparing the water-air mixture in vortex mixers was on 8–10% lower than in flotation plants using a saturator.
Key words: industrial wastewater, oil-containing impurities, suspended substances, reagent flotation purification, vortex mixing devices.

1. Introduction

One of the most efficient methods for cleaning industrial oil-containing wastewater is the pressure flotation method, using apparatus and plants having various designs [1-7]. An essential factor influencing the flotation treatment degree is the method of gas-liquid mixture preparation. In pressure flotation, high-pressure air is introduced into the saturator and dissolved in water. When a supersaturated aqueous solution enters the flotator, air is released from the water in the form of bubbles. The disadvantage of pressure flotation is the insignificant gas content of the supersaturated solution (not greater than 4-4.5%), with the increase of which air bubbles coalesce and the quality of purification deteriorates. The development of a new type mixers that would create a finely dispersed water-air mixture with high gas saturation without preliminary dissolution of air in water is an urgent task.

Tubular compact vortex mixing devices (VMDs) can be used as such mixers; their constructions was discussed in sufficient detail in [8]. The first-stage VMD (I VMD) is a vortex ejector, whereas the second-stage VMD (II VMD) is a coaxial "pipe-in-pipe" system with a rotational-translational motion of the water-air mixture. Such a system allows to increase quantity of bubbles contacts with pipe walls at high values of tangential stresses generated by the velocity gradient of the gas-liquid mixture, which intensifies the water-air mixture dispersion [9].

To determine the effect of the energy dissipation degree in a tubular mixer on the average diameter of dispersible air bubbles, formula [10] is used

$$d_b \approx 0.43\left(\frac{\sigma}{\rho_l}\right)^{0.6} \cdot \varepsilon^{-0.4}, \quad (1)$$

where ε_0 is the average value of the power dissipated in the unit mass of the mixed medium, W/kg or m^2/s^3; ρ_l is density of the liquid phase, kg/m^3; σ is surface tension at the air-liquid interface, n / m.

The value ε_0 is related to the velocity gradient G, s^{-1} in the mixer by the relation [11]

$$\varepsilon_0 = G^2 \cdot v_m, \quad (2)$$

where v_m is the kinematic coefficient of viscosity of water-air mixture, m^2/s.

The velocity gradient G, s^{-1}, is determined by the formula

$$G = \left(\frac{E}{t_m \cdot V_m \cdot \rho_m \cdot \nu_m}\right)^{0.5}, \quad (3)$$

where E is the energy expended on mixing, J; t_m is the mixing time, s; V_m is the volume of the mixer, m^3; ρ_m is density of water-air mixture, kg/m^3.

The energy value for tubular mixers is found from expression

$$E = Q_m \cdot \Delta\rho \cdot t_m, \quad (4)$$

where Δp is pressure losses in the mixer, Pa; Q_m is the flow rate of the water-air mixture, m^3/s.

Taking into account (2) and (4), the formula (1) for determining the mean diameter of air bubbles in the hydraulic fluid will have the following form:

$$d_b = 0.43\left(\frac{\sigma}{\rho_l}\right)^{0.6} \cdot \left(G^2 \cdot \nu_m\right)^{-0.4} = 0.43\left(\frac{\sigma}{\rho_l}\right)^{0.6} \cdot \left(\frac{\Delta\rho}{\rho_l}\right)^{-0.4}. \quad (5)$$

The value ρ_m can be determined with sufficient accuracy by the formula

$$\rho_m = \rho_l\left(1 - \varphi/100\right), \quad (6)$$

where φ is volumetric gas content in the mixture, %.

Thus, by setting the structural dimensions of stage II tubular mixer and the technological parameters of treating the water-air mixture, it is possible to determine the degree of dispersion of the air bubbles obtained in a certain range. In this article, the task was set of determining the average size of air bubbles in the flotator and comparing the efficiency of flotation treatment of oily wastewater using water-air mixtures obtained in the VMD and in the saturator.

2. Materials and methods

The object of research was the industrial wastewater of a machine-building plant entering, after an oil separator, the flotation treatment plant. Industrial wastewater was characterized by the following composition:

- oil products C_{oo} = 12 - 31 mg/l;
- suspended substances C_{so} = 28 – 55 mg/l;
- pH 6.8 – 7.1.

In the experiments on flotation treatment of effluents, polyaluminum chloride (PAC) was used as a coagulant with an active part content of 18%, used in industrial flotators as the main treating agent. To enhance the effect of removing contaminants, 4-5 mg/l of K-555 cationic flocculant is dosed to the wastewater after adding coagulant.

The oil-containing wastewater flotation treatment plant was installed at the industrial effluents treatment plant of the machine-building plant.

The experimental flotation plant (figure 1) consisted of a flotation column 1 with a total height of 2.2 m, a saturator 2 with a capacity of 0.017 m^3, first 3 and second 4 stage vortex mixing devices (VMDs), a receiving tank 5, a centrifugal pump 6, a compressor 8, coagulant dosing pumps 17 and flocculant 18. The installation also comprised pipelines, instrumentation and shut-off valves, auxiliary tanks.

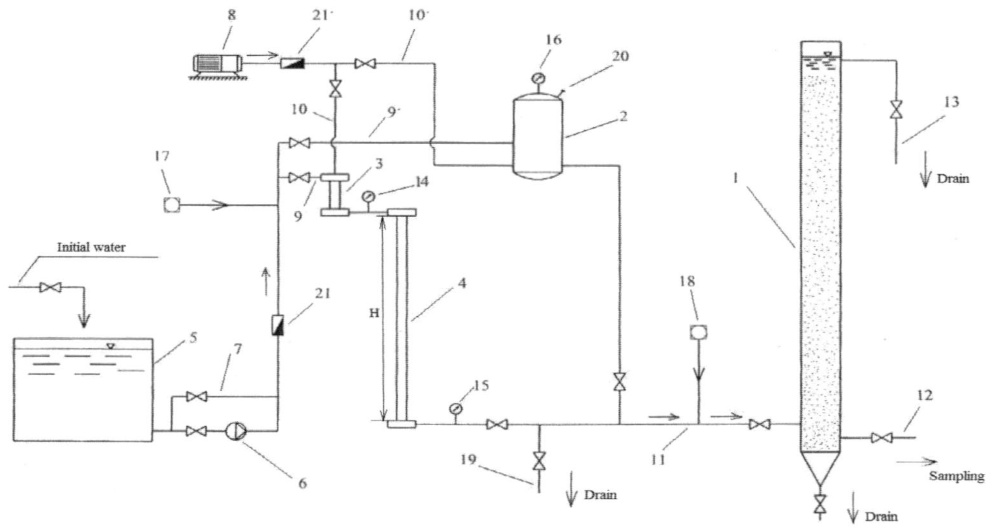

Figure 1. Layout of laboratory flotation plant: 1 – flotation column, 2 – saturator, 3, 4 – stage I and II VMDs, 5 – receiving tanks, 6 – pump, 7 – bypass line, 8 – compressor, 9, 9′ – pressure pipes of the original effluent, 10, 10′ – compressed air pipelines, 11 – air-water mixture pipeline, 12 –clarified water draw-off pipe, 13 – flotation sludge discharge pipeline, 14, 15, 16 – pressure gauges, 17 – coagulant dosing pump, 18 – flocculant dosing pump, 19 – air-water mixture discharge pipe, 20 – air vent, 21, 21′ – flow meters.

The initial wastewater was drawn off to the receiving tank 5 from the pipeline supplying oily wastewater to industrial flotators. The wastewater was then fed by pump 6 to the water-air mixture preparation devices – saturator 2 or VMDs 3 and 4. These devices worked alternately by closing the valves on the pressure pipes 9 and 9′. The flow rate of wastewater supplied to the saturator 2, was 0.1 m^3/hr due to the opening of the valve on the bypass line 7 and the circulation of most of the flow. Simultaneously, from compressor 8 through pipeline 10′air was supplied at a pressure of $4.5 - 4.6 \cdot 10^5$ Pa to saturator 2. The time of saturation of water with air was within 5 - 6 minutes. Excess compressed air was removed through valve 20, the pressure in the saturator was monitored by pressure gauge 16. The water-air mixture was throttled by a valve installed on pipeline 11 in the immediate vicinity of flotation column 1. The degree of water aeration (gas saturation) φ was 3.8 – 4%. The coagulant solution was fed by the dosing pump 17 to the original effluent feed pipe upstream of the water-air mixture preparation devices.

During the examination of the VMDs operation, a gate valve was opened on pipeline 9 and with the gate valve on the bypass line 7 closed, the sewage water from the pump 6 was fed to stage 3. The compressed air from compressor 8 through pipeline 10 was also entered there. The (gas) air saturation was from 6% to 8% and was controlled by flow meters 21 и 21′. After the primary mixing, the water-air mixture was fed to stage II VMD 4 and then flowed through the pipeline 11 to the flotation column. Excess of the air-water mixture flow leaving the VMD was discharged into the sewage system through the branch pipe 19. The pressure differential across VMD was maintained within $3.3 - 4.2 \cdot 10^5$ Pa due to the valves installed on the pipelines 9, 11 and was monitored by manometers 14 and 15. The height of the barrel of stage II VMD H in the experiments varied from 1.5 to 2.0 m. The wastewater-air mixing time in the VMD was under 2 s. The general view of the VMD is shown in figure 2.

IOP Conf. Series: Materials Science and Engineering **828** (2020) 012009 doi:10.1088/1757-899X/828/1/012009

The water-air mixture, obtained in the saturator or the VMD, was mixed immediately upstream the flotation column with the flocculant K-555 solution, fed by dosing pump 18 (see figure 1). The flotation time was 15–20 min. The flow rate reduced to the flotator's separation surface was 5.6 – 5.7 $m^3/m^2 \cdot hr$. The sludge formed during the operation of column 1 through the branch pipe 13 was discharged to the sewer, the clarified water through the branch pipe 12 was drawn off for chemical analysis.

Figure 2. General view of vortex mixing devices (VMD): 1– recycle pump, 2 – first stage VMD (I VMD), 3 – second stage VMD (II VMD), 4 – compressed air supply.

The experimental program provided for the addition of a coagulant solution into the original water, water aeration in the saturator or in stage I and II VMDs, treatment of the water-air mixture with flocculant at a constant dose of 5 mg/l, and flotation purification of water from contamination in the laboratory column. In each experiment, purified water samples were taken to determine residual concentrations of pollutants. The average size of the air bubbles in the flotation column was determined from the rate of the collective ascent of the bubbles in accordance with the recommendations [3, 12].

3. Results and discussion

The results of experiments to determine the average diameters of air bubbles in a flotation column are shown in figure 3.

The average size of the air bubbles d_b in the flotation column downstream of the saturator was 70-75 μm (Diagram 1, figure 3). The diameter of the air bubbles depended on the height of the VMD's barrel of in the flotation column downstream of the VMD and air saturation of wastewater. At H = 1.5 m and $\varphi = 6\%$, the average bubble diameter was 120-125 μm, and at $\varphi = 8\%$ the bubble size varied from 150 to 155 μm (diagram 2, figure 3). The pressure losses Δp in the VMD's barrel for H=1.5 m varied from 3.6·105 to 3.8·105 Pa. At H = 2.0 m and $\varphi = 8\%$, the average bubble diameter db was 95-100 μm, and at $\varphi = 6\%$ it was equal to db = 80 - 85 μm (diagram 3, figure 3). The pressure losses Δp в in the VMD's barrel for H=2.0 м varied from 4.0·105 до 4.2·105 Pa. It should be noted that the power consumption for the pump and compressor during water aeration using the VMDs were 8-20% lower than those when a saturator was used.

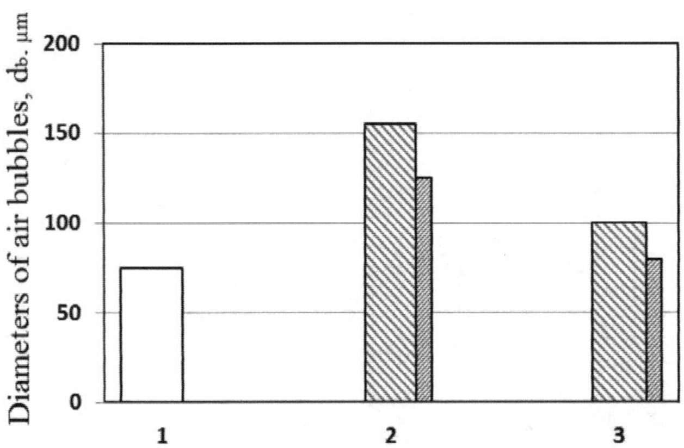

Figure3. Diagrams of average diameters of dispersed air bubbles in the flotator downstream saturator (1), downstream second stage VMD with H=1.5 m (2), downstream second stage VMD with H=2.0 m (3)

– at φ=6%; – at φ=8%.

The dependences of residual concentrations of oil products and suspended substances on the doses of the PAC coagulant following a 20 min flotation treatment using the water-air mixture having gas saturation of 3.8 - 4% obtained in the saturator are shown in figure. 4. The dose of the K-555 flocculant in all the experiments was D_f=5 mg/l. The purification efficiency was determined by the formula

$$E = \frac{C_{oo}(C_{so}) - C_0(C_s)}{C_{oo}(C_{so})} \cdot 100\%,. \tag{7}$$

where C_{oo} (C_{so}) and $C_o(C_s)$ are the concentrations of contaminants, mg/l, in the original and purified effluent respectively.

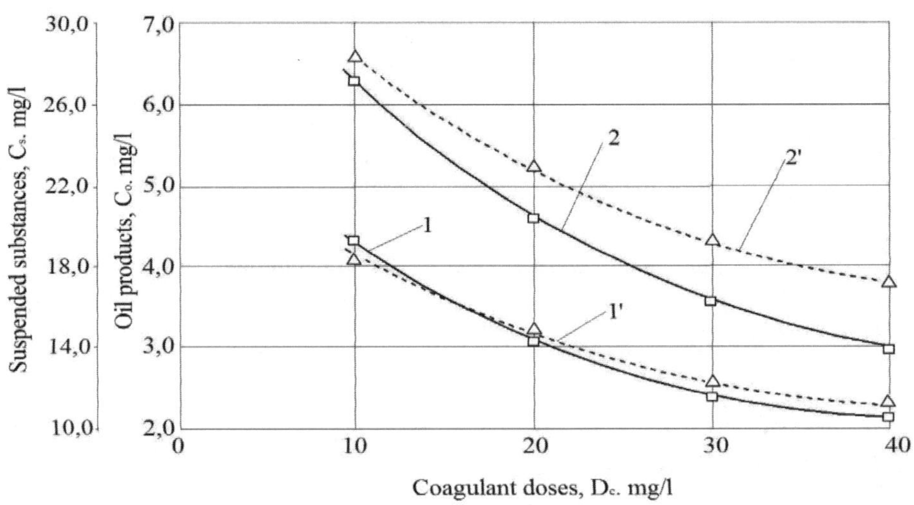

Figure 4. The dependences of residual concentrations of oil products C_o and suspended substances C_s in the effluents on the coagulant doses D_c after flotation with water-air mixture obtained in the saturator at initial concentrations C_{oo}=15 mg/l (1), C_{oo}=30 mg/ l (2), C_{so}=30 mg/l (1′), C_{so}=50 mg/l (2′),

——— oil products; - - - - - - suspended substances.

The results of experiments on flotation treatment of effluent using a water-air mixture obtained in the VMD with a barrel height of H = 1.5 m are shown in figures 5 and 6.

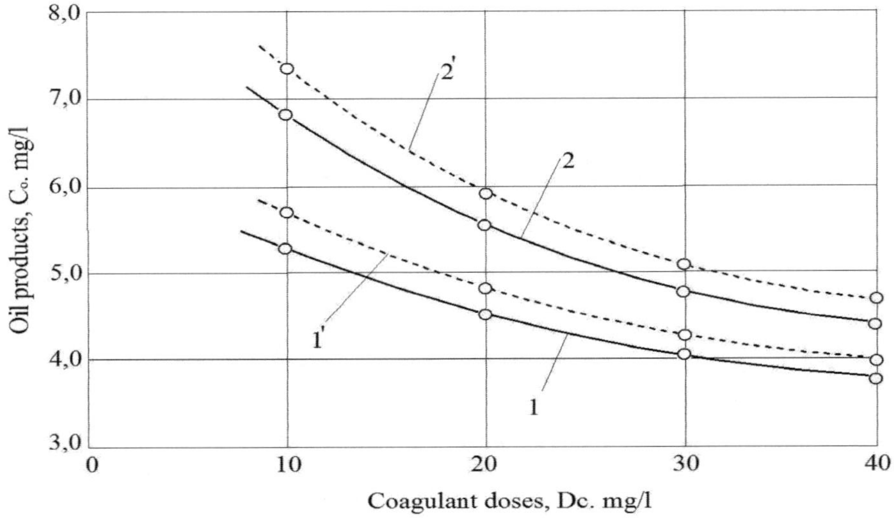

Figure 5. The dependences of residual concentrations of oil products C_o on the coagulant doses D_c after flotation with water-air mixture obtained in the VMD with a barrel height of H=1.5 m at initial concentrations C_{oo}=15 mg/l (1, 1′), C_{oo}=30 mg/l (2, 2′)
- - - - - - at φ=6%; ———— at φ=8% .

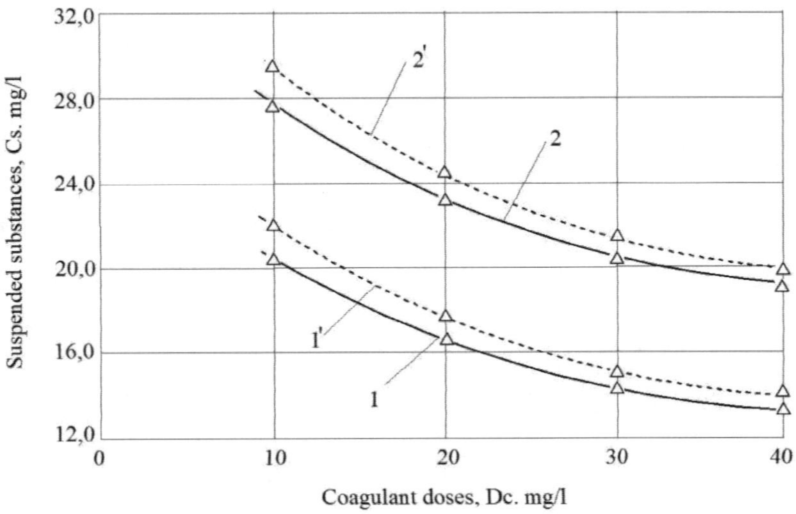

Figure 6. The dependences of residual concentrations of suspended substances C_s on the coagulant doses D_c after flotation with water-air mixture obtained in the VMD with a barrel height of H=1.5 m at initial concentrations C_s=30 mg/l (1, 1′), C_s=50 mg/l (2, 2′)
- - - - - - at φ=6%; ————at φ=8%.

A comparative analysis of the graphs in Fig. 4-6 showed that the efficiency of flotation treatment of wastewater from oil products and suspended substances with the use of the air-water mixture produced downstream the VMD with the barrel height H = 1.5 m was somewhat lower than the treatment efficiency using a saturator.

The results of experiments with the use of a water-air mixture produced in the VMD with a barrel height H=2.0 m for flotation treatment of wastewater are shown in figures 7-8.

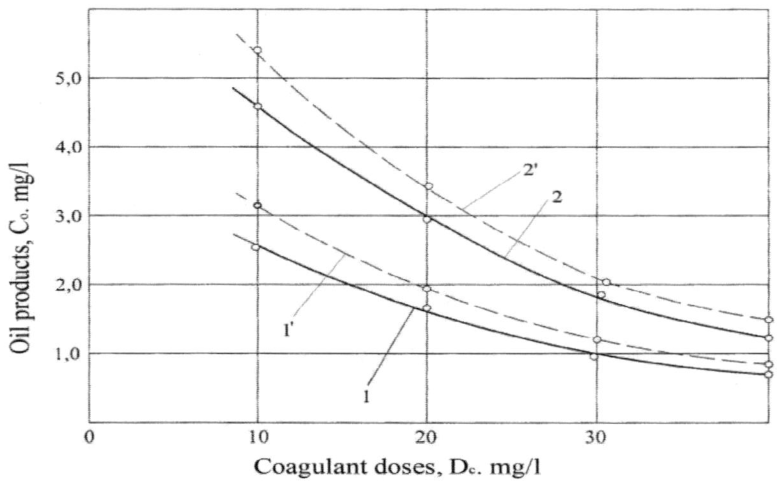

Figure 7. The dependences of residual concentrations of oil products C_o on the coagulant doses D_c after flotation with the water-air mixture obtained in the VMD with a barrel height of H=2.0 m at initial concentrations C_{oo}=15 mg/l (1, 1'), C_{oo}=30 mg/l (2, 2')

- - - - - - at φ=6 %; ———— at φ=8 %.

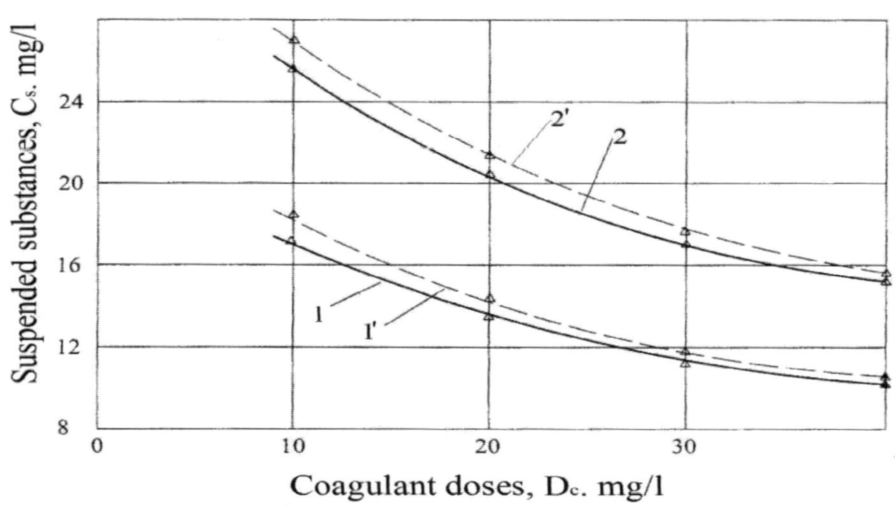

Figure 8. The dependences of residual concentrations of suspended substances C_s on the coagulant doses D_c after flotation with water-air mixture obtained in the VMD with a barrel height of H=2.0 m at initial concentrations C_{so}=30 mg/l (1, 1'), C_{so}=50 mg/l (2, 2')

- - - - - - -at φ=6 %; ———— at φ=8 %.

Analysis of the graphs in figures 4, 7 and 8 showed that due to the higher gas saturation of the water-air mixture in the flotation column with the use of the VMD (barrel height H=2.0 м and in-barrel pressure losses Δp=(4 ÷4.2)· 10^5 Pa) a higher cleaning effect on oil products and suspended solids was achieved in comparison with the experiments in which a saturator was used, with lower power consumption (by 8-10%). The effect of purification of industrial effluents with the use of water-air mixture from the saturator with the PAC doses of D_c =40 mg/l was the same as in the case of using in the flotation column of water-air mixture obtained downstream of the VMD (H=2.0 m) with coagulant doses of D_c =18-22 mg/l at φ=6 % and D_c =16-20 mg/l at φ=8 %. The effect of purification of industrial effluents from suspended substances with the use of water-air mixture from the saturator with the PAC doses of D_c =40 mg/l was equal to the effect o purification with the use in the flotator of water-air mixture obtained downstream of the VMD with coagulant doses of D_c =33-35 mg/l at φ=6 % and D_c =30-33 mg/l at φ=8 %.

4. Conclusions

1. The time required for retention of wastewater in the VMD's barrel for obtaining a highly disperse water-air mixture is reduced by more than 150 times compared with the water aeration time in the saturator. The pump and the compressor power common consumption during water aeration using VMDs was 8-10% lower than those using the saturator.

2. With the increase in gas saturation in the VMDs from 6 % to 8 % an increase of the average size of the air bubbles is observed from 120–125 μm to 150 - 155 μm with the height of the stage II VMD H=1,5 m and c 80–85 μm to 95–100 μm at H=2.0 m. The pressure losses in the VMDs are $(3.6 \div 3.8)\cdot 10^5$ Pa and $(4.0 \div 4.2)\cdot 10^5$ Pa respectively.

3. Vortex mixing devices (VMD), by providing a higher gas saturation of the air-water mixture in comparison with the saturator allow for a 5-7% increase on average in the efficiency of flotation treatment of wastewater from oil products and suspended substances at lower power consumption and coagulant doses.

References

[1] Glembotskiy V A and Klassen V I 1973 *Flotation* (Moscow: Nedra) p. 384
[2] Karelin Ya A 1982 *Oil Refineries Wastewater Treatment* (Moscow: Stroyizdat) p. 183
[3] Pokrovskiy V N and Arakcheyev Ye P 1980 *Thermal Power Plants Wastewater Treatment* (Moscow: Energia) p. 256
[4] Ponomaryov V N, Yoakimis E G and Mongayt I L 1985 *Oil Refineries Wastewater Treatment* (Moscow: Khimia) p. 254
[5] Rozentsvaig A K and Pergushev L P 1981 *Engineering and Physics Journal* **40** (**6**) 1013-18
[6] Tchobanoglous G, Burton F and Stensel H D 2013 *Wastewater Engineering: Treatment and Resource Recovery* (New York: McGraw-Hill) p. 1856
[7] Rao D G et al 2013 *Wastewater Treatment: Advanced Processes and Technologies* (Boca Raton: IWA Publishing and CRC Press) p. 365
[8] Grishin B M et al 2016 *Regional Architecture and Construction* **2**(**27**) 112- 17
[9] Adelshin A B 1996 *Flow Energy in the Processes of Intensification of Oil- Containing Wastewater Purification* (Kazan: KGASA) p. 200
[10] Yevilevich M A and Braginskiy L N 1979 *Optimization of Biochemical Wastewater Treatment* (Leningrad: Stroyizdat) p. 160
[11] Levich V G 1959 *Physicochemical Hydrodynamics* (Moscow: Fizmatgiz) p. 700
[12] Kafarov V V and Glebov M E 1991 *Mathematical Modeling of the Main Processes of Chemical Production* (Moscow: Vysshaya Shkola) p. 400

IOP Conf. Series: Materials Science and Engineering **828** (2020) 012010 doi:10.1088/1757-899X/828/1/012010

Conservation and rational use of natural resources: methods of circular economy assessment

Maria Andreevna Gureva

FSBEI HE «Industrial University of Tyumen»

E-mail: gurevama@tyuiu.ru

Abstract. Issues of resource conservation in terms of inclusion of raw materials and advanced technologies of environmental management are indisputably important in modern production and consumption development. The scientific community focuses on the development of technological methods for cyclic use of raw materials, construction of logistics supply chains, introduction and modernization of circular processes in production. The next logical stage of the study is to analyse various tools for assessment of circular economy development, promoting conservation and sustainable use of various natural resources; the article defines the main differences and results of the analysis. Carried out with the financial support of the Grant of the President of the Russian Federation to support young scientists Candidates of Sciences in the study MK-587.2019.6 " Development of theoretical and methodological foundations of the concept of circular economy as a new trend in the formation of sustainable socio-economic space."

1. Introduction.

The circularity concept has undeniable relevance, it is considered as an oriented process of modernization for the linear model of the economy to achieve qualitative development of various socio-economic phenomena. A historical review was previously presented, it covers circular economy emergence and formation and its relationship with new industrialization phenomenon; a variety of interpretations of "circular economy" concept was studied and main approaches to its formation were determined [5]. The authors described in detail using the circular economy model by countries on specific examples of enterprises characterizing introduction of main business models of the circular economy [2, 3]. The next logical stage is to study methodological aspects of assessing circular economy development.

The purpose of the article is to analyse various methodological approaches to the assessment of circular economy development in the socio-economic space, necessary for further development and testing a new assessment tool. Poor and insufficient elaboration of a methodological approach to the concept of circularity in Russian-language sources is indisputable fact testifying to scientific novelty.

Statistical information accumulated in various areas of socio-economic development, as a rule, is not informative and requires a thorough competent analysis. Indication is applied for quantitative and qualitative analysis of various indicators. This methodological approach has gained wide relevance in natural sciences, providing further insight into the causal relationship between the studied phenomena. The indicators used during the assessment offer a clearer view of the system state under study, which is influenced by a set of internal and external factors. The basic principles of indicators construction include aggregation and creation of indicator systems; at the same time, classical formation of an indicator is focused on presentation of the degree of achieving the reference value [4].

2. Main part.

In 2019, a group of scientists, after studying more than fifty-five different approaches to circular economy assessment "... that are developed and used by scientists, companies, environmental organizations, government agencies, created their taxonomy based on the needs arising from application of such indicators, including ten categories for differentiation and definition of C-indicators driven by the principles of the circular economy..." [25].

The categories/criteria identified by authors for proposed taxonomy of C-indicators include:

1) level (micro, meso, macro);
2) cycle (conservation, reuse / recovery, recycling);
3) performance (internal, impact);
4) perspective (actual, potential);
5) usage (for example: improvement, comparison, communication);
6) transversality (general, branch);
7) size (one, several);
8) units of measurement (quantitative, qualitative);
9) format (for example: a web-based tool, Excel-formula);
10) sources (scientists, companies, agencies) [25].

C-indicators can be considered as extraordinary incentives to move to a more advanced level of circularity.

A detailed analysis of methods for assessing circular economy development is beyond the scope of this article, so it briefly presents only a few.

In 2007, in China the scientific literature gives the earliest references to the indicators of circular economy evaluation. The first system of indicators "... is based on the 3R principles and is designed to promote the application of circular economy, assess the overall effectiveness of practices and support processes..." [17]. In 2011, the Regional Circular Economy Index System was introduced, based on three indicators, according to the 3R imperatives: pollution reduction, waste recycling and recycling of materials [22].

In 2011, researcher J. Guo-gang presented an index system for assessing the level of circular economy development, consisting of 16 indicators grouped into four groups:

1) consumption (water consumption per million GDP, water consumption per capita, elasticity in water use, energy consumption per million GDP);
2) environmental violations (norms of industrial wastewater discharge, level of harmlessness of household garbage, application of chemical fertilizers per unit of acreage);
3) waste management (urban wastewater treatment per capita, integrated solid industrial waste utilization rate, resource utilization network coverage, "three waste" utilization rate);
4) social development (GDP per capita, urbanization rate, unemployment rate, Engel coefficient, GDP growth) [19].

In 2012, a group of scientists led by Yu. Geng suggested a system of indicators structured into two groups for evaluation:

- macro-level (22 indicators);
- industrial park (12 indicators).

This system is designed to facilitate methodological processes of introduction of circular economy and increase attention to environmental problems (Geng Y., Fu J., Sarkis J., Xue B., 2012) [17].

In 2013, in China, the Ministry of Environmental Protection (MEP) drafted indicators system for circular economy assessment at the meso-level based on 21 indicators divided into four groups: economic development, waste management, pollution control, administration and management [20].

In 2015, C. Ruiter (2015) proposed "the Circular Economy Performance Index", acting as a useful tool for assessing the level of business circularity. The system consists of 25 key performance indicators (KPI) of the circular economy, classified into three levels according to impact degree: high (red), medium (orange) and low (green). According to the obtained calculated values the analyzed object can be assigned to one of five categories: "non-compliance", "compliance", "beyond compliance", "integrated strategy", "goal/mission", providing an opportunity to quickly realize shortcomings in the circular economy concept implementation [23].

In 2015, "EU Resource Efficiency Scoreboard 2015" presents "... a system for assessing the circularity of economy on 32 indicators formed into a three-level system: a common leading indicator of "resource productivity"; a second-level dashboard of additional macro indicators for materials, land, water and carbon; and a third-level of thematic indicators to measure progress towards key thematic goals, as well as actions and milestones set out in the road map..." [16, 28]. They are grouped into main topics and sub-topics:

- resource productivity (main indicator);
- dashboard indicators (materials, earth, water, carbon);
- transforming the economy (turning waste into resources, supporting research and innovation; pricing correctly);
- nature and ecosystems (biodiversity; clean air; land and soil);
- key areas (solving the food problem; improving buildings; ensuring effective mobility) [16, 28].

Resource productivity is the main indicator of the evaluation system under consideration. It is used as a measure of resource efficiency, i.e. how effectively an economy uses material resources to produce products and services available in the market. It is an absolute indicator, measured in euros per kg and monitors the dynamics of changes (2000 is considered as a base year). The efficiency of resources use by the state largely depends on the structure of national economy, the size and structure of international trade, while the economy is able to create more wealth without a proportional increase in resource consumption [16, 28].

In 2015, Ellen MacArthur Foundation, GRANTA Design and LIFE, developed an indicator "The Material Circularity Indicator" (MCI) presented in "Circularity Indicators: An Approach to Measuring Circularity. Methodology" [14].

Along with the MCI indicator in the Circular Metrics Landscape Analysis report 2018, The World Business Council for Sustainable Development (WBSCD) identifies three other important tools for assessing the circular economy:

1) the Life Cycle Assessment (LCA) is a method of assessing the environmental impact associated with all stages of the product life cycle from extraction of raw materials to disposal; it is an indirect tool used in the circular economy, but, nevertheless, so-called "life cycle analysis" (or "ecobalance") is extremely important;

2) the Circular Economy Toolkit (CET) is an assessment method that identifies and evaluates the potential improvement in circularity of products, i.e. it is also associated with life cycle analysis. This is an online test without score points, which includes 33 questions, developed at the University of Cambridge. Questions are divided into 7 subcategories according to the stages of product/service life cycle. It is presented as a web page and gives a qualitative assessment, structured into three categories (low, medium, high);

3) circular economy Indicator (the Circular Economy Indicator Prototype, CEIP) evaluates the performance of cyclic products. Developed on the basis of calculation functions of MS Office Excel, using a questionnaire with a score system of assessment, consisting of fifteen questions divided into 5 stages of a life cycle, namely: design or redesign; manufacture; commercialization; use and end of life. Presented as a spreadsheet in an Excel file, gives a quantitative estimate, measured in % [12, 18, 23, 24, 27].

In 2019, a group of researchers (Z. Steinmanna Z. J. N., Huijbregtsa M. A. J., Reijndersbb L., 2019) proposed the material quality circularity index (Qc), where the numerator expresses the net energy savings due to processing of primary material (MJ / kg) and the denominator is the embodied energy of 1 kg of primary material (MJ/kg), which can be attributed to the evaluation indicators of the circular economy at the micro-level [26].

In 2019, A. Avdiushchenko A., Zając P. proposed a system of evaluation "Circular Economy Indicators" which includes 25 indicators divided into 7 groups:

- economic development (GDP per capita, average life expectancy at birth for men, registered unemployment rate, poverty risk level);
- zero economy (municipal waste collected selectively in relation to the total amount of municipal waste collected; municipal waste collected per inhabitant; industrial and municipal waste water requiring treatment; expenditures on fixed assets serving environmental protection and water resources management related to waste processing and disposal);
- innovation economy (research and development expenditures per capita, fixed prices; average share of innovative enterprises in the total number of enterprises; adults involved in education and training; patent applications per 1 million inhabitants);
- energy efficiency and renewable energy (share of renewable energy sources in total electricity production; expenditures on fixed assets serving environmental protection and water resources management related to energy saving per capita; electricity consumption);
- low-carbon economy (emissions of carbon dioxide from plants particularly harmful to air purity; emissions of particles; cars; pollutants remaining or neutralized in pollutant reduction systems in common pollutants produced from plants particularly harmful to air purity; costs of fixed assets serving environmental protection and water management related to air and climate protection);
- smart economy (households with a personal computer with broadband Internet connection; businesses with broadband Internet access);
- spatially efficient economy (forest cover indicator; urban greenery and the share of parks, lawns and green areas in residential areas in the total area; urbanization coefficient) [9].

The table provides a comparative analysis of the considered tools for assessing the circular economy.

Table 1. Comparative analysis of circular economy assessment methods*

Year of development	Title** Full	Short	Authorship / country	Quantity: assessment subgroups	Quantity: indicators for assessing	Graphical shell	Level of assessment
1	2	3	4	5	6	7	8
2006	The Life Cycle Assessment	LCA	-	-	-	+	Micro
2007	The Regional Circular Economy Index System	RCI	China	2	3	-	Macro
2011	Regional Circular Economy Development index	-	J. Guo-gang / China	4	16	-	Macro
2012	Material flow analysis to evaluate Circular economy	MFA	Geng Y., Fu J., Sarkis J., Xue B.	2	34	-	Macro Micro
2013	Ministry of Environmental Protection CE	MEP	Su B. W. et al. / China	4	21	-	Meso

IOP Conf. Series: Materials Science and Engineering **828** (2020) 012010 doi:10.1088/1757-899X/828/1/012010

Year of development	Title** Full	Short	Authorship / country	Quantity assessment subgroups	Quantity indicators for assessing	Graphical shell	Level of assessment
	indicator system						
2015	The Circular Economy Performance Index	-	Ruiter C.	-	25	+	Micro Meso
2015	The Resource Efficiency Scoreboard	-	Eurostat European Commission	5	32	-	Macro
2015	The Material Circularity Indicator	MCI	Ellen MacArthur Foundation, Granta Design и LIFE	-	2	+	Macro
2016	Circular Economy Index for the consumer goods sector	-	L.H. Verbeek	-	25	+	Micro
2016***	The Circular Economy Toolkit	CET	Cambridge University	7	33	+	Micro
2016	The Circular Economy Indicator Prototype	CEIP	Griffiths, P. Cayzer, S.	5	15	+	Micro Meso
2018	Raw Materials Scoreboard 2018	-	EIP	5	26	-	Macro
2018	Monitoring framework for the circular economy	-	European Commission and Eurostat	4	10	-	Macro
2018	Adaptation of Monitoring framework for the circular economy for practice in the Republic of Belarus	-	BEROC / the Republic of Belarus	4	14	-	Macro
2018	Indicators for a Circular Economy	-	Vercalsteren A., Christis M., Van Hoof V.	-	18	-	Micro Meso Macro
2018	Circularity Rating Indicator	CL	O. I. Sergienko, E. S. Smaznova, D. V. Razumova	1	1	-	Micro Macro
2018	Indicators for evaluation of clusters' competitiveness through circular economy	-	K. Razminiene M. Tvaronaviciene	4	15	+	Meso Macro
2018	Circular Economy Development Index	CEDI	Vetrova M. A., Pakhomova N. V., Richter K. K.	1	1	-	Meso

Year of development	Title** Full	Short	Authorship / country	Quantity assessment subgroups	indicators for assessing	Graphical shell	Level of assessment
2019	Indicator for the circularity of material quality	-	Steinmanna Z.J.N., Huijbregtsa M.A.J., Reijndersb L.	1	1	-	Micro
2019	Circular Economy Indicators	-	Avdiushchenko A., Zając P.	7	25	-	Macro

* the table is compiled by the author using the cited sources in the bibliography list [1-30].
** there are more evaluation tools than are presented in this table.
*** the reliability of this data is in doubt, as the official data source has no date [27].

3. Conclusion

The bibliometric method of literature analysis showed that the assessment of circular economy development takes place at three levels: micro, meso and macro. Macro-level indicators are needed for evaluation and monitoring in order to improve various programs at the state level. Micro-level indicators tend to cover 3R imperatives without reflecting the full range of distinctive features of the circular economy, analyzing the environmental friendliness and waste-free production and consumption. At the same time, there is no clear distinction in the set of indicators, in this regard, they can be repeatedly used at the same time. Most of the studied indicators are focused only on one and/or several specific environmental problems than contribute to the estimated subjectivity.

The research carried out in this article has identified several problem areas regarding the methodological approach to the concept of circularity: insufficient statistical coverage of baseline data; lack of in-depth justification of metrics and evaluation systems; lack of a uniform tool for assessing circular economy development. Therefore, it is necessary to develop methodological assessment tools with wider coverage of the spectrum of circular economy paradigm than existing tools, providing relevant integration with real business practices.

An extremely important condition for activating the intensification of transition to the circular economy model is creation of a comprehensive methodological tool for assessing its development, which is required for effective strategic and program planning at the state level.

The continuation of the study, based on the conceptual, theoretical and methodological provisions of the circular economy, should be aimed at improving the assessment tools, which will contribute to the qualitative achievement of the goals of society sustainable development.

References

[1] Vetrova M A 2018 *Justification of strategic and operational decisions of enterprises in the transition to a circular economy : 08.00.05 25.00.17: Dis. ... Candidate of Economic Sciences* (St. Petersburg State University) 432 p. URL: https://disser.spbu.ru/files/phd_spsu/vetrova_disser.pdf
[2] Guryeva M A 2019 *Questions of innovative economy 9* **(4).** DOI: 10.18334/vinec.9.4.41236
[3] Guryeva M A and Butko V V 2019 *Economic relations* **9 (4).** DOI: 10.18334/eo.9.4.40991
[4] Guryeva M A and Simarova I S 2016 *Global scientific potential* **10 (67)** 78-86
[5] Guryeva M A 2019 *Economic relations* **9 (3).** DOI: 10.18334/eo.9.3.40990
[6] Razminiene K and Tvaronaviciene M 2018 *TERRA ECONOMICUS* **16 (4)** 50-65. DOI: 10.23683/2073-6606-2018-16-4-50-65
[7] Sergienko O I, Smaznova E S and Razumova D V 2018 *Bulletin of PNRPU. Applied ecology. Urban studies* **4** 80-92
[8] Shershunovich E and Tochitskaya I 2018 *BEROC Green Economy Policy Paper Series, PP GE3*. URL: http://www.beroc.by/webroot/delivery/files/PP_GE_3.pdf

[9] Avdiushchenko Anna and Zając Paweł 2019 *Sustainability* **11(11)** 3025. DOI: https://doi.org/10.3390/su11113025

[10] Banaitė D and Tamošiūnienė R 2016 *Journal of Security and Sustainability Issues* **6(2)** 315–323. DOI: http://dx.doi.org/10.9770/jssi.2016.6.2(10)

[11] Banaitė D 2016 *Social Transformations in Contemporary Society* **4** 142-150. URL: http://stics.mruni.eu/wp-content/uploads/2016/07/STICS_2016_4_142-150.pdf

[12] Cayzer Steve 2016 *Sustainable Design and Manufacturing.* DOI: 10.1007/978-3-319-32098-4_27

[13] *Circular Metrics Landscape Analysis. A joint report on the current landscape of circular metrics use and recommendations for a common measurement framework* May 2018. URL: https://docs.wbcsd.org/2018/06/Circular_Metrics-Landscape_analysis.pdf

[14] *Circularity Indicators: An Approach to Measuring Circularity. Methodology.* URL: https://www.ellenmacarthurfoundation.org/assets/downloads/insight/Circularity-Indicators_Project-Overview_May2015.pdf

[15] *Communication from the Commission to the European Parliament, the Council, the European Economic and Social Committee and the Committee of the Regions on a monitoring framework for the circular economy.* URL: http://ec.europa.eu/environment/circular-economy/pdf/monitoring-framework.pdf.

[16] *EU Resource Efficiency Scoreboard* 2015. URL: https://ec.europa.eu/environment/resource_efficiency/targets_indicators/scoreboard/pdf/EU%20Resource%20Efficiency%20Scoreboard%202015.pdf

[17] Geng Y, Fu J, Sarkis J and Xue B 2012 *Journal of Cleaner Production* **23** 216-224

[18] Griffiths P and Cayzer S 2016 *3rd International Conference on Sustainable Design and Manufacturing, SDM 2016; Springer Science and Business Media Deutschland* (GmbH: Berlin, Germany) pp. 307–321

[19] Guo-gang J 2011 *Energy Procedia* **5** 125–129

[20] Heshmati Almas 2015 *A Review of the Circular Economy and its Implementation* URL: http://ftp.iza.org/dp9611.pdf

[21] *Indicators for a circular economy. EASAC.* 2016. URL: https://www.easac.eu/fileadmin/PDF_s/reports_statements/Circular_Economy/EASAC_Indicators_web_complete.pdf

[22] Jia C and Zhang J 2011 *Procedia Environmental Sciences* **11** 637– 64.

[23] Ruiter C 2015 *VU University Amsterdam.* URL: http://dspace.library.uu.nl/handle/1874/337188/

[24] Saidani Michael, Yannou Bernard, Leroy Yann and Cluzel François 2017 *How to Assess Product Performance in the Circular Economy? Proposed Requirements for the Design of a Circularity Measurement Framework Recycling* **2** 6. DOI: 10.3390/recycling2010006

[25] Saidani Michael, Yannou Bernard, Leroy Yann, Cluzel François and Kendall Alissa 2019 *Journal of Cleaner Production* (Elsevier) **207** 542-559.

[26] Steinmanna Z J N, Huijbregtsa M A J and Reijndersb L 2019 *Resources, Conservation and Recycling* **141** 362-363. DOI: https://doi.org/10.1016/j.resconrec.2018.10.040

[27] *The Circular Economy Toolkit.* URL: http://circulareconomytoolkit.org/Toolkit.html

[28] *The Raw Materials Scoreboard is part of the monitoring and evaluation strategy for the European Innovation Partnership (EIP) on Raw Materials.* URL: https://op.europa.eu/en/publication-detail/-/publication/117c8d9b-e3d3-11e8-b690-01aa75ed71a1

[29] Verbeek L H 2016 *A Circular Economy Index for the consumer goods sector.* URL: https://dspace.library.uu.n

[30] Vercalsteren An, Christis Maarten and Van Hoof 2018 *Veronique Indicators for a Circular Economy.* URL: https://vlaanderen-circulair.be/en/summa-ce-centre/publications/indicators-for-a-circular-economy

IOP Conf. Series: Materials Science and Engineering **828** (2020) 012011 doi:10.1088/1757-899X/828/1/012011

Climate control in road-building machinery

E L Iovleva[1] and P A Stepanov[1]

[1] Department of "Machine Science", Road Faculty, M.K. Ammosov North-Eastern Federal University, 58, Belinsky ave., Yakutsk 677000, Russia

E-mail: elizaveta-iovleva@yandex.ru

Abstract. Climate control is an advanced air conditioning system that allows controlling the temperature in the machinery cabin more effectively in any weather conditions. Today, almost all new road-building machinery is equipped with this system. This system allows adapting the temperature in the cabin according to your needs more effectively. Climate control allows controlling the climate in the cabin regardless of the weather conditions outside the machine. Climate control works just like home air conditioners, thanks to the automatic control of the heating, ventilation and air conditioning systems. With the help of electronics, the temperature is automatically maintained at the level set by the driver. This system is necessary not only to maintain a stable temperature in the cab, but also to maintain the efficiency and concentration of the driver. However, there is no air conditioning system in the old road construction machines. In the Republic of Sakha (Yakutia) the material and technological foundations in many road enterprises are outdated for now. And in order to keep the driver operating and protect his health, this article discusses the possibility to equip road-building machinery with climate control. THE Volvo BL61B forklift was chosen as an example. This machine initially lacked a climate control system, but thanks to a temperature control device, it was possible to achieve the optimum temperature in the cabin.

1. Introduction

The health of the driver is an important factor that determines his readiness to control a car, which ultimately affects traffic safety. A person feels comfortable if certain ambient temperature is reached.

"Climate in the car" directly affects the driver's condition, the safe driving of a car, and overall road safety.

Comfortable temperature in the car is determined by the outdoor ambient temperature and the amount of air exchange in the cabin:

- at low outdoor temperatures, e.g. -20 °C

high cabin temperature 28°C
high air exchange 8 kg / min.

- at high outdoor temperatures, e.g. 40°C

moderate cabin temperature 23°C
high air exchange 10kg / min.

- at moderate outdoor temperatures, e.g. 10°C

moderate cabin temperature 21.5°C law air exchange 4kg/min.

Scientific studies conducted by the World Health Organization have shown that the degree of concentration and rapid reaction of a driver under adverse stress on his body are significantly reduced.

Such an unfavorable factor is heat. For the driver to feel comfortable, the ambient temperature in the cabin of the car should lie in the range from 20 to 22 °C. [5,6]

This problem is most urgent for drivers of road-building equipment, as the enterprises of the road industry often operate the equipment manufactured at a time when they did not focus on the air conditioning system. Driver's health and traffic safety are fundamental to any industrial work. The construction season, almost in any city of Russia, falls on the summer period and therefore the system for maintaining a favorable temperature in the driver's cabin is an important component. This article shows the possibility of installing a climate control system in road construction equipment of the old model, but operated in many road enterprises of the Republic of Sakha (Yakutia).

2. Results and Discussion

STC-1000 is a ready-made device with two relays, one for cooling and heating. One of the advantages of the thermostat is the calibration and programming of the temperature received from the sensor. Intelligent control digital thermostat controls the temperature of cooling, heating, automatic STC-1000 temperature controller products, refrigeration, heating is automatically converted, using temperature hysteresis. It is a cost-effective, versatile product. Table 1 shows the technical characteristics of the thermostat. [7-9]

Table 1. Characteristics of the STC-1000

Type of thermostat	electronic
Supply voltage	220 V
Method of installation	embeddable
Load switching	relay
Minimum temperature	-50°C
Maximum temperature	100°C
Maximum power	2.2 kW
Hysteresis	1-30°C
The division value of the thermostat	0.1°C
The accuracy of the thermostat	1°C
Calibration	Is available
Controlled minimum temperature	-50°C
Controlled minimum temperature	100°C
On delay	Is available
Type of thermostat	electronic
Supply voltage	220 V
Method of installation	embeddable

This STC-1000 thermostat was set in the VOLVO BL61B excavator loader (figure 1).

Figure 1. Volvo BL61B excavator loader

Figure 2. The location of the temperature sensor

Connect the temperature sensor to the power supply of the excavator loader, the air conditioner and the heating system, using tape and a flat screwdriver, as close as possible to the driver (figure 2).

After connecting the temperature sensor, measure the temperature in the cabin (figure 3). Next, adjust the controller in the cab of the excavator loader by connecting to the air conditioner and heating system.

Figure 3. The temperature inside the cab of the excavator

IOP Conf. Series: Materials Science and Engineering **828** (2020) 012011 doi:10.1088/1757-899X/828/1/012011

The display of the STC-1000 thermostat shows two modes of operation: heating and cooling. There are four control buttons: turn on / off, enter programming mode (S) and two up and down arrows. Turn on and off by holding the power button for at least 3 seconds. After the first switch on the thermostat shows the temperature of the medium in which the sensor is located.

To view the temperature that the controller is programmed to maintain, press "up" (Can be set in the range from -50 to 99.9 degrees with an accuracy of one-tenth. The default of 10.0 Sec).

Pressing the down arrow displays the hysteresis that the controller is configured for (It can be set from 0.3 to 10 degrees to the nearest tenth. In factory settings 0.5 degrees).

To enter the programming mode, press and hold the S button for at least 3 seconds, F1 will appear on the display.

Press the S button and hold. The set temperature is displayed. Without releasing the button S arrows "up" or "down" set the required temperature.

Release the S button and press the up arrow. The display shows F2. Press the S button and the display will show the set temperature delta. Hold down the S arrow "up" and "down" set the required Delta.

Release the S button and press the up arrow. The display shows F3. Press the S button and the display will show the delay to turn on (default 3 minutes, the range of available values from 1 to 10 minutes). Release the S button and press the up arrow. The display shows F4. Here you can set the sensor calibration value. (Values range from -10 to +10°C in 0.1 increments. The default value is 0).

To exit the programming mode while saving the settings, press the on/off button briefly.

If you need to exit without saving the settings, the controller will automatically enter the operating mode if you do not press the buttons for 30 seconds.

If the display shows Er (error), then press any button and hold until the controller reboots. After that, all settings will be reset to the factory settings. [10]

Like any technical system, climate control has its advantages and disadvantages. Advantages of such a system are:

- in providing comfortable conditions for staying in the cabin of vehicles for driving;
- automatic operation reset;
- in a large number of regulated parameters;
- in minimizing the participation of a human in the work of the system;
- in a simple, convenient and intuitive interface that provides quick learning how to work with the system.

The disadvantages of climate control are:

- high cost of maintenance and repair;
- increased consumption of fuel in case of running the system
- self-repair of climate control is not possible. [11], [12]

However, despite the fact that there are some disadvantages, machines equipped with a climate control system increases the comfort of the driver.

3. Conclusion

During the construction of roads, drivers of road vehicles constantly expose their lives to the risk of harmful vapors and gases of asphalt aerosols, increased dust, ionizing radiation, temperature fluctuations, various aspirator diseases, microorganisms and many other negative factors [13].

The climate control system allows the driver to create a comfortable atmosphere in the cab for work.

This article has shown the possibility of setting the SCT-1000 thermostat in the VOLVO BL61B excavator loader to improve the climate control in the cabin.

This system creates comfort for the driver, and secondly, minimizes participation in operating the system.

References

[1] *Behr Hella Service GmbH, Dr.-Manfred-Behr-Straße 174523 Schwäbisch Hall.* URL: www.behrhellaservice.com
[2] Ambartsumyan B, Nosov B, and Tagasov B 1999 *Environmental safety of road transport* (M.: LLC publishing house "Nauchtekhlitizdat") 208 p
[3] Bogdanov A V 2007 *Health and Safety* **2** 2-3
[4] Zakharov E A, and Shumsky S N 2007 *Environmental problems of road transport: textbook* (VSTU. Volgograd) 107 p.
[5] Bogdanov A V 2007 *Health and Safety* **2** 2-3
[6] Balovnev V.I. 2006 *Multipurpose road-building and technological machines: textbook (*Omsk-Moscow: OJSC "Omsk printing house") 320 p
[7] *"Air conditioners". Directory.* 1981 ("CIITElektromash", M.)
[8] "Electronics in the car" 2012 *Renovation* (M.: SOLON PRESS) **123** 83-85.
[9] Mitin M A, and Pchelintsev N I 2013 Climate system in a modern car *Repair* (M.: SOLON PRESS) **127** 72 p.
[10] Pchelintsev N 2012 *Repair & Service* **7** 47-54.
[11] Ergunova O, Lizunkov V, Blaginin, V, Politsinskaya E, and Shaykina O 2018 *International Journal of Energy Economics and Policy* **8(1)** 122-127
[12] *Self-study program №208. Automotive climate control systems. Device and mode of functioning-Volkswagen AG* 1998 (Wolfsburg) 74 p.
[13] Lizunkov V, Politsinskaya E, Malushko E, Kindaev A, and Minin M 2018 *International journal of energy economics and policy* **8 (3)** 250-257

IOP Conf. Series: Materials Science and Engineering **828** (2020) 012012 doi:10.1088/1757-899X/828/1/012012

Environmental Rehabilitation of Volga-Akhtuba Floodplain

V V Kalinin[1], V F Loboyko[2], O A Solovyova[2], M A Likhomanova[2], A Y Ovcharova[2] and A V Loboyko[2]

[1] ZAO «Volgovodproject»
[2] Volgograd State Agrarian University, 26, University Ave, Volgograd, 400002, Russia

E-mail: ovcharova_82@list.ru

Abstract. The research considers the water bodies of the Volga-Akhtuba floodplain requiring an optimal hydrological regime, aimed at preserving floodplain oak groves, soil fertility, spawning sites and at the development of the territory agricultural potential. New solutions are presented, aimed at complex approach to floodplain territory rehabilitation, at minimization of floodplain degradation and at its restoration, instead of traditional hydro-mechanized work on cleaning tracts, eriks and lakes of the floodplain. A phased restoration of the floodplain natural state and creation of balanced ecosystem with an operating self-purification mechanism are planned as a result of the proposed measures being implemented. These measures enable to increase the acreage of irrigated lands for vegetable crops by accumulating spring flood waters in the water bodies of Volga-Akhtuba floodplain.

1. Introduction

The Volga-Akhtuba floodplain within the Volgograd region has a high biological and landscape diversity. It contains numerous water bodies: the river Akhtuba, stretching for 90 km within Volgograd region; Krasnoslobodsky tract (33 km long); Kashirinsky tract (99.3 km); 8 backwaters; 6 large ducts; 173 eriks, over 200 large lakes (unique accumulators of liquid and solid river runoff) and 2 marshes. The total area amounts to 206.8 ha [1].

In total, within the Volgograd region the Volga-Akhtuba floodplain has over 400 water bodies and valuable natural complexes: wetlands; spawning grounds of valuable commercial fish, including sturgeon; places of wetland birds density, including globally rare species; nesting of the white-tailed eagle; floodplain oak groves of a pedunculate oak special ecotype (an outpost of oak in the extreme southeast of Russia); habitats of relict plants (ferns); a unique combination of wetlands with gallery oak groves.

Today, being a specially protected natural area the Volga-Akhtuba floodplain requires an optimal hydrological regime. It is aimed both at preserving floodplain oak groves, soil fertility, spawning sites and at developing the agricultural potential of the territory.

The environmental parameters of water bodies are affected by the artificial regulation of the hydrological regime of the floodplain, as well as by the expansion of residential areas. Thus, they are transformed and lose numerous ecosystem properties that are significant for humans. Intense and unregulated anthropogenic impacts result in violations of the natural physicochemical parameters of water bodies, loss of biodiversity, degradation of the structure and functions of biological communities. It causes reduction in the range of potential ecosystem services - the benefits "that people get from ecosystems". The restoration of the lost properties of aquatic ecosystems is a key factor in increasing the sustainability of developed landscapes, creating comfortable conditions for living and managing the population [2].

The main factor that has caused the need for rehabilitation of the Volga-Akhtuba floodplain eriks and lakes is the change in the hydrological and environmental situation. The situation resulted from changes in the hydrological regime by regulating the river Volga flow after the construction of the Volzhskaya hydroelectric power plant.

Eriks are one of the main waterways of the Volga-Akhtuba floodplain. Numerous sub-parameters of floodplain water bodies depend on their water content. In this regard, of practical importance is work aimed at improving the state of water supply to degraded water bodies. The degradation of water bodies can include siltation, overgrowth of alien shrub and woody vegetation, overlapping by unauthorized enclosures that result in losing most of ecosystem functions. The hydrological regime of floodplain ecological systems depends on the functioning of the main elements of natural water bodies.

An important part optimizing the hydrological regime of the network of tracts, eriks and floodplain lakes is played by both the volume of special release and the duration of maximum discharge through the Volgograd hydroelectric complex. The optimal flooding of the floodplain territory is achieved at maximum flow rates – 27,000 cu m/s, its duration being 14 days at least [3, 4].

The State Program of the Volgograd Region "The use and protection of water bodies, prevention of negative impact of water on the territory of the Volgograd Region for 2014-2020", approved by the Government of the Volgograd Region dated August 30, 2013 No. 453-п, provides for the implementation of environmental rehabilitation of the Volga-Akhtuba floodplain.

2. Materials and Methods

In order to minimize the degradation of the floodplain, and also to restore it, a fundamentally new solution was adopted instead of the traditional hydro mechanized works on clearing tracts, eriks and lakes of the Volga-Akhtuba floodplain. It is aimed at a comprehensive approach to the rehabilitation of floodplain territories proper, namely: clearing water bodies from reed vegetation and the removal of alien species of woody-shrubby vegetation that oppress the local floodplain flora; the formation of a complex bottom topography with the necessary depths and shallow waters, planting of trees and shrubs of a typical floodplain flora; formation of recreational qualities of the coastline.

InordertoachievethegoalssetintheframeworkofcooperationwithVolgograd Regional Committee of Natural Resources, Forestry and Ecology, the functioning of water bodies located in the Volga-Akhtuba floodplain against the background of environmental protection measures was reviewed and analyzed.

Sites for cleaning water bodies were chosen based on route and topographic surveys. Areas with noticeable rifts and tree blockages constricting the watercourse were visually identified. Based on the materials of topographic surveys, shallow areas of lakes and areas requiring an increase in the cross section of the rivers and channels were determined.

3. Ecological rehabilitation of water bodies in Volga-Akhtuba floodplain

Dredging and clearing of water bodies have a negative impact on biocenoses, since there is a direct destruction of tree-shrub and grass vegetation in the work area, storage of soil, and placement of construction sites. Without active recovery, plant complexes are either restored for decades, or are not restored at all. In the clearing zone, perennials are almost completely replaced by ruderal annuals. Alien (invasive) plant species easily naturalize disturbed habitats. They occupy habitats typical for the floodplain of black poplar stands, willows and osier beds, changing the properties and functions of natural ecosystems. It results in the change in forest-forming species. Studies show that the restoration of the normal functioning of aquatic ecosystems in a natural way without disturbing the bottom surface takes place in the second or the third year. Studies conducted in 2013 on the Krasnoslobodsky water tract, where large-scale dredging was carried out, showed that the biocenoses restoration processes are very slow. After the hydraulic works, the riverbed was a ditch filled with water. They

IOP Conf. Series: Materials Science and Engineering **828** (2020) 012012 doi:10.1088/1757-899X/828/1/012012

completely destroyed coastline biotopes which are characterized by the highest rates of species richness and species diversity in the water body were overgrown with macrophytes. The restoration of this biotope and overgrown biocenosis was not observed even a year after the work completion. The species composition of the bottom fauna was characterized by extremely low rates. In cleared areas, only 7 zoobenthos species were detected [5].

So, on the Krasnoslobodsky waterway, measures were taken to rehabilitate the eriks Degtyarny, Dudarev, Nareznoy and Zhernova — more than 16 km long, and erik Averkin — 4.8 km long.

In 2019, the environmental rehabilitation measures of erik Sudomoyka, Sakharny and Lake Zapornoye were started. More than 5,500 trees and shrubs are planned to plant on the area of more than 57 hectares.

On the Kashirinsky waterway, measures were taken to rehabilitate erik Kashirin and duct from erik Kashirin to Lake Proklyatoe, over the area of 29 hectares.

Figure1. Riverbed woody vegetation of the erik Sudomoyka

In 2019 rehabilitation measures the eriks Chaika, Old Kashirin, the lakes Proklyatoe, Kamyshistoe and Kruzhnoe started: more than 13,000 trees and shrubs were planted on the area of over 68 hectares; rehabilitation of 19 km long erik Shumrovaty: planting over 10,000 trees and shrubs on the area of 220 hectares; and rehabilitation of Lake Shirokogorloye: planting over 5,100 trees and shrubs on the area of 128 hectares.

As a result of implementing these measures, it is planned to restore gradually the natural state of the floodplain territories, to create a balanced ecosystem with an effective self-cleaning mechanism, which will increase the area of irrigated land for vegetable crops by accumulating spring flood waters in the previously created reservoirs of water bodies of the Volga-Akhtuba floodplain.

Figure 2. Lake Malaya Nevidimka

Figure 3. Lake Bolshaya Nevidimka

IOP Conf. Series: Materials Science and Engineering **828** (2020) 012012 doi:10.1088/1757-899X/828/1/012012

Based on the materials of visual route surveys of sites with the probe bottom examination in certain sections of water bodies, the need for the following activities is considered:

- clearing the channels of eriks, the ducts and lakes with dredging;
- clearing trees fallen into water bodies;
- clearing the bottom of household waste in areas adjacent to settlements.

The formation of hydraulic parameters that increase the throughput and capacity within one water body, without going beyond the boundaries of coastlines, enables to improve the sanitary condition of water bodies in the low-water period.

Technical rehabilitation is carried out on the formed, stable (in the lay-out) channel without cutoff.

These measures contribute to improving the throughput capacity of water bodies, both during the period of maximum discharges and during the phase of the flood rise and fall, besides, the network of existing eriks and the channels provides hydraulic connection to numerous lakes of the Volga-Akhtuba floodplain.

Currently, the water bodies of the Volga-Akhtuba floodplain are mainly a typical example of natural systems degradation due to high anthropogenic impact: water bodies are located close to highways, surrounded by villages, dammed up to ensure the safety of settlements in floods; lands adjacent to water bodies are used for agriculture. The built-up area increasing and highways being constructed, lakes and rivers lost their flowage.

In this regard, there is a justified need for a comprehensive cleaning and deepening of water bodies of the Volga-Akhtuba floodplain as of flood water receivers.

General parameters and dynamics of the model territory (landscape), including the lake Zapornoe:

- The total area of the model plot is 101.5 sq km;
- The build-up area (%): 1940 - 6.298%; 1984 - 9.271%; 2012 - 16.157%. Thus, in the period from 1940-2012 the build-up area increased by 156.5%;
- The population size at the model plot in 2010 amounted to 21601 people;
- Density of population 213 people/sq km;
- Density of road and path network (km/sq km): 1940 - 0.835; 1984 - 0.834; 2012 - 2.089.

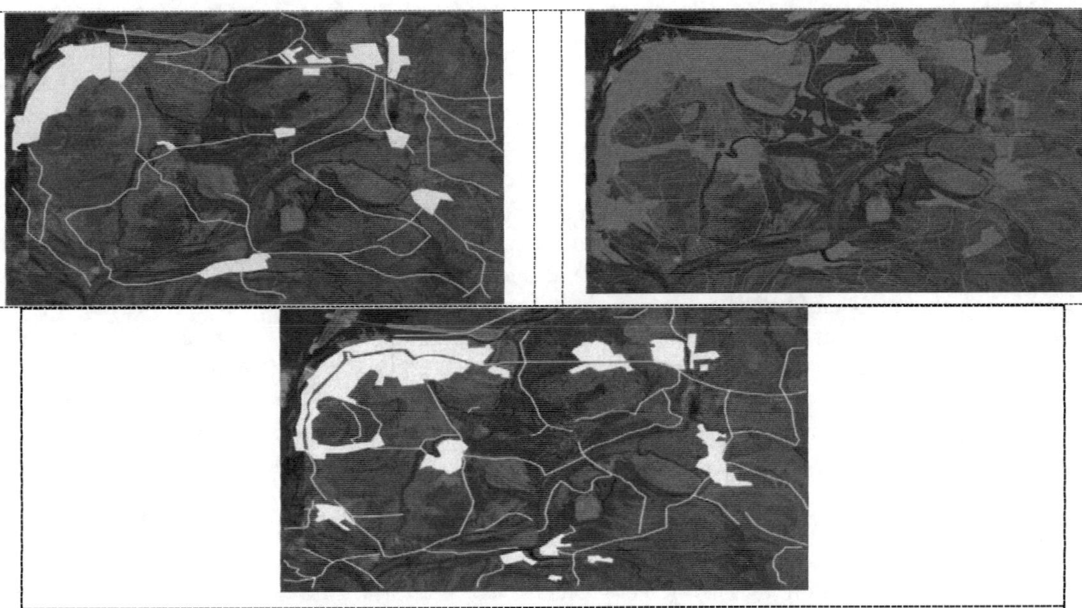

Figure 4. Comparison of the residential area near the lake Zapornoye in 1940, 1984 and 2012: the state of building-up and of the road-path network (based on data from topographic maps of 1940, 1984 and a space photograph of 2012 near the lake Zapornoye)

Flood scale and duration decreasing, siltation processes accelerate, eriks and lakes are quickly overgrown with air-water plants. Overgrowth accelerates the siltation of shallow waters and enhances waterlogging, slows down the water flow during floods, and contributes to a decrease in flora and fauna diversity both in shallow water and in adjacent meadows. Due to the multiple mechanical dredging of the water bodies beds, the hydrobiont fauna and the aquatic flora, which should maintain water quality, are depleted. Fish productivity is steadily declining. Ground vegetation is highly transformed as a result of high anthropogenic impact. The animal population corresponds to the parameters of the anthropogenically disturbed landscape. Habitats of aquatic plants, of bottom invertebrates, as well as nursery grounds for hatchlings were lost. Thus, the water bodies have almost completely lost most of their ecosystem functions and need to be restored through the implementation of special rehabilitation measures [6].

Figure 5. Common reed thickets on the lake Zapornoye

Having analyzed the state of water bodies, we believe, that environmental rehabilitation of the Volga-Akhtuba floodplain should be based on ecosystem restoration measures. They include increasing water availability in the discharge flow of the Volgograd hydroelectric complex, provided for by the operation regulations (at the level of 25000-27000 cu m/s, duration 5 - 7 days); maintaining the design levels after the flood decline, restoring typical biodiversity groups in the formed habitats (tree-shrub and grassy vegetation) and improving water quality [7].

Bed formation in eriks and ducts enables to increase throughput both during the peak of the flood and during the phases of ups and downs. This measure enables the lakes to be filled in the flood for 5-7 days to the marks of the normal headwater elevation (hereinafter NWL).

Improving the quality of water, maintaining design levels after the flood decline and the formation of recreational qualities of the coastline are achieved by removing thickets of aquatic vegetation. It is done in a part of the area between the boundaries of the water surface at the NWL and the top of dead storage (TDS). Besides, the bottom of the lakes is cleared of household clogging garbage, abandoned nets and other ownerless fishing gear. Works are also underway in the water protection area of the channels to the lakes; alien trees being eliminated and local trees being planted.

Ecological rehabilitation of water bodies is an environmental protection measure aimed at optimizing the hydraulic regime and restoring the normal functioning of the main elements of the ecological systems of water bodies.

Considering the above provisions the ecological rehabilitation of the water bodies is carried out in 2 steps:

- Technical phase:

- littoral formation with depths 0.2 to 0.7 m at least 20% of the lake area, and with depths 0.8 to 1.5 m 20% at least.

- the formation of a pelagic zone with depths of 1.6 to 2.0 m at least 20% of the lake and with depths of 2.1 to 3.0 m at least 10% of the lake.

- Biological phase:

- According to the ecological conditions of the landscape under rehabilitation, primarily, the level of water flooding, it is essential to restore tree-shrub vegetation, meadow floodplain communities, submerged littoral vegetation and to reconstruct of the helophyte community.

Considering the real conditions, based on the specific local conditions on the area to be cleared (mainly on the depths, the width of the clearing and the type of soil), clearing can be carried out by two types of mechanisms: an excavator and a dredger.

Clearing and dredging of water bodies on cramped and built-up plots of the coast are supposed to be done by the dredger. In the areas where excavation equipment and vehicles are accessible, excavation is carried out.

Figure 6. Excavator clearing the erik

Figure 7. Dredger clearing the erik

The channels are cleared by an excavator up to the design level with the subsequent loading of soil onto vehicles and transportation to the permanent dump for the allocated places for their placement.

Water availability and cross section dimensions determine channel formation works with the dredger. This eliminates the need to remove trees and shrubs along the banks of water bodies.

When clearing water bodies with a dredger, the extracted soil is transported via floating and onshore soil pipelines to the alluvium map. It is washed in a non-cascading manner into the alluvial map with a ring levee. The pulp is supplied through a welded main pulp line. Soil on the map is laid by slip-joint pipes or quick-disconnect pipes. The clear water is discharged from the map through the discharge chandler well to erik over the discharge pipeline through the local treatment plant.

Before the works start, the fertile soil layer is removed from the surface of the designed alluvial maps and the base of the map dipping dams for subsequent reclamation of alluvial maps.

Fallen trees during the clearing of water bodies of the Volga-Akhtuba floodplain are removed either manually or from a motorboat by sawing and towing to the point of extraction to the shore. Trunks are subject to sawing and transferring to the population as firewood, unwanted branches are transported to a household waste dump.

The implementation of these measures results in the gradual restoration of the natural state of the floodplain area, the creation of a balanced ecosystem with an operating self-cleaning mechanism. Besides, after the spring flood, it will enable to ensure the use of previously created storage tanks of water bodies to increase the irrigated areas during the cultivation of vegetable crops.

However, the close location of the Volga-Akhtuba floodplain to Volgograd and Volzhsky, as well as the historical features of the floodplain territory use, namely the development of housing construction, recreation, agriculture, leave their imprint on the preservation of the unique and typical natural complexes of the floodplain.

IOP Conf. Series: Materials Science and Engineering **828** (2020) 012012 doi:10.1088/1757-899X/828/1/012012

Figure 8. Removing fallen trees while clearing erik

4. Conclusions

Studies on the ecological rehabilitation of floodplain water bodies provide for an increase in the receiving capacity of lake basins, water availability, flow rate and depth of water bodies. Rehabilitation will enable to increase the filling rate Zapornoye lake with water and, accordingly, to reduce the introduction of sediment deposits from fields and meadows. Therefore, the effects of Volga flood regulation will be mitigated.

Water quality as a result of ecological rehabilitation will be improved due to restoration of aquatic vegetation developing in the littoral zone of lakes. There is a large amount of scientific data indicating the existence of a close relationship between the vegetation of higher aquatic plants, self-cleaning of water bodies, the circulation of substances and the processes of formation of water quality in them.

- higher aquatic plants, due to their morphological (stem structure, location of organs, etc.) and environmental (density of thickets, etc.) features, can serve as a biological filter when suspended and poorly soluble contaminants enter the water body [9];
- higher aquatic plants can serve in a water body not only as mechanical suspension filterers, they are also able to absorb and accumulate mineral and organic substances, to exhibit a certain activity in the processing of suspensions using them in the course of metabolism, thereby they affect the quality of water, being the main factor in regulating water quality [10];
- ions absorbed by higher aquatic plants are removed from substances cycle in a water body not for a short period of time, as is the case with the development of phytoplankton, but for a longer period - 5 to 6 months;
- biogenic elements and organic compounds obtained by plants do not remain ballast, but are used by them to maintain functional activity aimed at the formation of biomass containing a significant amount of proteins, amino acids, carbohydrates, etc. and can be used for business purposes [11];
- coastal-aquatic plants inhibit the development of phytoplankton and "blooming of water" with a biomass of 1.5 kg / sq m [9].

An important direction in planning and implementing environmental rehabilitation of disturbed landscapes is the restoration of a typical vegetation cover, since it is plants that form the habitat structure for various groups of animals on a landscape basis [12].

The restoration of full-fledged bottom community's characteristic of the floodplain should be expected after the formation of thickets biocenosis. In this case, the restoration of zoobenthos, in the presence of nearby non-drying water bodies, a kind of refugiums, occurs quite quickly. This is facilitated by the high adaptation of the taxonomic and ecological complex of benthic fauna of the floodplain to habitat in stressful situations. Species richness and biodiversity of the bottom fauna of

IOP Conf. Series: Materials Science and Engineering **828** (2020) 012012 doi:10.1088/1757-899X/828/1/012012

the cleared channel to Proklyatoye lake will be restored. Erik Kashirinis the refugial water body from which hydrobionts will penetrate and populate.

The process of fish fauna restoration in disturbed areas is determined by the characteristics of the water bodies of the northern part of the Volga-Akhtuba floodplain. During floods, floodplain water bodies, as well as the Volga and Akhtuba for a period form a single water system. Fish are given the opportunity to move within it, and they actively use it. Thus, during high water there is a redistribution of fish between floodplain water bodies, the Volga and the Akhtuba [4]. The increase in depth in some parts of the water body to 2.5 - 3 m creates more favorable conditions for fish wintering. In addition, the extraction of excess bottom sediments saturated with detritus from water bodies improves the hydro-chemical regime, thus reducing the possibility of clogging phenomena.

To maintain the desired structure of natural complexes in the landscape, it is necessary to formulate these parameters and requirements and to use management tools to maintain the described structure or to change it in the direction desired.

References

[1] Shoda M E, Sprague L A, Murphy J C and Riskin M L 2019 *Sci Total Environ* **650** 2314–2324. DOI:10.1016/j.citotenv.2018.09.377.

[2] Hoagland B, Schmidt C, Russo TA, Adams R and Kaye J 2019 *Sci Total Environ* **649** 979–994. DOI:10.1016/j.scitotenv.2018.08.379.

[3] Wilk-Woźniak E, Walusiak E, Burchardt L, Cerbin S, Chmura D, Gąbka M and Żbikowski J 2019 *Ecological Indicators* **98** 736–742. DOI:10.1016/j.ecolind.2018.11.025.

[4] Chen J, Zhu W, Zheng Y, Tian Y Q and Yu Q 2019 Data *Water Science & Technology: Water Supply* **19 (1)** 274–281. DOI:10.2166/ws.2018.077.

[5] Brylev V A and Ovcharova A Yu 2013 *Izvestia Universities. North Caucasus region. Natural sciences* **5** 72–75.

[6] Loboyko V F, Ovcharova A Yu and Nikitina NS 2018 *Izvestia of the Lower Volga agro-university complex: Science and higher professional education* **4 (52)** 89–96.

[7] Ovcharova A Yu 2013 *Izvestia Universities. North Caucasus region. Natural sciences* **1** 77–80.

[8] Tyutyuma N V, Tuz R K and Konev S V 2017 *Izvestiya Nizhnevolzhskiy agrouniversity complex: Science and higher professional education* **1 (45)** 175–181.

[9] Ovchinnikov A S, Loboyko V F, Ovcharova A Yu, Kosenkova S V and Agisheva NR 2018 *Small rivers of the Volgograd region* (Volgograd: "Printing yard") 208 p.

[10] Ovchinnikov A S, Borodichev V V, Dedova E B and Sazanov M A 2015 *Izvestiya Nizhnevolzhskiy agrouniversity complex: Science and higher professional education* **3 (39)** 9–19.

[11] Lizunkov V, Politsinskaya E, Malushko E, Kindaev A, and Minin M 2018 *International journal of energy economics and policy* **8 (3)** 250-257

[12] Kruzhilin I P, Ganiev M A, Melikhov V V, Rodin K A, Dubenok N N, Ovchinnikov A S, Fomin S D and Abdou N M 2017 *J. of Eng. and Applied Sci.* **12 (24)** 7118–7123.

Lexical analysis of technical terms in the field of materials science

O O Kandrashkina[1] and E V Revina[1]

[1] Samara State Technical University, 244 Molodogvardeyskaya street, Samara, 443100, Russia

E-mail:petergrifon@mail.ru

Abstract. The article presents lexico-morphological analysis of terminology in the research articles on materials science. The article considers lexical features distinguished in scientific texts. The main purpose of the paper is to reveal the most frequent and productive types of word formation used in making specialized words and terms in technical English. For carrying out the linguistic study the analysis of the immediate constituents has been applied. Special attention is given to affixation, composition and contraction as the means of forming new words in the materials science engineering. Compounding or composition has been regarded as an important source of forming new words. Compound adjectives functioning as attributes are found to be primarily used for describing materials physical and chemical properties, conditions and stages of the experiments. Derivational units include borrowed and native productive or non-productive frequent affixes connected with different scientific notions. The study has been carried out on the technical terms and vocabulary from the journals "Advanced engineering materials" and "Advanced composite materials".

1. Introduction

Language used in science describes and interprets the world, projecting the objective character of scientific knowledge. Scientific and technical communication can be defined as a process of gathering, organizing, presenting and refining information. It is also a process inevitably shaped by its contexts, and which is improved when it recognizes its contexts.

The language of science is governed by the aim of the functional style of scientific prose. The main aims of any scientific work are to prove a hypothesis, to create new concepts, to disclose the internal laws of existence, development, relations between different phenomena. The language means used in science tend to be objective, precise, unemotional, and impersonal.

Any scientific text employs a special technical vocabulary which consists of two main groups: words associated with professional communication and a less exclusive group of so-called learned words.

Scientific English generally includes terms and specialized vocabulary. Terminology may refer to three concepts: terminology as the study of terms, terminology as the practical aspect of doing terminographical work, and terminology as a set of specialized terms. In this article we consider terminology as a set of specialized terms [1]. Terminologies as the lexical components of specialized languages emerge from theoretical and technological innovation: new scientific insights and novel tools enrich the conceptual and practical environment of the specialists, and in the process expand

IOP Conf. Series: Materials Science and Engineering **828** (2020) 012013 doi:10.1088/1757-899X/828/1/012013

their vocabularies. A term is a dynamic phenomenon that is born, formulated and delves into the process of cognition, the transition from a concept (mental category) to a verbalised concept associated with some theory to conceptualise a particular field of knowledge or activity [2]. This understanding of the term is inextricably linked with explaining connections between the structures of language and knowledge structures.

The written form of scientific discourse is characterized by the following lexical features:

a. lexicalization of *-ing* and *-ed* verb forms and their transition into adjectives and nouns;
b. a great number of words derived from one radical stem. The word-terms of the one root are quite abundant even within one given article;
c. the use of contracted or shortened words and acronyms connected with some specific experimental processes or names of an element;
d. the usage of compound words related to the properties of materials, characteristics, physical conditions of reactions and experiments.

Morphologically, the terms are mainly nouns or verbs semantically correlated with nouns and adjectives derived from nouns: absorption (n) — absorb (v), activation (n) — activate (v); anion (n) — anionic (adj), corrosion (n) — corrosive (adj), circle (n) — circular (adj). It should be noted that rows of the same root word-terms function in scientific discourse: adsorb—adsorption—adsorbent—adsorptive, corrode—corrosion—corrosive, diffuse—diffusion— diffused—diffusible [3].

So, lexical and morphological analysis of terminology in scientific articles presents special interest because they contain a great number of specialized terms which are characterized by the different types of word-formation.

2. Word-building types

We consider only those productive word-building types that are of importance for lexical analysis of terms from research articles, that is affixation, composition and contraction.

2.1. Affixation

Affixation is the process of making new when one or more affixes are added to a root. Affix is a derivational morpheme added to the stem and forming a new derivative belonging to a different part of speech or different word class. The meanings of affixes refer to the concept conveyed by the whole word to a certain category.

Affixation is generally divided into prefixation and suffixation depending if the morpheme precedes or follows the root.

There exists an essential difference between suffixal and prefixal derivatives. In the English language suffixation is a characteristic of noun and adjective formation, while prefixation is typical for verb formation. The other difference lies in the semantic structure of the suffix and the prefix [4]. The part-of-speech meaning has a much greater significance in suffixes compared to prefixes. Due to this, a prefix may be confined to one part of speech as or may function in more than one part of speech. Suffixes, on the contrary, function in one part of speech and often form a derived stem of a different part of speech as compared with that of the base.

2.1.1. Prefixation

Prefixation is the formation of words with the help of prefixes.

Prefixes may be classified on different principles. Diachronically there is a division into native and borrowed prefixes. Synchronically prefixes may be classified into:

a) the type of lexical-grammatical character of the base: de-verbal, de-nominal, de-adjectival.
b) the generic denotational meaning prefixes
   — negative (un-, non-, in-, dis-)

— reversative or privative (un-, de-, dis-)
— pejorative (mis-, mal-, pseudo-)
— prefixes of time and order (fore-, pre-, post-, ex-)
— prefix of repetition (re-)
— locative (super-, sub-, inter-, trans-)

Prefix of repetition *re-* is observed in the research papers in hyphenated way in order to distinguish the meaning. One of the lexical features of scientific writing is the abundance of words composed of Greek prefixes.

These derived lexical units are mainly adjectives and nouns and their semantics is directly connected with materials names that are used in conducting research or experiments.

Consider some examples: *bioceramics, biocompatibility, biomaterial*. All these words are the prefixal nouns. Prefix *bio-* is a borrowed one of Greek origin. The derived nouns, therefore, possess the meaning connected with living things or someone`s life or as in the case with *bioceramics* — the material with natural properties and that is not hazardous.

The other prefixes that are frequent in the articles are *-micro* and *-micro*. There are numerous examples of nouns formed with them: *microstructure, microlevel, microscale, micro-tomography, micro-hardness, macrostructure, macro-architectural, macroscale, macro-level.* The borrowed prefix of Greek origin *macro-* refers to the meaning of size ("large"), and words with the prefix of opposite meaning *micro-*, meaning "small in size" or "extremely small".

It should be noted that above-mentioned prefixes usually form so-called neoclassical compounds because of their connection with Greek or Latin. The only difference between the neoclassical forms and native compounds is that the non-native elements are obligatorily bound. They use the so-called combining forms, which are bound morphemes that in principle differ from bound roots and affixes, however, this distinction is difficult to draw in many cases [5]. Scholars consider these neoclassical formations as compounds and do not treat them as cases of affixation.

Other examples of widespread use of borrowed prefixes are *post-processing, post-sintering, post-fabrication*. The prefix *post-* has a temporal meaning of "after". In all the words it emphasizes sequence or consequence of processes expressed by the verb bases.

The non-native prefix of Latin origin *inter-* is found to show the relations inside or between processes or elements: *intermetallic, interdistance, interchangeable, interconnected*.

More examples of derivational units having non-native word-building elements in their form are words with prefix *nano-*: *nanoporous, nanoindentation, nano-interface.* The meaning of the prefix refers to connection with science and technology, describing the newest achievements in the particular field of study. It is also associated with measurements and has been coined to define a one billionth part of the whole unit.

As for negative prefixes some examples of non- have been observed: *non-oxide-based, non-dimensional, non-comnensurate, non-homogenity, nonionic, non-monotonic.* In contrast to *un-* and *in-*, negation with this prefix does not carry evaluative force. Furthermore, contradictory and complementary opposites of the concepts are formed via this morpheme.

2.1.2. Suffixation

Suffixation is the process of creating words with the help of suffixes. Suffixes usually modify the lexical meaning of the base and transfer words to a different part of speech. They can be classified variously using the word class of their base or using the word class newly formed, or both:

a) noun-forming suffixes: -er, -dom, -ness, -ation, -ing.
b) adjective-forming suffixes: -able, -less, -ful, -ic, -ous;
c) verb-forming suffixes: -en, -fy, -ise (-ize);
d) adverb-forming suffixes: - ly [6].

In the analyzed research articles the most common suffix added to the base is noun-forming frequent suffix of Latin origin -(t)ion: characterization, tension, compression, nucleation, dispersion, teration, solution, reduction, discretization, elongation, delamination, galvanization.

Noun-forming suffix -ty is mainly applied for describing the properties of a given material or element in the articles under analysis: resistivity, porosity, conductivity, dispersability, diffusitivity. In the given examples one can see that the derived noun form is made up of verb bases mainly (resist, conduct, disperse, fuse).

There are some examples of adjectives derived with a borrowed suffix of Latin origin *-al*: *quadrilateral, computational, mictostructural, multidirectional, colloidal, sacrificial, topological.*

To describe properties or states of the materials derivational adjectives are formed with the help of the productive frequent suffix *-ed* added to the verb or noun bases are used: *required, solidified, scaled, dealloyed, structured, boronized, prefabricated, infrared, heated, processed.*

According to the lexico-grammatical character of the radical suffixes are classified into the following types:

— de-verbal suffixes (added to the verbal base): *-er, -ing, -ment, -able*,
— de-nominal suffixes (added to the noun base): *-less, -ish, -ful, -ist, -some*
— de-adjectival suffixes (affixed to the adjective base): *-en, -ly, -ish, -ness.*

In the research articles under analysis there is a considerable number of de-verbal suffixes which make nouns. The derived nouns with productive *-ing* suffix possess the meaning of process implied in the verb base: *sintering, handling, leaching, cutting, grinding, processing, acidizing, strengthening, backtracking.*

There are also semi-affixes, which can either be attached to the base to form a word of can function as independent lexemes.

3. Composition

Composition is a process of word-formation where two or more lexemes combine into a single new word. Compound words may be represented by one word or by two words joined together with a hyphen. The syntactic relation between the words can be traced and shown. Compounds are made up of two immediate constituents (according to morphological analysis) which are called derivational bases. Compound words are inseparable vocabulary units.

Structurally compound words are characterized by the specific order and arrangement of the bases following one another. The order in which the two bases are placed within a compound is rigidly fixed in Modern English and it is the second immediate constituent that makes the head-member of the word, in other words, forms its structural and semantic centre [7]. The term head is generally used to refer to the most important unit in complex linguistic structures.

The classifications of compounds are based upon the type of composition, the linking element, the part of speech and upon the structural pattern.

As for presence or absence of connecting elements, compounds are divided into those without connecting elements or hyphened, compounds with a vowel or consonant as a linking element and composed words with linking elements represented by preposition or conjunction stems

According to the structure, compounds can be neutral (without linking elements). There are three subtypes of neutral compounds depending on the structure of the constituent stems: simple or simplexes consisting of simple affixless stems; derivational compounds or complexes having affixes in their structure; contracted compounds which have a shortened (contracted) stem in their structure.

The structural type of compound words and the word-building type of composition have certain advantages for communication purposes. The structural meaning of the derivational pattern of compounds may be abstracted and described through the interrelation of its immediate constituents.

Compound words usually belong to adjectives as a part of speech. Lexical analysis of technical vocabulary from the research papers revealed the following types of compounds. The first largest group are presented by hyphenated derivational or simple compounds formed to the pattern *Adj + Adj (adjective+adjective), Adj+N (adjective + noun), N+N (noun+noun), N+Adj (noun+adjective)*:

Adj+Adj: *hexagonal-prismatic, linear-elastic, electro-active, hot-extruded, electrically-actuated, electroless-deposited, multi-layered, electro-deposition, mixed-made, inter-tube;*

IOP Conf. Series: Materials Science and Engineering **828** (2020) 012013 doi:10.1088/1757-899X/828/1/012013

Adj+N: *cohezive-zone, high-purity low-voltage high-ductility high-strength, hot-pressing, reduced-weight*;

N+Adj: *load-transfer load-carrying, center-notched, node-release, size-dependent, element-based, energy-efficient, size-dependent, bulk-nanostructured, load-applied*;

N+N: *sress-strain, displacement-delamination, traction-separation, shape-memory, scale-removal, core-shell, liquid-phase.*

Adjectival compounds formed by the structural pattern *N+Adj* can have nouns or other adjectives as non-heads. The interpretation of noun-adjective compounds follows the principles that the non-head element can function either as a modifier as an argument of the head.

Within the structural pattern *N+Adj* in some cases we can observe with the derivational suffix *-ed*, whose derivatives can be paraphrased as "having some quality" or "provided with some quality or feature". Besides, compounds with adjectival heads that are based on past participles also can imply a passive interpretation.

The above-written compound words have the meanings connected with materials processing, the inner structure of element and its properties or for the description of physical and chemical conditions during the experimental studies.

The next group of hyphenated compounds are implied in technical writing to define the mixture of substances or elements under analysis: *graphene-oxide, ceramic-polymer, alumina-polymer, polymer — only, boride-based, single-crystal, ceramic-metal.*

Morphological compounds are quite few in number: *thermoelectric, hydrofluoric, morphometric, thermodynamics, lithiographic, electrochemical, flexographic, petrochemical.*

Compounds of different structural types are numerous in the articles under investigation. A very common type of compounds is syntactic ones. It can be illustrated by the following examples: *push-to-pull, length-to-diameter, coupling-to-tubing, thermal-to-electrical strength-to-density, roll-to-roll, smaller-is-stronger.*

All these composed lexical units have linking elements represented by preposition stems. These derivational adjectives are based on the semantic relations of sequence or consequence or manner. They are usually formed according to the pattern N+to+N or Adj+to+Adj.

We have divided composed words from the articles into several groups according to the pattern they are built upon.

4. Shortening (Contraction)

Contracted or curtailed words are produced in two different ways. The first is to make a new word from a syllable or two of the original word. The second way of shortening is to make a new word from the initial letters of a word group. This type is called initial shortenings. The new form acquires some linguistic value of its own.

Abbreviation refers to any shortened form of a word or a phrase. An acronym is an abbreviation of a special type that abounds in technical and scientific writing. Acronyms imply forming new elements not by combining existing morphemes and proceeding from sound forms to their graphic representation but coining new words from the initial letters of phrasal terms originating in texts.

Abbreviations can be a useful way to make writing and speaking more concise, less repetitive, and more accessible to nonscientists [8].

Abbreviation does not change the part-of-speech meaning - it produces words belonging to the same part of speech as the primary word. Initial abbreviation in which the first element is a letter and the second a complete word. Shortening may be regarded as a type of root creation because the resulting new morphemes are capable of being used as free forms and combine with bound forms.

There are few classifications of shortened words. The generally accepted division is based on the position of the clipped part. According to whether it is the final, initial or middle part of the word that is cut off we distinguish: 1)final clipping (or apocope) 2) initial clipping (apheresis) 3) medial (or syncope) [9].

Graphical abbreviations are the result of shortening of words and word-groups only in written speech while in spoken language full forms are used [10].

In technical writing abbreviations are used to make long names of experimental processes and long technical terms easier to remember and less tedious to refer to repeatedly in a scientific article. In such contexts the long name or technical term first is often given in full form with the abbreviation in brackets after it. After it has been mentioned, just the abbreviation is used. There are a lot of examples of this kind in scientific papers on materials science:

additive manifacturing — AM
computer-aided design — CAM
selective laser sintering - SLS,
dielectric elastomer actuator - DEA
electroactive hydraulic sar - EAHS
lamellar dislocation substructured - LDS
electron channeling contrast imagine - ECCI
electron backscatter diffraction - EBSD
misorientation histograms - MDH
surface mechanical grinding treatment - SMGT
sream-assisted gravity drainage - SAGD
thermal and cyclic steam stimulation - TSS and CSS
physical vapor deposition —PVD
chemical vapor deposition - CVD
carbon nanotube - CNT
resistance pressure sintering- RPS
SEM - scanning electron microscopy.

They are primarily used for the economy of space and effort in writing.

5. Conclusion

Scientists and researches make use of different terms and vocabulary units to convey the results of their research. Technical writing has its own lexical features, such as nominalization, acronyms, usage of Greek and Latin derivational word-building elements, affixation and composition as the main productive ways of word building [11]. The aim of the article was to define the most productive ways of creating new words in the field of materials science. It has been stated that composition is the primary source of word formations for creating new adjectives. It can be explained by the domain of research articles [12]. Materials science deals with new ways and technologies of improving materials properties and performance as well as with creation of new materials types. Therefore, compound adjectives of different structural patterns serve to describe all the core concepts and describe physical and chemical properties, conditions and stages of the experiment. The study of technical vocabulary shows that the hyphenated simple or derived compounds of structural patterns like *N+Adj, Adj+N, Adj+Adj, Adj+N* are the most common ones. More complex terms are sometimes expressed by noun phrases consisting of three or more words. These nominal groups are contracted to abbreviations in scientific writing. It saves the space and makes the long term shorter and clearer to the reader. Another type of word formation - affixation - is also proved to be a productive way of forming technical words in the field of materials science. The study has revealed the most frequent adjective-forming, noun-forming and adverb-forming suffixes. Prefixes that are mainly of Greek origin connected with concepts of size and nature (*micro-, macro-, bio-*). They are used to make derivational compounds like *microstructure, biomaterial.* Moreover, words derived with these combining initial forms are called neoclassical compounds and they present an active source of vocabulary extension in the English language. Other derivational affixes observed in the articles are negative prefix *non-*, locative prefix *inter-*, adjectival suffix *-al*, noun-forming suffix *-ion* and adjectival suffix *-ed* and

The above-mentioned features tend to cluster together as lexical characteristic of scientific writing. The result of the current study will be immensely useful for the researchers who regularly present their findings because it can help them to understand the technical vocabulary implied and to create new words which are connected with materials, their chemical and physical properties, procedures and stages of the experiments involved. Moreover, being aware of lexical features in scientific articles it is of great importance to scientists who are not native English speakers, as it helps them to get the main point of the article and to understand the text more clearly.

References

[1] Kockaert H and Steurs F 2015 Handbook of Terminology vol 1 (Amsterdam: John Benjamins Publishing Co.) pp. 507-508
[2] Leychik V 2007 Terminology: Subject, Methodology, Structure (Moscow: Librokom) pp. 21-22
[3] Rusko T 2014 COACTIVITY: Philology, Educology 22(1) 82-83
[4] Ginsburg R 1979 A Course in Modern English Lexicology (Moscow: High School) pp.110-112
[5] Diaz-Negrillo An 2014 Linguistik Online 68 5-6
[6] Nikolenko A 2007 English Lexicology – Theory and Practice (Vinnitsa: New Book) pp. 51-53
[7] Jackson H and Ze Amvela E 2007 Words, Meaning, and Vocabulary: An Introduction to Modern English Lexicology (London: Continuum) pp. 92-93
[8] Day R and Sakaduski N 2011 Scientific English: A Guide for Scientists and Other Professionals (3rd Edition, Greenwood) pp. 90-91
[9] Antrushina G, Afanasieva O and Morozova N 2019 English Lexicology (Moscow: URSS) pp. 115-116
[10] Malushko E, Bolsunovskaya L, and Martyushev N 2019 Asian EFL Journal 23 (3.2) 315-328.
[11] Minin M G, Politsinskaya E V, and Lizunkov V G 2019 Vysshee Obrazovanie v Rossi 28(10) 83-95.
[12] Shamne N L, Milovanova M V, and Malushko E Yu 2019 IOP Conference Series: Materials Science and Engineering 483(**1**) 012081. DOI: 10.1088/1757-899X/483/1/012081

IOP Conf. Series: Materials Science and Engineering **828** (2020) 012014 doi:10.1088/1757-899X/828/1/012014

Using Russian Federation taxes information and analytical system to monitor economic condition of the manufacturing industry

A A Ksenofontov, A Sh Kamaletdinov, I V Trifonov, P V Trifonov and N A Cherepovskaya

Financial University under the Government of the Russian Federation, 49 Leningrad Ave., Moscow, 125993, Russia

e-mail: a.ksenofontov@mail.ru

Abstract. This paper describes a study of the financial and economic state and functioning of economic sectors of the constituent subjects of the Russian Federation based on the application of advanced information technologies. This study aims to analyse the performance of the Russian Federation national economy and the economy of 85 Russian Federation constituent subjects with a breakdown by different economic activities based on the Russian Federation Taxes information & analytical system of regional tax revenues designed. Basic calculations were made using the index method of economic activity evaluation developed by the authors. The index construction technique makes it possible to analyse dynamic trends in general and individual index development. Based on the method proposed, the constituent subjects of the Russian Federation have been ranked in accordance with their tax revenue indices for all kinds of economic activities in 2016. The efficiency of tax revenues by individual indicators included in the index has been reviewed. The constituent subjects of the Russian Federation have been ranked by the "Manufacturing Activity" indicator. Index distribution indicators have been calculated for each constituent subject of the Russian Federation.

1. Introduction

For the time being, the situation in the Russian Federation national economy is rather difficult. The economic development of the country is affected by both external and internal factors. External factors hampering the economic development include geopolitical situation aggravation, continuous expansion of sanction lists aimed to impede business activities of Russian companies, world market fluctuations, energy price fall, etc. The list of internal factors slowing the pace of economic development of the country is even broader. It includes prolonged financial and economic crisis, technological backwardness of domestic companies and industries, inefficient use of human and natural resources, unfavorable conditions for granting loan facilities to companies in the domestic market, aging and low commission rates of fixed assets.

One of the ways out of this situation is a detailed financial and economic analysis of the Russian national economy and economy of the 85 Russian Federation constituent subjects. This kind of study will assist in searching for those constituent entities, which require development support, and will provide the keys to identifying the industries requiring improvements. The proposed methodology was implemented in the analytical module of intellectual studies of the Russian Federation Taxes information & analytical system.

2. Methodological foundations of the study

The subject of this study is the economy of the Russian Federation constituent subjects and the industries running in their territory. For the study, 85 Russian Federation constituent subjects and 13 types of economic activity were selected. The purpose of the study is the financial and economic analysis of 85 constituent subjects of the Russian Federation and the types of economic activities operating on their territory. The task of the study is to develop and apply methods for evaluating the performance of economic activities in the administrative entities under study.

These studies were realized using economic & statistical methods, system analysis and general scientific comparison and correlation methods. The main calculations were made using the index method of evaluation types of economic activity, developed by the authors. The index method refers to the statistical analysis methods and is applied in many science and technology sectors. This method was specifically applied in economic studies.

To date, this approach is the most popular in macroeconomic studies. For example, at the World Economic Forum in Davos in 2018, it was suggested to use the country's Inclusive Development Index (IDI). The application of the index method in the economic studies is described in several papers issued by domestic [1] and foreign [2] authors. Particular attention is paid to the development of methods for calculating the stable economic well-being index known as Genuine Progress Indicator (GPI). GPI was developed by Herman Daley and John Cobb in 1989. The scientists suggested this index as the "economy assessment method that gives the best benchmarks for those who are interested in the economic well-being improvement" [3]. Establishing the GPI was one of the successful attempts to synthesize an aggregated monetary index. At present, the following papers [4], [5], [6] are known in the field of GPI research.

The environmental economy often uses the Environmental Sustainability Index (ISE), developed by the Columbia and Yale universities for the World Economic Forum in Davos. At present, the applied aspect of using the ISE is described in the papers [7] and [8]. The index method may also be used, for example, in logistics [9], in analyzing the green production development [10] or in analyzing the efficiency of production activities in various sectors of the economy [11], [12], etc. The index method can also be used to assess the development of the Russian Federation territories.

3. Results of the study

To ensure an objective assessment of the economic activity of the Russian Federation constituent entities, the authors suggest the method that is based on the integral indicator of tax revenues from all kinds of economic activity types (tax revenue efficiency index). To evaluate the effectiveness of tax revenues in the constituent entities of the Russian Federation using the index method is suggested. Statistical data for 2015–2016, presented by the Federal Tax Service of the Russian Federation, and statistical data on the main socio-economic indicators, provided by the Federal State Statistics Service of the Russian Federation, were used to compute the index values. To simplify subsequent analysis, the acquired data was consolidated in the database of the Russian Federation Taxes information & analytical system of regional tax receipts.

Let's introduce the relative performance indicator that is the ratio of tax income value to the average annual number of employed populations of a constituent entity for a certain type of economic activity. The Ministry of Economic Development of the Russian Federation and the Federal Agency for Technical Regulation and Metrology developed the "All-Russian Classifier of Types of Economic Activity - OK 029-2014". This document provides in-depth details of all the economic sectors of the Russian Federation, which are enlarged into 13 groups:

- Agriculture, hunting and forestry; fishing, fish farming;
- Mining of mineral resource;

- Manufacturing activity;
- Production and distribution of electricity, gas and water
- Construction
- Wholesale and retail trade; repair of motor vehicles, motorcycles, household goods and personal items
- Hotels and restaurants
- Transport and communication
- Real estate transactions, lease and provision of services
- Education
- Health and social services
- Providing other public, social and personal services
- Other activities.

Let's call the suggested indicator as called the relative tax revenue efficiency index (RTREI) by the type of economic activity. Then, for each constituent entity, thirteen such indicators *RTREIij* can be defined according to the number of economic activity types, where the index i takes values from 1 to 85 (according to the number of the Russian Federation constituent entities, n = 85), and the index j corresponds to the economic activity type (m = 13) and varies from 1 to 13. This average figure for an entity characterizes the tax income, brought to the country's budget by one employee engaged in the concerned type of economic activity. In the future, let's denote this indicator for simplicity of designation and use as Xi,j.

Then, let's calculate for each type of economic activity the average value of the relative tax revenue efficiency index for all constituent entities. As a result, we will obtain thirteen average values, which were determined by the simple average determination formula $\bar{X}_j = \frac{\sum_{i=1}^{n} X_{ij}}{n}$.

Then, for each tax, let's determine the variance D_j and root-mean-square deviation σ_j of the relative tax revenue efficiency index for all constituent entities. These variation indicators were calculated using the following formulas $D_j = \frac{\sum_{i=1}^{n}(X_{ij}-\bar{X}_j)^2}{n-1}$ and $\sigma_j = \sqrt{D_j}$. By using calculated average values, let's center all relative indicators X_{ij} according to the formula $\dot{X}_{ij} = X_{ij} - \bar{X}_j$. Now, if for all computed centered relative indicators $\dot{X}_{ij}$ we use a normalization operation with the help of the root-mean-square deviations σ_j, we will receive standardized relative tax revenue efficiency indices (indicators) for each type of economic activity for all constituent entities $U_{ij} = \frac{\dot{X}_{ij}}{\sigma_j}$. Note that the values U_{ij} are normalized and are dimensionless. In addition, all average values of standardized relative tax revenue efficiency indices for each type of economic activity are equal to zero, i.e. $\bar{U}_{ij} = 0$ for all j , whereas the all the variances of these indicators $D_j = \frac{\sum_{i=1}^{n}(U_{ij}-\bar{U}_j)^2}{n-1}$ are equal to one.

Since tax revenues from different economic activities are affected by many factors of a different nature, it can be assumed that standardized values U_{ij} are subject to different laws of distribution, but with the same averages and variances. Thus, each constituent entity of the Russian Federation may be characterized by a system of dimensionless indicators with the same average, values equal to zero, and variances, equal to one.

If we summarize the indicators obtained by all types of economic activity for each constituent entity, then we will receive a certain aggregate indicator that reflects the efficiency of tax revenues in the constituent entity for all types of economic activity: $I_i = \sum_{j=1}^{m} U_{ij}$. The resultant tax revenue efficiency index is based on 13 indicators, and each indicator corresponds to the specific economic activity type and evaluates the level of economic development of the constituent entity (index value in the i-th constituent entity is I_i).

The suggested tax revenue efficiency index is an integral indicator that can be used to measure economic development level of regions, their comparison and assess tax revenue efficiency in various economic activities. The index constructing method makes it possible to analyze its dynamics and variations of its individual indicators, which allows evaluating changes in the economic activities of constituent entities dependent on the redistribution of the role of economic activities in terms of tax revenue efficiency. The calculation method suggested allows the use of a variety of different measurements in any socio-economic system (evaluating the resource, economic, social and other system potentials). This makes it possible to obtain a comprehensive indicator of the economic status and development prospects of the system as well as to assess its potential, define goals, identify infrastructural problems and economy diversification shortcomings, and assess investment risks and threats.

Below are some examples, which demonstrate possible use of the tax revenue efficiency index while analyzing the economic activities of constituent entities of the Russian Federation. Fig. 1 shows the ranking of the Russian Federation constituent entities in accordance with the tax revenue efficiency index values I. The constituent entities are fairly strong stratified by this indicator. 28 constituent entities have a positive value of this index (above the average level), and 57 constituent entities have a below-zero value of this index. From this picture it can be concluded that the economic development of the Russian Federation constituent entities is very uneven.

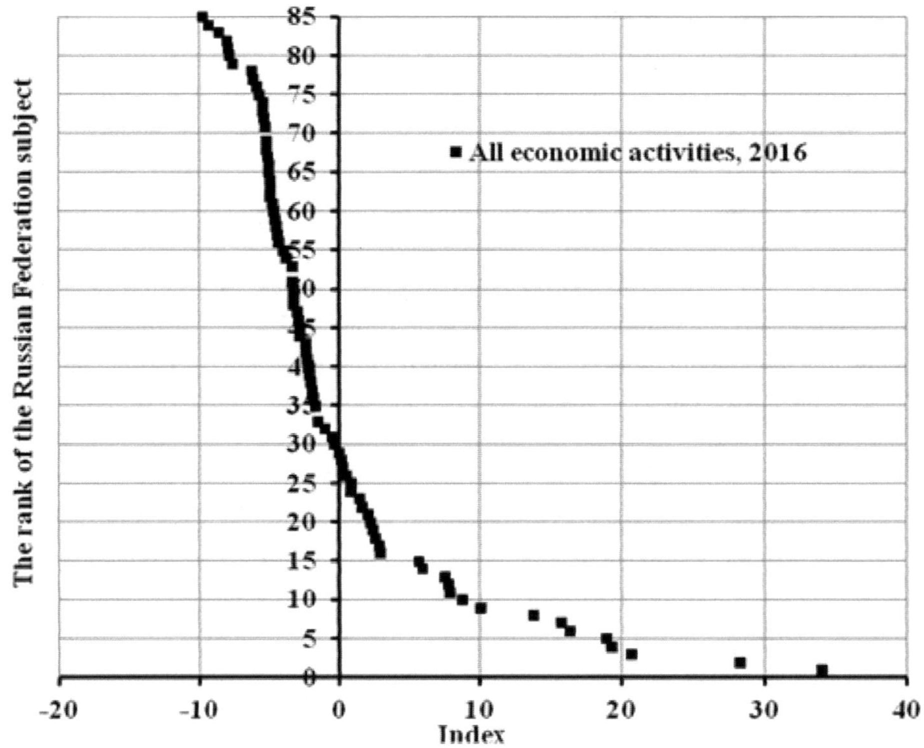

Figure 1. Distribution of the Russian Federation subjects in accordance with the index of tax revenues for all types of economic activity in 2016

Further, tax revenue efficiencies for specific types of economic activity, i.e. for individual indicators included in the index, may be analyzed. Fig. 2 is a graph, visually showing the distribution of the Russian Federation constituent entities by the Manufacturing Activity indicator. In total, 25 constituent entities are above the average level (zero mark), and 60 constituent entities are below this mark. Yamal-Nenets Autonomous District, Leningrad Region, St. Petersburg, Kaliningrad Region, Komi Republic, Omsk Region, Ryazan Region, Yaroslavl Region, Khanty-Mansi Autonomous Area, Khabarovsk Region and Saratov Region are considered the most developed constituent entities by this indicator. Chechen Republic, Republic of Ingushetia, Republic of Kalmykia, Republic of Tyva, Astrakhan Region, Jewish Autonomous Region, Republic of Buryatia, Amur Region, Ivanovo Region and Republic of Sakha (Yakutia) are the weakest ones.

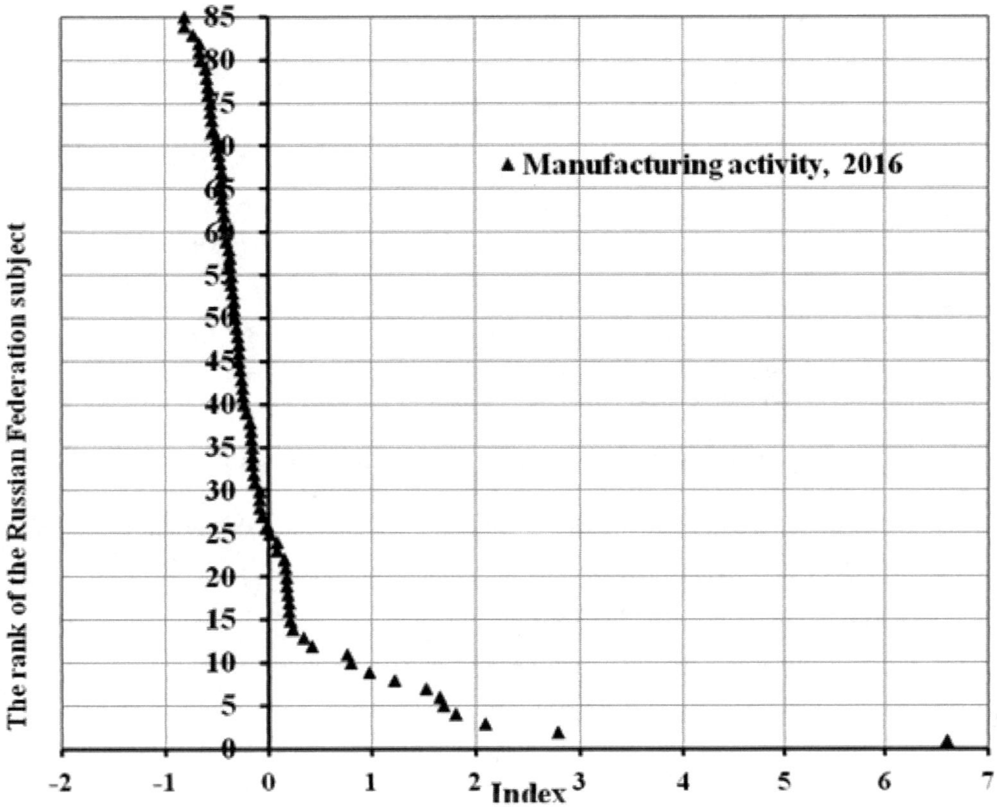

Figure 2. Distribution of the Russian Federation subjects in accordance with the index of tax revenues by type of economic activity "Production and distribution of electricity, gas and water" in 2016

4. Conclusions

To conclude, we note that this article provides a review of scientific papers of domestic and foreign authors describing the application of the index method and its indicators. It has been revealed that this method is widely used in various fields of economics and management. The tax revenue efficiency index, which is based on 13 indicators, each of which corresponds to the specific type of economic activity and evaluates the economic development level of the concerned constituent entity of the Russian Federation, has been suggested. The method developed has been embedded in the analytical module of the Russian Federation Taxes information & analytical system. Based on the method suggested, the constituent entities of the Russian Federation have been ranked in accordance with the value of the tax revenue efficiency index for all foreign economic activities in 2016. The efficiency of tax revenues has been reviewed by individual indicators included in the index. Using this approach, the constituent entities of the Russian Federation have been ranked by the Manufacturing Activity indicator.

It should be noted that the index method presented in this paper makes it possible to evaluate the efficiency (productivity) of tax revenues in the constituent entities of the Russian Federation simultaneously for all types of economic activity, compare constituent entities with each other according to the value and structure of this indicator, classify the constituent entities of the Russian Federation in terms of the efficiency of tax revenues, analyze dynamic index changes and predict its values in the future. The suggested method of computing the index allows the use of a variety of factors of different nature, which affect the socio-economic system (economic, socio-demographic, technological, etc.). As result, a comprehensive indicator of the activities and development of the system may be obtained, system potential can be assessed, goals can be defined, infrastructure problems can be identified, shortcomings in the economy diversification can be revealed, and investment risks and threats can be assessed.

References

[1] Sharafutdinov R et al 2018 *National Academy of Managerial Staff of Culture and Arts Herald* **2(2)**
[2] Bisht SS et al 2010 *Social Indicators Research* **98(1)** 167-81
[3] Daly H and Cobb J 1989 *For the Common Good* (Boston: Beacon Press)
[4] Hamilton C 1999 *Ecological economics* **30(1)** 13-28
[5] Bleys B and Whitby A 2015 *Environmental Economics* **117** 162-72
[6] Bagstad KJ 2012 *Ecological indicators* **18** 330-41
[7] Sales B et al 2018 *Revista ambiente* **10 (2)** 1-22
[8] Lopez MV et al 2007 *Journal of business ethics* **75 (3)** 285-300
[9] Rezaei J et al 2018 *Transport policy* **68** 158-169
[10] Meng B and Chi G 2018 *Singapore economic review* **63(2)** 229-48
[11] Azapagic A 2004 *Journal of cleaner production* **12(6)** 639-62
[12] Ksenofontov A A, Ksenofontov A S et al 2019 *Conf. Series: Materials Sci. and Eng.* **125** 012114

Research of equipment for pulsed heating supply

A P Levtsev[1], E S Lapin[1], B M Grishin[2], E G Ezhov[2] and Yu V Rodionov[2]

[1] National Research Mordovia State University, 68 Bolshevik st., Saransk, 430005, Russia
[2] Penza State University of Architecture and Construction, 28 Germana Titova, Penza, 440028, Russia

e-mail: *bgrishin@ rambler.ru*

Abstract. Proposed scheme of individual heating unit (IHU) contains a mixing device based on energy-efficient two-flow membrane pump operating in pulse mode at available head of the heating network. The problem was solved of finding the optimal regimes of membrane pump work, which provided qualitative and economical efficacy of heating devices. Experimental graphs illustrate the pressure changes in the working chambers of the pump equipped with a pulse heating medium flow distributor. It is shown that implementation of the membrane pump in the IHU circuit allows distributing temperature more evenly throughout the heating devices surface due to pulsating mode of the heating media movement.
Key words: individual heating unit, mixing device, two-flow membrane pump, pulse flow distributor, heating medium temperature distribution.

1. Introduction

To date, most of the buildings in urban settlements are connected to district's heat networks. In such cases, heating medium is mixed mainly by means of mixing devices based on the water-jet elevators. Water-jet elevators use the available head in the heat network and do not require additional drive, they are quite reliable and easy to operate, but they are highly dependent on the available head in the heat network. Head required for normal operation of the water jet elevator is 20-30 m [1]. If this condition is not met, operation of the heating unit based on a water jet elevator is partially or completely disturbed [2]. This phenomenon became especially frequent with expansion of the effective range of heat networks which, in some Russian cities, exceeds 9 kilometers. It is very difficult to provide a proper hydraulic mode in these conditions, therefore new buildings are connected to heating networks using an independent scheme, through a heat exchanger. It is mainly not possible to connect old buildings using an independent connection scheme for heating and ventilation systems of the building because of multiple inputs and old pipeline circuits in basements. Energy surveys performed in 2012 in the budget sector revealed main drawbacks of traditional dependent schemes for buildings connection to heating networks and main of them is overheating of returned network water at the change of available head in the heating network and reduction of heating efficiency. Technical solutions were proposed consisting in installation of additional monoblock pumps of Wilo, Grundfos type, etc. in the feed-up line. In some cases, in places where energy saving potential was revealed in buildings, weather regulators were installed in the heating medium supply pipe of the building individual heating units (IHU). However, work of the latter has a significant impact on hydraulic equipment of neighboring buildings [3]. There is an opinion among experts that mass use of weather regulators can completely disturb hydraulic mode of a heating system [4]. On the other hand, use of additional mixing pumps and weather regulation in the IHU, especially in the autumn-spring period, reduces efficiency of heating devices. Adding of such a membrane pump into the IHU circuit provides high-quality mixing of heating medium, which will allow more evenly distributing temperature throughout the heating device surface due to the pulsating mode of the heating medium movement [5,6,7]. In this regard, it is very important and practical to search for constructive solutions that could solve these problems with minimal energy consumption.

2. Methods and equipment

This paper describes a scheme of an individual heating unit with energy-efficient mixing device based on special two-flow membrane pump operating at the available head in the heating system. Adding of such a membrane pump into the IHU circuit provides high-quality mixing of heating medium which will distribute temperature more evenly over the surface of the heating devices due to the pulsating mode of the heating media movement [8, 9, 10].

Figure 1 shows a diagram of individual calorific unit including membrane pump which comprises: feed 1 and reverse 2 pipe-lines, left 3 and right 4 sections of membrane pump; pulse flow distributor 5 with 6 (right) and 7 (left) valves; right 8 and left 9 stems of pulse flow distributor; cam of pulse flow distributor 10 connected to electric drive 11, left 12 and right 13 working chambers of the membrane pump; left 14 and right 15 nozzles of the membrane pump; stem 16; mechanism for switching of dart valve 17; left 18 and right 19 dart valves; heating devices of the right 20 and left 21 sections; pumping chambers 22 and 24; discharge recirculation check valves of right 23 and left 25 sections; discharge recirculation check valves 26, 27 and suction check valves 28, 29.

Individual heating unit with membrane pump works as follows. Suppose that the left dart valve 18 is closed, and the right dart valve 19 is open, cam of the pulse flow distributor 10 is in position I (figure 2), at that the right valve of the pulse flow distributor 6 is open and the left valve of the pulse flow distributor 7 is closed. When the heating medium is fed into the supply pipeline 1 it begins moving along the following circuit: supply pipeline 1, pulse flow distributor 5, supply nozzle of the right section of the membrane pump 15, right working chamber 13 of the right section 4 of the membrane pump, right dart valve 19, heating device of the right section 20 and further into the return pipeline 2. When the steady-state mode of the heating medium is reached, the electric drive 11 turns on and begins to rotate the cam 10 and the left rod of the pulse flow distributor 9 jumps into the slot of the cam 10, this will lead to subsequent sharp closing of the left valve 7, and the right rod 8 will ride over the cam and smoothly open the right valve 6. If to rotate the cam further, the left valve 7 will be closed, and the right valve 6 will open for some time, and then will be kept open until position II of the cam. When the right valve 6 reaches fully open position, the stem 16 will be in a left extreme position, the switching mechanism of the dart valves 17 will operate that will cause sharp switching of the dart valves: the right dart valve 19 will close, and the left dart valve 18 will open. When the right dart valve 19 is abruptly closed, a water hammer will occur and further reverse wave appeared will create a pressure pulse that will reach the right working chamber 13 of the right section 4 and will give acceleration at the beginning of the movement of the rod 16. Further movement of the rod will occur under the influence of the pressure difference between the direct and reverse pipeline 2. The membrane of the right section of the membrane pump 4 will move from left to right, dragging the rod 16, and heating medium will be fed from the pump chamber 22, through the discharge check valve of the right section 23 to the heating device of the right section 20 and further into the return pipeline 2. When the heating media passes through the heating device, secondary heat of the heating medium is transferred of the right section 20 and accumulated primary heat of the heating media is transferred to ambient air. At this time, heating medium from the left working chamber 12 of the left section of the membrane pump 3, through the discharge recirculation check valve of the left section 26 and suction check valve of the left section 28, will flow through the open dart valve 18 and the heating device of the left section 21 into the left pump chamber 24, under the action of the pressure difference in them. When the heating medium passes through the heating device of the left section 21, heat energy is accumulated and transferred.

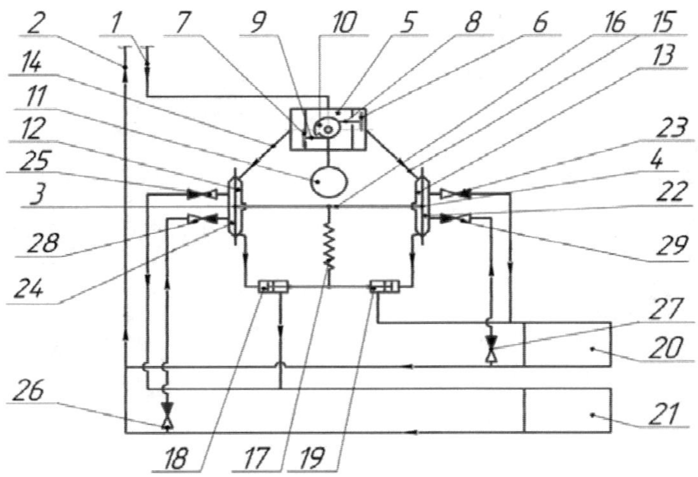

Figure 1. Scheme of individual heating unit with membrane pump

When cam of the pulse flow distributor 10 is switched to position II, valves of the pulse flow distributor will be switched and the process will occur in the right section that similar to that in left.

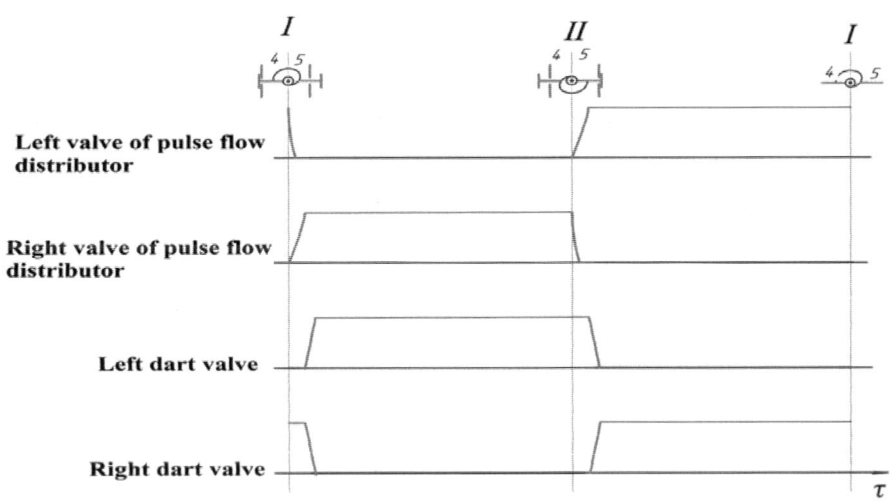

Figure 2. Timing diagram of valve operation.

Reliable operation of the device is provided at frequency of heating medium flow interruption from 0.4 to 0.8 Hz. At that, heating medium flow rate is reduced to 30 % at increase in the flow interruption frequency. Ratio of primary and secondary heating medium mixing is 2.

The scheme is implemented at the Department of heat and power systems of FSBEI of Higher Education "Mordovia State University named after N.P. Ogarev" in the form of IHU prototype with consumption of 3 t/h. The IHU prototype includes a pulse flow distributor of heating medium with the electric drive (Figure 3) and a two-flow membrane pump (Figure 4).

IOP Conf. Series: Materials Science and Engineering **828** (2020) 012015 doi:10.1088/1757-899X/828/1/012015

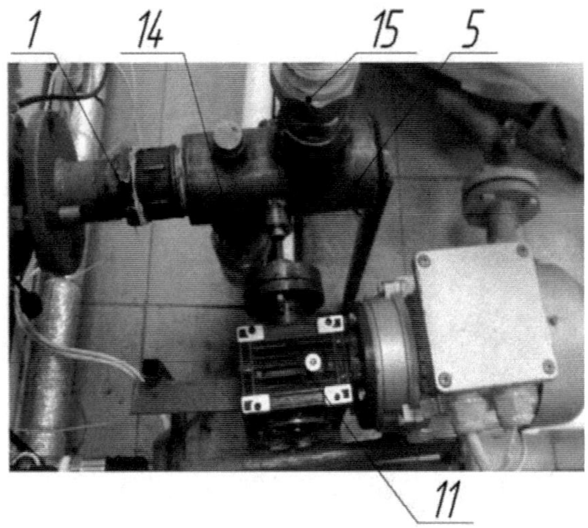

Figure 3. Pulse flow distributor of heating medium with the electric drive.

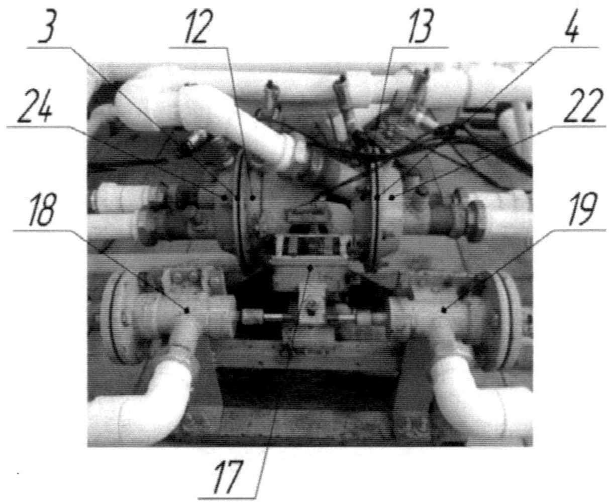

Figure 4. General view of individual heating unit with energy-independent membrane pump.

3. Results and discussion

The main task of testing IHU with a two-flow membrane pump is defining the following data: dependence of pressure in the working and pumping chambers on flow interruption frequency; dependence of flow rate on flow interruption frequency; temperature distribution pattern throughout the surface of heating devices. Graphs of pressure in the chambers at two frequencies - 0.4 and 0.8 Hz - are shown in Figures 5 and 6.

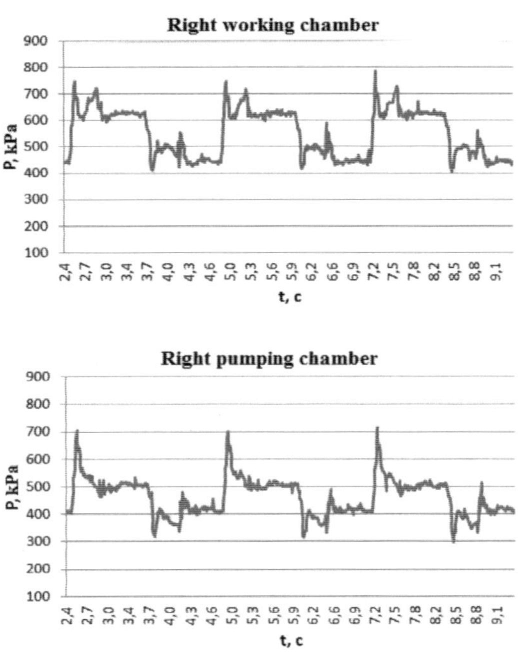

Figure 5. Graph of pressure at a frequency of 0.4 Hz.

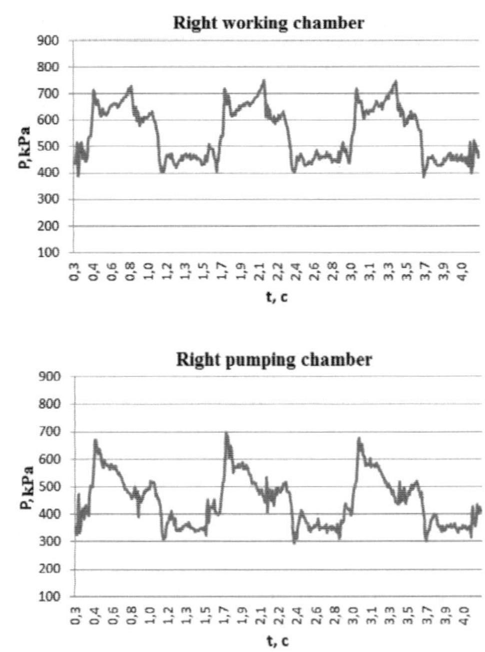

Figure 6. Graph of pressure at a frequency of 0.8 Hz.

It can be seen from the graphs that when the flow interruption frequency changes from 0.4 Hz to 0.8 Hz, pressure amplitude in the working and pumping chambers almost does not change.

Hydraulic characteristics of the IHU prototype depends on frequency of the pulse heating medium flow distributor (figure 7). As can be seen in this Figure, a significant change in the flow rate is observed in the frequency band from 0.5 to 0.6 Hz. At frequencies less than 0.5 Hz and more than 0.6 Hz, heating medium flow rate varies slightly; this is due to the design features of the pump piping.

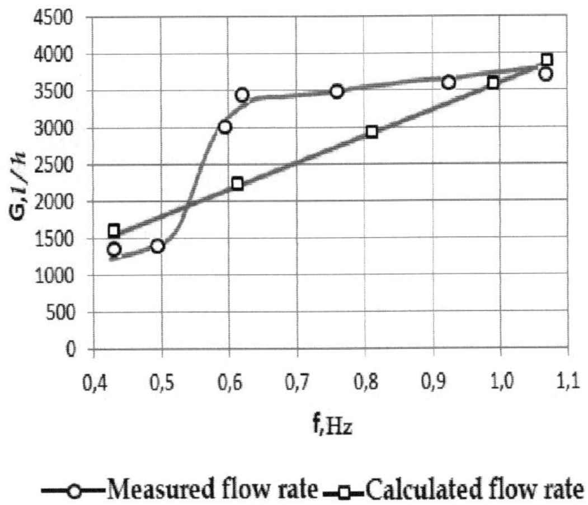

Figure 7. Flow rate of the membrane pump versus frequency.

The IHU prototype was loaded with BM RBS-500 bimetallic radiators. Temperature distribution throughout the surface of one of the heating devices, at static mode and at pulse mode with flow interruption frequency of 0.45 Hz, is shown in figure 8. As can be seen from this Figure, the pulsating flow of the heating medium allows increasing about 30% the convective heat exchange in the working units with low temperatures (43-47.6^{0}C) on the radiator surface.

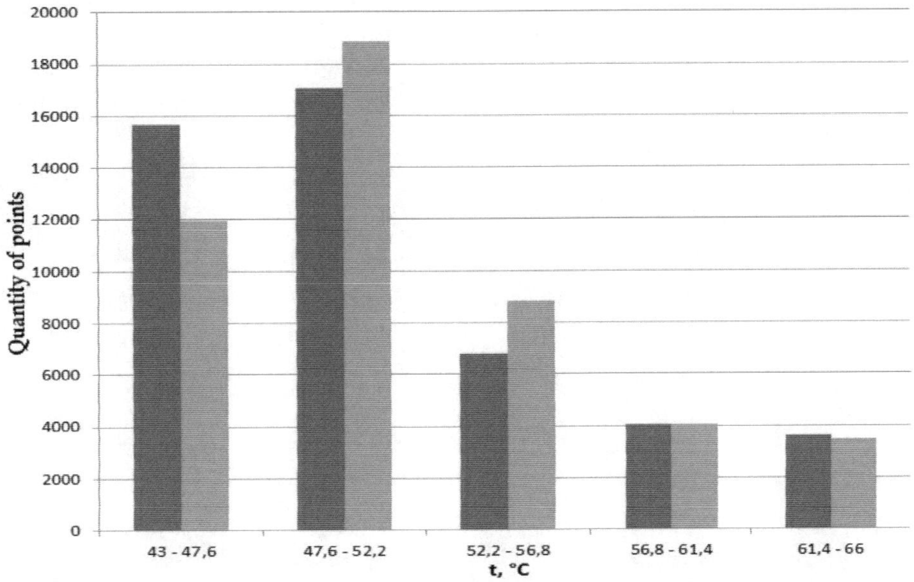

Figure 8. Temperature distribution diagrams in the working units of the sectional bimetallic radiator
- heating medium supply in pulse mode;
heating medium supply in static mode.

IOP Conf. Series: Materials Science and Engineering **828** (2020) 012015 doi:10.1088/1757-899X/828/1/012015

4. Conclusions

1. The practice of IHU operation makes it possible to conclude that use of additional mixing pumps and weather regulation, especially in the autumn-spring period, reduces efficiency of heating devices. In this regard, it is very important and practical to search for design solutions that could solve these problems with minimal energy consumption. One of the possible solutions is addition to the IHU scheme a pump-mixing device based on a special two-flow membrane pump operating at the available head in the heating network. Adding of such a membrane pump into the IHU circuit provides high-quality mixing of heating medium which will distribute temperature more evenly throughout the surface of the heating devices due to the pulsating mode of the heating medium movement.

2. As a result of the IHU testing with energy-independent pump-mixing device based on a special two-flow membrane pump operating at available head of the heating system, it was found out that, when flow interruption frequency changes from 0.4 Hz to 0.8 Hz, pressure amplitude in the working and pumping chambers practically does not change.

3. Hydraulic characteristics of the IHU prototype depends on frequency of the heating medium pulse flow distributor. At the same time, a significant change in the flow rate is observed in the frequency band from 0.5 to 0.6 Hz. At frequencies less than 0.5 Hz and more than 0.6 Hz, heating medium flow rate varies slightly, this is due to the design features of the pump piping.

4. Temperature distribution throughout the surface of one of the BM RBS-500 heating devices at static mode and at pulse mode at flow interruption frequency of 0.45 Hz shows that pulsating flow of the heating medium increases the convective heat exchange by 30 % in working units with low temperatures.

References

[1] Sokolov E Ya 2001 *Heating and Heating Networks* (Moscow: Publishing house of MEI) p. 472

[2] Sergeeva S K 1998 *Energy Audit and Rationing of Energy Costs: Collection of Methodical Materials* (N Novgorod: NSTU) p. 130

[3] Levtsev A P, Eniwatov A V and Artemov I N 2016 *Intercollege Collection of Academic Papers* (Saransk: NRMSU)

[4] Lapin E S, Levtsev A P and Kudashev S F 2016 *Energy-Efficient and Resource-Saving Technologies and Systems: Proceedings of Research and Practice Conference Dedicated to the Memory of Professor F. Kh. Burumkulov* (Saransk: NRMSU)

[5] Miladze V 2015 *Plumbing, Heating, Air Conditioning* **9** (**165**) 44-45

[6] Ovsepyan VM 1968 *Hydraulic Ram and Ram Installations* (Moscow: Mechanical Engineering) p. 124

[7] Galitseysky M B 1977 Thermal and Hydrodynamic Processes in Oscillating Flows (Moscow: Mechanical Engineering) p. 318

[8] Popov I A, Yakovlev A B and Shchelkov A V 2011 *Power Engineering of Tatarstan* **1** 44-7

[9] Valueva E P 2006 *Thermal Physics of High Temperatures* **44** **(10)** 118-26

[10] Fedotkin I M 1972 *Intensification of Heat Exchange in Food Production Apparatus* (Moscow: Food industry) p. 212

Organizational and methodological aspects of improving the organization and operation of production processes

M G Mustafaev [1], D G Mustafaeva [1] and G A Mustafaev [1]

[1] North Caucasian Institute of Mining and Metallurgy (State Technological University), 44 Nikolaeva Street, Vladikavkaz, 362021, Russia

E-mail: dzhamilya79@yandex.ru

Abstract. The issues of improving the efficiency of the operation and quality of the organization of production processes in the enterprise are considered. The company's success is ensured by the creation of organizational and technical solutions at all levels of the organization of production processes to create competitive products, management, compliance with established requirements of production processes covering all stages of the product lifecycle and ensuring a level of product quality that can fully meet all consumer needs and expectations, have confidence that the quality of the products delivered will be stable and sustainable. The company's quality system includes a set of organizational structure, processes, procedures and resources needed to implement overall quality management, with documented performance rules, forms and quality management practices that ensure that products meet the requirements, and its effective functioning ensures that problems are prevented rather than identified after they occur. Process management is systemic in nature by creating and understanding the network of processes, their consistency and interaction. Management of interconnected processes as a system contributes to the efficiency and efficiency of the company in achieving the intended results, and allows you to manage the relationships and interdependencies between the processes of the system.

1. Introduction

In today's environment, in enterprises, the relevance of the quality of the organization and operation of production processes is determined by its focus on ensuring a level of product quality that can fully meet all requests consumers. High quality products are the main component and determining their competitiveness. The priority of the operation of enterprises among other areas of development is to ensure the stable quality of products [1-4], relevant to the requirements of consumers.

In a competitive environment, continuous, systematic and focused quality-enhancing work is needed to ensure the production of competitive products [5, 6], the enterprise has a quality system that is an organizational structure that clearly distributes responsibility, procedures and resources to manage quality.

2. Improving the organization and operation of production processes

Increased market requirements for products encourage enterprises to change the approach to the problem of management and the quality of the organization of production processes. It is not enough to control the quality of products and the production process. A prerequisite for the successful operation of the company in the market is the creation of methodological and systematic principles to improve the

IOP Conf. Series: Materials Science and Engineering **828** (2020) 012016 doi:10.1088/1757-899X/828/1/012016

efficiency of the organization of production processes, a quality system covering all stages of life cycle products.

In market conditions, the company is forced to pay more and more attention to management and quality problems, as quality is the main criterion for consumers, leads to the strengthening of the company's position and to operate successfully in the market. Consumers of products are becoming more demanding and expect a high level of quality, they need confirmation and assurance that the quality is respected. And this kind of guarantee becomes methodological and system-technical principles of the organization of production processes, organizational and technical solutions at all levels of the organization of competitive product creation processes, the presence of on quality system enterprise.

The operation of management and quality system extends to all stages of the product life cycle. The quality system is designed to take into account the specific activities of the enterprise and includes a set of organizational structure, processes, procedures and resources necessary to implement a common quality management containing documented rules of performance, forms and quality management practices that ensure that products meet the requirements.

The quality system functions in a way that ensures that:

- the system is properly understood, applied, provided with the necessary support and has the necessary efficiency;
- manufactured products really meet the needs and expectations of the consumer;
- the needs of society and the demands of environmental protection are considered;
- the focus is on preventing problems rather than identifying them after problems.

The main target is to build quality systems and certify products that meet the requirements of consumers and, accordingly, the consumer-oriented ability of the company to consistently produce products of a certain level of quality.

In order to ensure the efficient operation and organization of production processes in enterprises, and to ensure a high level of product quality, appropriate methodological support is needed, including these activities in the systems quality. Methodological provision is seen as a set of interconnected processes united by one goal of achieving the required quality of measurements, establishing a rational range of measurable parameters and optimal standards for measurement accuracy in product quality control and process management.

The efficiency of the organization of production processes is also ensured on the basis of metrological support, which is a systematic, strictly defined set of tools and methods to obtain measuring information, have the properties necessary to make decisions to bring the control object to a target state. The main provisions of the metrological provision are carried out in the development, production, testing and operation of products.

The main objectives of the efficiency of the operation and the quality of the organization of production processes are:

- achievement, support and commitment to continuous improvement in the quality of work;
- continuous satisfaction of all established and anticipated requirements of consumers and other stakeholders;
- ensuring to internal management and other employees that quality requirements are met and maintained, and that quality is improving;
- ensuring the confidence of consumers and other stakeholders in meeting quality requirements;
- ensuring that the quality requirements of the system are met.

The quality system considers the company's features and minimizes the cost of product development and implementation. The consumer, with a quality and certification system, has confidence that the quality of the products delivered will be stable and stable.

The modern market economy has fundamentally different requirements for the quality of its products. The quality is an important tool in the fight for markets and ensures the competitiveness of products. It consists of the technical level of the product and its usefulness to the consumer. Competitiveness is determined by a combination of quality and value features of products that can meet the needs of the consumer, and the cost of purchasing and consuming the products involved.

The quality system of production based on the interrelated categories: object, objective factor, subject, methods, functions, means, principle, and others.

The quality system is considered effective when the products produced meet the requirements of the consumer and the presence of an effective system of product quality management is recognized by the consumer.

In the process of managing the quality of products, they carry out:

- develop a program to manage, plan and improve product quality;
- collecting and analysing information about the object that affects the quality;
- management solutions, site impact preparation and quality management solutions;
- analysis of information about changes in the quality of the object.

In the management of product quality, the direct objects of management are the processes on which the quality of the products depends. They flow both before the production, production and after the production stage of the product life cycle.

Management solutions are made by comparing information on the actual state of the managed process with the characteristics set by the management program.

In order to effectively organize the management process, a quality strategy is developed tailored to the product lifecycle, which is closely linked to the overall strategy of the company and provides for next steps to achieve goals and achieve goals.

With a strategy on the quality of the organization of production processes, a short-term management tactic is also developed, which represents a targeted activity that ensures that the parameters are constantly approaching the specified parameters. process quality.

The system of organizing production processes and management includes functions:

- strategic, tactical and operational management;
- decision-making, impact management, analysis and accounting, information-control;
- specialized and common to all stages of the product lifecycle;
- science, technology, production, economic and social factors and conditions.

The tools that ensure the quality of the organization of production processes include:

- regulatory bank, which regulates product quality;
- metrological means, including physical benchmarks, exemplary and/or working means of measurement;
- state system for ensuring the unity of measurements;
- public service standard reference data on the properties of substances and materials.

The quality assurance system for the organization of production processes is designed to consider the company's activities and to implement certain policies in achieving the goals set. The scale of the quality system must be consistent with quality assurance objectives and objectives.

Understanding and managing interconnected processes as a system contributes to efficiency and efficiency in achieving the intended results. This approach allows you to manage the relationships and interdependencies between the processes of the system, so that the overall results of the enterprise will be improved.

The quality of the production process organization includes:

- the documented information required by the standard;
- documented information identified by the organization as necessary to ensure the performance of the quality system.

Where the company determines the need for changes in the system of organization and management of the quality of production processes, these changes should be carried out on a planned basis.

The company plans, implements the processes necessary to meet the requirements for delivery and action, and to manage these processes through:

- determining product requirements;

- setting criteria for processes and acceptance of products;
- identify the resources needed to meet product requirements;
- process management in accordance with established criteria;
- identify, develop, update and apply, as well as register and preserve documented information to the extent necessary for:
- to ensure that the processes were carried out as planned;
- to demonstrate product compliance requirements.

The results of such planning should be appropriate for the enterprise. The company monitors data on consumers' perception of their needs and expectations and determines how to obtain, monitor and analyse this information [7].

The quality of the company's production processes should be carefully documented. All elements of the system, requirements and regulations adopted for quality assurance must also be documented. In this case, the documentation of the quality system provides a unified understanding of the policies, quality objectives, allows for a clear distribution of responsibilities, rights and responsibilities on quality, establish how units and performers interact in the performance of their functions and ensure efficient operation of production processes.

The most important advantage of the monitoring system is its ability to identify problems and develop action to systematically prevent its recurrence in the future [8]. Corrective and preventive actions must be conducted and analysed, which must be planned and implemented. All this leads to a continuous process of improvement.

A quality system is a management system for managing and managing an organization in relation to quality [9]. At the same time, it is a set of organizational structure, techniques, interconnected processes and resources necessary for the overall management and management of the enterprise in relation to quality.

Sustainability management helps the company make highly effective decisions that are continuous and successful for a long time. Sustainable development refers to the way the enterprise is managed, which provides long-term economic efficiency in meeting the requirements of environmental protection and social development.

The success of an enterprise depends on the degree to which the economic, environmental and social aspects of the business are integrated [10]. The main goal of the company is to ensure efficiency. Applying the principles of sustainable development, contributes to the continuous improvement of the company's activities. Improvement can be achieved in the following ways:

- strengthening relationships with partners by demonstrating to the parties involved responsible for the economic, environmental and social activities in the short and long term;
- strengthening the internal unity of the team based on a deep understanding of the company's work by the staff, continuous improvement of the activities and decision-making processes;
- developing trust among the parties involved by applying transparency and accountability principles;
- encouraging learning and innovation that provides new opportunities in relationships with the parties involved;
- understanding of possible hazards and risk management through systematic identification, ranking of significance and risk management.

The sustainable success of the company is achieved through its ability to meet the needs and expectations of its consumers and other stakeholders in a long-term and balanced manner [11]. Sustainable success is achieved through effective management, through awareness of the environment of its existence, through learning and the proper application of improvements and innovations.

The company is developing a management system to ensure:

- making good use of resources;
- decision-making based on facts;
- focusing on meeting consumer needs as well as the needs and expectations of other stakeholders.

IOP Conf. Series: Materials Science and Engineering **828** (2020) 012016 doi:10.1088/1757-899X/828/1/012016

The enterprise can achieve sustainable success by consistently meeting the needs and expectations of all stakeholders in a balanced way on a long-term basis.

In order to develop, adopt and maintain an effective strategy and policy, the enterprise should ensure that processes are in place to:

- continuous monitoring and regular analysis of the environment, including the needs and expectations of its consumers, the competitive situation, new technologies, political changes, economic forecasts or sociological factors;
- identify and identify the needs and expectations of other stakeholders;
- assessing the capabilities of existing processes and available resources;
- identify future resource and technology needs;
- updating your strategy and policies;
- determining the outputs of processes needed to meet the needs and expectations of stakeholders.

Such processes must be established in a timely manner and supported by the necessary plans and resources.

The strategy considers activities such as analysing consumer needs or regulatory requirements, its products, strengths and weaknesses, opportunities and threats. There is a clear process of developing and analyzing the company's strategy.

The organization processes are specific and vary depending on the type, size and level of the enterprise. Operations in each process must be defined and matched by the size and distinctiveness of the enterprise. Process management is systemic in nature by creating and understanding the network of processes, their consistency and interaction. The company establishes and plans processes, as well as identifies the functional units needed to produce products that are able to meet the needs and expectations of consumers and other stakeholders at a constant basis.

Processes are planned in accordance with the company's strategy and relate to management activities, resources, product lifecycles, monitoring, measurement and analysis. Process planning helps determine an enterprise's need to develop new technologies or master new technologies, develop new products, or bring new product characteristics to the consumer to create added value.

To achieve sustainable success in an ever-changing and unsustainable environment, the company regularly monitors, measures, studies and analyzes the effectiveness of its operations.

The management of the company's sustainable development is based on principles that meet its values. Principle is a fundamental basis that must be matched by the decisions and activities of the company. Principles shape the social, economic and environmental aspects of sustainable development management.

The right organization and quality system in the enterprise successfully implements the tasks of improving the quality:

- improving the ideological foundations of quality management, including leadership and leadership responsibility;
- developing and improving the process;
- introduction of statistical technologies in the quality management system;
- automation of management within the quality management system;
- improving the core functions of the quality management system;
- improving consumer-supplier interactions;
- improving the technical control and audit system at the company;
- improving design and process manufacturing processes;
- improving production maintenance processes;
- improving the effectiveness of corrective and preventative actions in relation to inconsistencies;
- training of staff in the field of quality.

Maintaining process parameters within certain limits ensures stable reproduction of operations and quantifies their impact on product quality.

IOP Conf. Series: Materials Science and Engineering **828** (2020) 012016 doi:10.1088/1757-899X/828/1/012016

The use of different control methods in production allows to establish a statistical relationship between product parameters and process modes and patterns of distribution depending on technological, design, operational and other factors.

Materials, products and process are subject to control in the production process. To monitor and measure, collect and register data in all kinds of monitoring, creating a unified system of quality assurance, dependency and interconnection between the individual elements of the system, determining sources and ways to collect information, measures to prevent deviations of characteristics from established requirements.

3. Conclusion

Improving the organization and operation of production processes and improving the company's operations is achieved by applying a process approach, and a systematic approach to management - identifying, understanding and managing interconnected processes as systems contribute to a business's effectiveness and effectiveness in achieving its goals. Process management is systemic in nature by creating and understanding the network of processes, their consistency and interaction. Management of interconnected processes as a system contributes to the efficiency and efficiency of the company in achieving the intended results and allows you to manage the relationships and interdependencies between the processes of the system, and the overall performance of the enterprise improves.

References

[1] Herrmann A, Huber F and Braunstein C 2000 *International Journal of Production Economics* **66** 77–96

[2] Koilakuntla M, Patyal V S, Modgil S and Ekkuluri P 2012 *Economics and Finance* **3** 55–61

[3] Martınez-Costa M, Martınez-Lorente A R and Choi T Y 2008 *International Journal of Production Economics* **113** 23–39

[4] Sila I 2007 *Journal of Operations Management* **25** 83–109

[5] Weckenmann A, Akkasoglu G and Werner T 2015 *The TQM Journal* **27(3)** 281–93

[6] Aldrich H E and Martinez M A 2001 *Entrepreneurship Theory & Practice* **25** 41–56

[7] Ceci F and Lubatti D 2012 *Research Policy* **41** 565–79

[8] Lizunkov V, Politsinskaya E, Malushko E, Kindaev A, and Minin M 2018 *International journal of energy economics and policy* **8 (3)** 250-257

[9] Dewick P and Miozzo M *2004 R&D Management* **34** 323–33

[10] Mintzberg H 1983 *Structure in Fives: Designing Effective Organizations* (Upper Saddle River, NJ: Prentice Hall)

[11] Ergunova O, Lizunkov V, Blaginin, V, Politsinskaya E, and Shaykina O 2018 *International Journal of Energy Economics and Policy* **8(1)** 122-127

Substantiation of butane-based air-gas mixture using for facilities gasification

N N Osipova[1], B M Grishin[2], G I Greysukh[2] and O V Tarakanov[2]
[1] Yuri Gagarin State Technical University of Saratov, 77 Politekhnicheskaya str., Saratov, 410054, Russia,
[2] Penza State University of Architecture and Construction, 28 Germana Titova, Penza, 440028, Russia

e-mail: bgrishin@ramdler.ru

Abstract. The features of the use of technical butane in gas supply systems are considered, criteria are established to ensure the presence of excess pressure in the tank. A mathematical model of temperature distribution on the reservoir contour has been developed, the temperatures of technical butane vapor have been calculated, in the presence of heat exchange elements of gas supply systems with the surrounding soil. The thickness of thermal insulation is justified to ensure an overpressure in the tank. The composition of the gas-air mixture, taking into account the restrictions on the fluctuations in the Wobbe number, ensures the complete combustion of the mixture in gas-powered consumer installations.
Keywords: butane, air-gas mixture, mathematical model, overpressure, temperature distribu tion, Wobbe number.

1. Introduction

Social well-being of the population is one of the priority tasks of any developed country. This task can be resolved most simply in large urban areas with high population density, advanced energy and logistics infrastructure, mainly due to natural gas supply through a gas network. In areas with complicated terrain, with a low population density, separated individual farms and industrial production facilities, gasification with natural gas is, in most cases, economically impractical [1, 2]. In this case, alternative options for gasification based on hydrocarbon gases are considered [3, 4]. The significant cost of these systems limits their widespread implementation and reduces attractiveness for the potential consumer. Potential reduction of gas supply systems cost is possible by use of gas-air mixtures based on industrial butane.

As practice shows, autonomous gas supply systems in most cases are not designed for use of liquefied gas with a high content of butane fractions [5-7]. The most significant disadvantages of such gas supply systems include:

- low vapor capacity of tank installations;
- condensation of vapor phase in distribution pipelines;
- complexity of drain and filling operations due to low overpressure of vapor phase in the tanks in winter.

Adaptation of autonomous gas supply systems to work with butane fractions will improve the system efficiency and reduce the cost of gas fuel.

Industrial butane is a multi-component mixture, which, according to GOST R 52087-2003, is characterized by the following composition:

- methane, ethane, ethylene, propane, propylene in total not more than 40%;

- butane and butylene more than 60%, in total.

For such a mixture, overpressure in the tanks necessary for vapor phase extraction is formed mainly at positive temperatures (Table 1).

Table 1. Overpressure of industrial butane

Temperature, K	Pressure, MPa	Temperature, K	Pressure, MPa
250	0.06	290	0.25
260	0.095	300	0.345
270	0.13	310	0.458
280	0.19	320	0.6

Minimum overpressure required for normal regasification of butane and for supply of the vapor phase to the consumer is 0.1 MPa, which is provided at the ambient temperature of minus 12.8°C.

The simplest solution for overpressure maintaining is to install submersible pumps in tanks to supply liquid phase of gas to regasification plants. However, the pump use is reasonable when the pressure in the tank is in range from 0.1 to 0.3 MPa. When the pressure in the tank is more than 0.3 MPa, the pumps are not necessary since the vapor phase overpressure is sufficient to move gas through the pipeline system to the consumer. Thus, forced supply to regasification plants should be used only in winter, and this solution is not efficient for year-round use.

Additional research is needed to develop recommendations for overpressure creation in underground tanks with industrial butane.

2. Statement of the research problem

To solve this problem, we consider heat exchange processes between an underground tank with industrial butane and soil mass and define temperature on the tank outer surface, which determines temperature of the liquid phase for the gas in question.

Let us consider the theoretical background for the problem solution.

Estimated temperature of liquid butane t_b as it is stored is equal to natural value of temperature on the axis of the tank location t_{gr}. The tank dimensions are: radius of the tank R and length l.

Following assumptions were made to solve the problem:

- length of the tank exceeds its diameter;
- heat transferred from the ground is completely absorbed by the liquid phase of the gas (there is no resistance to heat exchange between the tank wall and the liquid and there is no resistance to heat transfer);
- soil is a homogeneous medium with thermal conductivity λ;
- resistance to heat transfer at the ground-air boundary is taken into account using an additional layer method.

Thus, the total depth of the tank is defined by the formula:

$$h = h_0 + \lambda\left(\frac{h_{sn}}{\lambda_{sn}} + \frac{1}{\alpha_{int}}\right), \tag{1}$$

where h_0 - depth of the tank relative to the ground surface, m; h_{sn} - height of snow cover, m; λ_{sn} - thermal conductivity of snow, W/(m K); α_{int} – heat transfer factor in outside air, W / (m^2 K).

Calculation model for the problem is given in Figure 1.

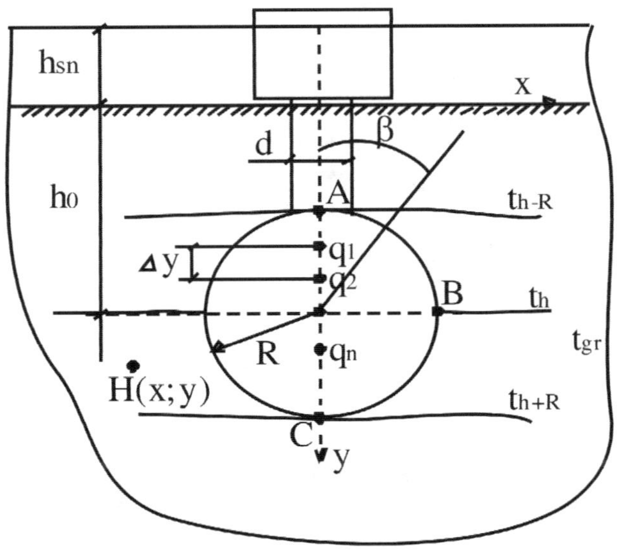

Figure 1. Calculation model for the problem

3. Results and discussion

The problem is solved by applying temperature fields. Let us select own temperature field of the soil which has a given temperature distribution by depth t (y). In this case, temperature on the cylindrical surface of the tank will have the following values:

- on the upper generator line t_{h-R};
- on the location axis t_h;
- on the lower generator line t_{h+R}.

The second temperature field is represented by heat sources of some intensity q_1, q_2, ...q_n arranged along y axis with some step Δy shown in figure1.

According to Forchheimer, temperature in point H (x; y) due to i-th heat source is defined by the formula

$$t_H = \frac{q_i}{2\pi\lambda} \ln\left(\frac{x^2 + (h_i + y)^2}{x^2 + (h_i - y)^2}\right)^{0.5}. \quad (2)$$

If there are several sources i=1, 2,....*n*, their combined effect can be described by the formula

$$t_H = t_{H_1} + t_{H_2} + \ldots + t_{H_i} + t_{H_n} \quad (3)$$

$$t_H = \frac{1}{2\pi} \sum_{i=1}^{n} \frac{q_i}{\lambda} \ln\left(\frac{x^2 + [(h-R) + i\Delta y + y]^2}{x^2 + [(h-R) + i\Delta y - y]^2}\right)^{0.5}. \quad (4)$$

A method for solving the system of equations for arbitrary point H located on the cylindrical part of the tank outer surface is described in detail in [6, 8]. The most accurate solution for the temperature distribution on the tank outer surface can be obtained when a number of equations $n \to \infty$. According to Gauss's theorem, total heat flow to a body with an isothermal surface located in soil mass is numerically equal to total intensity of heat sources located inside the body. Thus, the resulting value of the heat flow on the tank outer surface is zero.

Total effect of heat sources q_1, q_2, ... q_n is defined by the formula

$$q_1 + q_2 + \ldots + q_n = 0. \quad (5)$$

IOP Conf. Series: Materials Science and Engineering **828** (2020) 012017 doi:10.1088/1757-899X/828/1/012017

Formulas (1-5), taking into account mentioned assumptions, form a mathematical model of temperature distribution on the outer surface of underground tank filled with industrial butane. To implement the mathematical model, the following initial data were used:

- tank volume V=5 m^3 with radius R=0.7 m;
- tank depth in the soil (axis of the tank) h_0=1,3 m, height of snow cover h_{sn}=0.1 m;
- ground temperature on the location axis of the tank (cold climate zone) t_{gr}=- 3.6^{0}C;
- the study is carried out on the tank outer surface from A to C (figure 1);

Solution of the problem is given in Table 2

Table 2. Temperature on the tank outer surface

Angle β	0	45	90	135	180
Calculated temperature distribution, ^{0}C	-4.97	-5,16	-4.97	-4.78	-4.97
Average temperature, ^{0}C			-4.97		

As can be seen in Table 2, different depth of the tank outer surface points gives maximum deviation from natural temperature of the soil 1.56°C at the location axis. This effect also causes variations of overpressure 0.005-0.006 MPa which does not have effect on the performance of the tank units.

Thus, the underground tank filled with industrial butane provides stable vapor phase generation and formation of overpressure in all climatic zones, except for very cold ones.

At the same time, heat exchange between the tank neck and environment plays an important role in extraction of butane vapor phase from the tank and its movement through the elements of the gas supply system.

In accordance with the underground tank features, the neck is surrounded by soil mass, but a part exchanges heat with outside air. Thus, total heat exchange is described by the formula:

- for the underground part:

$$Q_{utn} = \frac{(h_{uth} - h_{sn})(t_b - t_{\min})\pi d}{\frac{h_{uth}}{\lambda} + \frac{h_{sn}}{\lambda_{sn}} + \frac{1}{\alpha_{\text{int}}}}; \quad (6)$$

- for surface part:

$$Q_{otn} = k_{\text{int}} \left[\pi d (h_{otn} - h_{sn}) + \frac{\pi d^2}{4} \right] (t_b - t_{\min}), \quad (7)$$

where h_{utn} is height of the underground part of the tank neck, m; h_{otn} is height of the surface part of the tank neck, m; k_{int} is heat transfer factor of vapor of the liquefied gas to outside air, W/(m^2 K); d is the diameter of the neck, m; t_{min} is temperature of the coldest day, °C

Following initial data were used to calculate heat loss:

- height of the underground part of the neck - 0.6 m;
- height of the above-ground part of the neck - 0.1 m;
- diameter of the neck - 0.4 m;
- temperature of the coldest day: minus 40°C.

After substitution in formulas (6) and (7) and summation of the results, heat loss through the neck was defined in the amount of 240 W, which, in accordance with formula (4), causes a temperature decrease of 7.1°C.

Thus, temperature in the area of the neck will be:

$$t=-4.97+(-7.1)=-12.07^0C.$$

At this temperature, the overpressure of the butane vapor phase is reduced to the minimum value, gas supply is stopped.

To prevent vapor phase temperature decrease, it is proposed to cover the neck and cap of the tank with thermal insulation. The insulation thickness is selected in accordance with a mathematical model and is equal 0.15 m.

One of the important criteria for butane use in gas supply to the consumer is completeness of the gas-air mixture combustion in domestic gas burners. If gas is not burned completely, efficiency of household gas appliances decreases and emission of carbon monoxide in gasified rooms increases. Adaptation of gas burners requires changes of gas combustion nozzles design.

Most simply, this problem is solved without constructive changes, just by application of gas-air mixtures (butane-air).

A criterion parameter determining interchangeability of combustible gases is Wobbe number [9, 10]. Interchangeability is defined as possibility of using alternative gas without disrupting normal operation of gas burners and their design solutions. Wobbe number of butane-air mixture is defined by the formula:

$$W = \frac{\left(Q_1^W\right)_b \cdot k_b^0}{\sqrt{\left(\frac{\rho_b k_b^o + \rho_{air} k_{air}^0}{\rho_{air}}\right)}}, \tag{8}$$

Where $(Q_1^W)_b$ is the lowest combustion heat of butane, MJ/m^3;

k_b^0 – volume concentration of butane in the gas-air mixture;

k_{air}^0 – volume concentration of air in the gas-air mixture;

ρ_{air} – specific density of air, kg / Nm^3;

ρ_b – specific density of butane, kg / Nm^3.

To substantiate composition of the butane–air gas mixture, calculations were carried out for various butane contents in gas-air mixture from 1 to 100%.

The calculation results are shown in figure 2.

As can be seen from the graph given in figure 2, butane content in the gas-air mixture should be not less than 48% and not more than 54%. In this case, the gas-air mixture fully meets the requirements of combustible gases and natural network gas interchangeability.

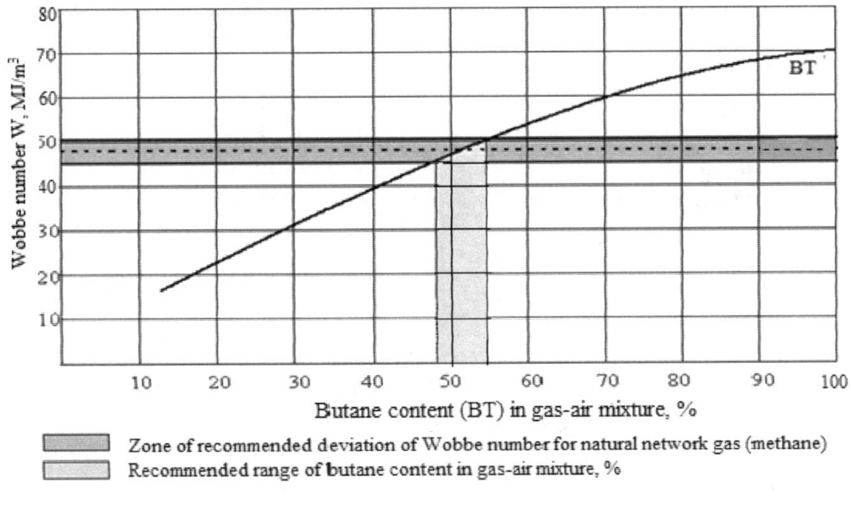

Figure 2. Defining butane-air mixture composition when using in substitution of natural network gas

As noted earlier, when using gas-air mixtures, dew point temperature decreases which favorably affects the fluid-free operation of gas supply systems. Table 3 shows values of dew point temperature of industrial butane and gas-air mixture based on butane.

It can be seen from Table 3 that use of air-gas mixture significantly reduces dew point temperature. For example, use of industrial butane causes condensation of water vapor and gaseous phase even at minus 1°C, and at pressure increase - in the range of positive temperatures that makes completely impossible using industrial butane even in the warm season.

Table 3. Dew point temperature of combustible gases based on butane

Pressure, MPa	Dew point temperature, °C		
	Butane content in gas-air mixture (recommended), %		Butane 100%
	Lower limit (48%)	Upper limit (54%)	
0.405	15	20	41
0.304	17	11	30
0.203	-3	1	18
0.182	-5	-2	15
0.162	-8	-5	12
0.142	-12	-9	8
0.122	-16	-12	3
0.101	-20	-18	-1

Use of the butane-air gas mixture displaces dew point temperature by more than 10°C over the entire pressure range of gas-air mixture; this eliminates condensation of the gas-air mixture vapors in the distribution pipelines and expands the scope of the mixture application for facilities gasification.

4. Conclusions

1) Application of gas-air mixtures based on industrial butane for gasification of facilities requires a reasonable approach to defining basic parameters of gas supply systems based on the solution of problems associated with heat exchange between industrial butane and soil through the tank walls;

2) To ensure overpressure of the butane vapor phase, it is necessary to exclude heat exchange with environment by applying thermal insulation to the elements of the gas supply system.

3) Mixing industrial butane vapors and air with butane content in the mixture of 48-54%, taking into account restrictions on variations of Wobbe number, will help to exclude changes in design of gas-burning devices in gas-using equipment when using the network natural gas;

4) Use of butane-air gas mixture provides lower dew point temperature values that has a positive effect on operation of distribution pipelines at negative temperatures.

References

[1] Medvedeva O N 2018 *Russian Journal of Building Construction and Architecture* **1** (**37**) 29– 41
[2] Spektor N Yu 2017 *Territory Neftegaz* **7 - 8** 118 –23
[3] Osipova NN, Bychkova I M and Ryapisova Yu S 2016 *Resource-Efficient Technologies in the Building Complex: Materials of the IV Intern. Scientific-Practical Conf.* (Saratov: SSTU)
[4] Osipova N N 2011 *Scientific and Technical Journal: Construction and Renovation* **5** (**37**) 39–45
[5] Staskevich N L 1984 *Gas Supply of Cities* vol 1(Leningrad: Nedra) p. 623
[6] Kuritsyn B N and Medvedeva O N 2011 *Technical and Economic Optimization of Heat Supply Systems* (Saratov: SSTU) p. 62
[7] Morge R P and Yerazunis S P 1967 *Heat Transfer with Phase Change* **63** 1–7
[8] Kuritsyn B N, Chirchinskaya G P and Pavlutin M V 2005 *Gas of Russia* **1** 31–35
[9] Worasaen A, Pannucharoenwong N, Benjapiyaporn C and Jongpluempiti J 2017 *Energy Procedia* **138** 278–81
[10] Costa F C and Moutinho dos Santos 2015 E *Brazilian Journal of Petroleum and Gas* **9** (**2**) 37–44

IOP Conf. Series: Materials Science and Engineering **828** (2020) 012018 doi:10.1088/1757-899X/828/1/012018

Approach to the Composing the Practicum on Safety Systems

O V Plitsyna and T N Rogova

Safety Management in the Technosphere, Russian University of Transport, Obraztsova st., 9, build. 9, Moscow, 127994, Russia

E-mail: plitsa@yandex.ru

Abstract. The practicum creates conditions for the controlled formation of important competences. Generalized practical lessons for master's programs are promising. This paper proposes practicum concept in which the interest in abilities of certain safety means stimulates attention to the terms that cause difficulties. Questioning the master's degree students in «Technosphere safety» revealed sustained interest in effective means of acoustic safety as well as difficulties with the terminology of fire safety. Prepared generalized practical lesson on acoustic safety contains the following: fire danger indicators introduced into reference data; automated calculations of sound insulation; developed assessment procedure. Results of the generalized practical lesson in the control group have confirmed both overcoming difficulties with the terminology of fire safety and increasing the degree of readiness for the expertise of projects

1. Introduction

Practicums create conditions for the controlled formation of important students' competences [1-4].

Assessment of competencies formed by means of discipline «Safety» for bachelor programs confirmed the wide opportunities of generalized practical lessons [5-7].

Obtained data indicated that generalized practical lessons are also justified for master programs.

Questioning the master's degree students in «Technosphere safety» showed:

- Materials to ensure acoustic safety are stable interest.
- The terminology of fire safety is the difficulty.

The noted results allow assuming that difficulty could be overcome if during the practical lesson on the discipline «Calculation and design of safety systems» to take into consideration the fire hazard class of the considered materials, as well as fire hazard class and fire resistance limit of the projecting constructions. In addition, there would be conditions for forming the competence's component «readiness».

The purpose of the paper is improving the practicum on safety systems due to the generalized practical lesson.

The problems were solved:

- Complementing the acoustic safety reference data.
- Automating the calculations.
- Developing the assessment procedure.
- Realizing the generalized practical lesson in the control group.

2. Composing the generalized practical lesson

2.1. Complementing the acoustic safety reference data

Information about structures and materials [8-11] was introduced into acoustic safety reference data (tables 1, 2).

Table 1. Options of walls with plates in a distance

Wall			Plate on distance			Air layer	
material	fire hazard class	thickness (mm)	material	fire hazard class	thickness (mm)	thickness (mm)	mineral wool's fire hazard class
Ceramsite concrete	K0	100	Drywall	KM1	6.5	40	KM0
Foam concrete	K0	180	PhoneStar	KM5	12.5	25	KM0

Table 2. Options of frame-sheath partitions

Type of partition	Structure	Fire resistance limit	Fire hazard class
C 111	mineral wool plate: density 37 kg/m^3, thickness 50 mm one layer of drywall: thickness 12.5 mm	EI 45	K0
	mineral wool plate: density 37 kg/m^3, thickness 50 mm one layer of fire-resistant drywall: thickness 12.5 mm	EI 60	K0
C112	mineral wool plate: density 37 kg/m^3, thickness 50 mm two layers of drywall: thickness 12.5 mm	EI 60	K0
	mineral wool plate: density 37 kg/m^3, thickness 50 mm two layers of fire-resistant drywall: thickness of 12.5 mm	EI 90	K0

Fire hazard classes and fire resistance limits correspond with normalized [12, 13] values.

2.2. Automating the calculations

Calculations were automated by Excel spreadsheets.
They give the following acoustic characteristics in the most acceptable form:

- Sound insulation of plate on distance (table 3).
- Sound insulation of single-layer wall (table 4).
- Sound insulation of single-layer wall with the plate on distance (table 5).
- Sound insulation of the multi-layer wall.
- Sound insulation index (tables 4, 5).

Table 3. Option of calculating the sound insulation of plate on distance

Parameter	Value	Values at geometric mean frequencies															
		100	125	160	200	250	315	400	500	630	800	1000	1250	1600	2000	2500	3150
t (m)	0.04																
E (Pa)	19000																
k (Pa/m)	4750000																
d (kg/m^3)	1100																
h (m)	0.01																
m (kg/m^2)	7.15																
f_0 (Hz)	129.72																
$3f_0$ (Hz)	389.17																
add_1	$f>f_0$	0	0	0.43	0.18	0.07	0.03	0.01	0.0	0.0	0.0	0.0	0.0	0.0	0.0	0.0	0.0
c_A (m/s^2)	340																
c_P (m/s^2)	2500																
f_{BOUN} (Hz)	3952.14																
l (m)	6.00																
s_L	0.01																
n	6																
add_2	0.06	0	0	0.06	0.06	0.06	0.06	0.06	0.06	0.06	0.06	0.06	0.06	0.06	0.06	0.06	0.06
add_3 add_2	f_0–$3f_0$ $f>3f_0$	0	0	0.49	0.24	0.13	0.09	0.06	0.06	0.06	0.06	0.06	0.06	0.06	0.06	0.06	0.06
Sound insulation R_{ADD} (dB)		0	0	3.0	6.5	9.0	10.5	12.5	12.5	12.5	12.5	12.5	12.5	12.5	12.5	12.5	12.5

IOP Conf. Series: Materials Science and Engineering **828** (2020) 012018 doi:10.1088/1757-899X/828/1/012018

Table 4. Option of calculating the sound insulation of the wall

Parameter	Value	Values at geometric mean frequencies															
		100	125	160	200	250	315	400	500	630	800	1000	1250	1600	2000	2500	3150
t (mm)	100																
f_B (Hz)	315																
d (kg/m^3)	1300																
m (kg/m^2)	130																
K	1.2																
m_{EQ} (kg/m^2)	156																
R_B (dB)	32																
Sound insulation R_1 (dB)		32	32	32	32	32	32	34	36	38	40	42	44	46	48	50	52
Sound insulation index (dB)	43																

Table 5. Option of calculating the sound insulation of the wall with the plate on distance

Parameter	Value	Values at geometric mean frequencies															
		100	125	160	200	250	315	400	500	630	800	1000	1250	1600	2000	2500	3150
R_1 (dB)		32	32	32	32	32	32	34	36	38	40	42	44	46	48	50	52
R_{ADD} (dB)		0	0	3.0	6.5	9.0	10.5	12.5	12.5	12.5	12.5	12.5	12.5	12.5	12.5	12.5	12.5
R (dB)		32	32	35	38.5	41	42,5	46,5	48,5	50.5	52.5	54.5	56.5	58.5	60.5	6.5	64.5
Estimative curve		33	36	39	42	45	48	51	52	53	54	55	56	56	56	56	56
Deviations down	34.5	1	4	4	3.5	4	5.5	4.5	3.5	2.5	1.5	0.5	0	0	0	0	0
Shifting the estimative curve	-1	32	35	38	41	44	47	50	51	52	53	54	55	55	55	55	55
Deviations down	23	0	3	3	2.5	3	4.5	2.5	2.5	1.5	0.5	0	0	0	0	0	0
Sound insulation index (dB)	51																

IOP Conf. Series: Materials Science and Engineering **828** (2020) 012018 doi:10.1088/1757-899X/828/1/012018

Sound insulation index is determined to compare with the normalized [14-17] value.

2.3. Developing the assessment procedure

Fuzzy Logic Toolbox was used for assessment procedures (figure 1).

The system has two inputs, one output, three rules «if ... then», three values for the centers of the membership functions of the inputs and output.

Assessment of the manifestation of the indicator:

- 0 points if the indicator is absent.
- 5 points, if the manifestation of the indicator is unstable.
- 10 points if the indicator is stable.

Degree of «readiness» competence's component formation:

- Low if there is no manifestation of any indicator.
- Average if the manifestation of the indicator of acoustic safety is unstable.
- High if fire safety indicator is stably manifested.

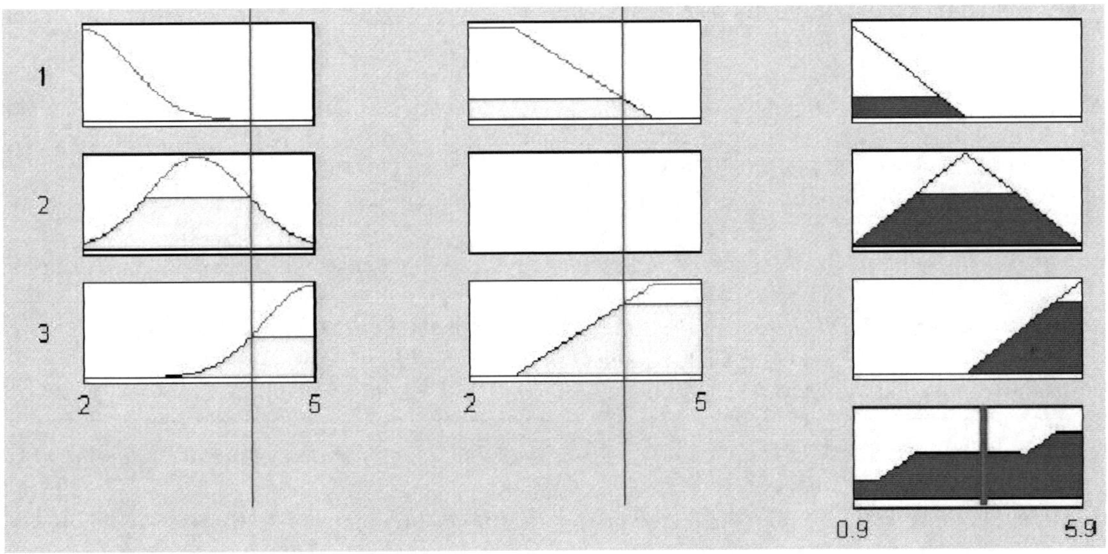

Figure 1. Option of assessment procedure

Self-assessment on acoustic and fire safety indicators is sufficient to fulfil a fuzzy expert system.

2.4. Realizing the generalized practical lesson in the control group

At the practical lesson control group master students worked on ensuring the acoustic comfort in the reconstructed room:

- The required index of sound insulation of walls and partitions was determined.
- The effectiveness of various sound insulation means was considered taking into account the permissible fire risk.
- Optimal design of the walls or partitions was calculated.
- Readiness to coordinate decisions and interact with other security experts was assessed in dynamics.

IOP Conf. Series: Materials Science and Engineering **828** (2020) 012018 doi:10.1088/1757-899X/828/1/012018

The following results have been obtained:

- Calculations and self-assessment automation ensure the sufficiency of study time.
- Interest in the characterization of new acoustic safety means stimulates attention to fire safety indicators.
- Proposed design solutions show increasing the readiness to coordinate design and interact with other safety professionals.

Thus, the suppositions of the appropriateness to clarify the practical lesson content on acoustic safety systems have been confirmed.

3. Conclusion

It is recommended to introduce into practicum on safety systems tasks that combine consideration of the fire hazard of materials and structures with the following calculations:

- Sound insulation of single-layer wall.
- Sound insulation of plate on distance.
- Sound insulation of single-layer wall with the plate on distance.
- Sound insulation of the multi-layer wall.
- Sound insulation index.

References

[1] Plutenko A D, Leyfa A V, Kozyr A V and Haletskaya T V 2018 *European Journal of Contemporary Education* **7(2)** 360–371
[2] Abramova I V, Shilova Z V, Varankina V I and Rubanova I V 2019 *European Journal of Contemporary Education* **8(1)** 187–200
[3] Pastushkova M A, Savateeva O V, Trotsenko A A and Savateev D A 2019 *European Journal of Contemporary Education* **8(2)** 328–337
[4] Gorbunova T N 2017 *European Journal of Contemporary Education* **6(2)** 254–263
[5] Lapteva M D and Pavlova A O 2018 *The Humanities and Education* **9(1)** 67–75
[6] Golovanova I I, Telegina N V and Donetskaya O I 2019 *Education and Self Development* **14(1)** 57–67
[7] Glinchikov D Y and Plitsyna O V 2019 *Proc. Int. Conf. on Preventin Rescue Aid* vol 16 (Khimki: ACP EMERCOM of Russia) pp 33–6
[8] Hummel H U 2014 *Proc. Int. Conf. on Acoustics in architecture* vol 1 (Moscow: RIA ARD) pp 4–6
[9] Rutling E V and Gurova E V 2017 *Bulletin of Volgograd university of architecture and civil engineering* **47(66)** 43–50
[10] Shchelokov Yu A 2015 *Noise Theory and Practice* **1(1)** 70–6
[11] Shchelokov Yu A 2016 *Noise Theory and Practice* **2(1)** 8–16
[12] *SP 2.13130.2012 Systems of fire protection. Fire-resistance security of protecting units*
[13] *ISO 717-1:2013 Acoustics – Rating of sound insulation in buildings and of building elements*
[14] *SP 51.13330.2011 Sound protection*
[15] *ISO 834-13:2019 Fire-resistance tests – Elements of building construction*
[16] *SP 23.103.2003 Projection of sound insulation of separating constructions in domestic and public buildings*
[17] *SP 254.1325800.2016 Buildings and territories. Rules for designing of industrial sound protection*

IOP Conf. Series: Materials Science and Engineering **828** (2020) 012019 doi:10.1088/1757-899X/828/1/012019

Improving the Fire-fighting Design of Silencers

O V Plitsyna and T N Rogova

Safety Management in the Technosphere, Russian University of Transport, Obraztsova st., 9, build. 9, Moscow, 127994, Russia

E-mail: plitsa@yandex.rux

Abstract. This paper presents fire-fighting design of silencer formed by stepped partitions. Providing the distance between the partitions not exceeding a quarter of the wavelengths and forming the set of ducts with alternating expansions and narrowings create the possibilities for noise reduction and spark arresting. Slit silencer model testing on the stand contained doubled reverberant chambers with removable elements and diaphragm as well as on the stand with internal combustion chamber demonstrates the following: noise reduction is not less than 7.5 dB in a wide frequency range; pressure losses are not more than 120 Pa; sparks don't pass. Correcting the silencer geometry allows forming the devices which acoustic and aerodynamic characteristics are similar to the characteristics of standard plate silencers. Calculating data show the dimensions of single and double silencer, ensuring spark arresting and requirement noise reduction in a gas flue.

1. Introduction

The advantages of silencers without sound-absorbing material include the stability of the characteristics in the aerosol flows and the possibility of dynamic spark arresting [1-4]. However, the prospects of fire-fighting design of silencers are often limited by the appearance of transverse resonant oscillations [5-7]. To ensure the required broadband noise reduction [8-10], it is necessary determining the dimensions of the silencer design [11-14].

Slit silencer construction is formed by means of stepped partitions as the set of ducts which one dimension is much smaller than the other two dimensions [15]. Each duct contains sequence sections of narrowing and expansion. The design is added by turning walls and filter cavities.

Slit silencer is described taking into account not less than one hundred forms of propagating waves [16-18] by the method of electroacoustic analogy. Acoustic calculations are realized using Simulink.

The purpose of the paper is detecting the silencer geometry ensuring the smooth throughout the normalized frequency range noise reduction.

The problems of calculated and experimental bringing out abilities of the slit silencers were solved.

2. Slit ducts geometry

2.1. Slit ducts cross-section

Silencer's slit ducts can be fulfilled vertical or horizontal (figure 1).

The silencer's slit ducts should have the smallest cross-section dimension no more than a quarter of the wavelengths propagating in the system.

IOP Conf. Series: Materials Science and Engineering **828** (2020) 012019 doi:10.1088/1757-899X/828/1/012019

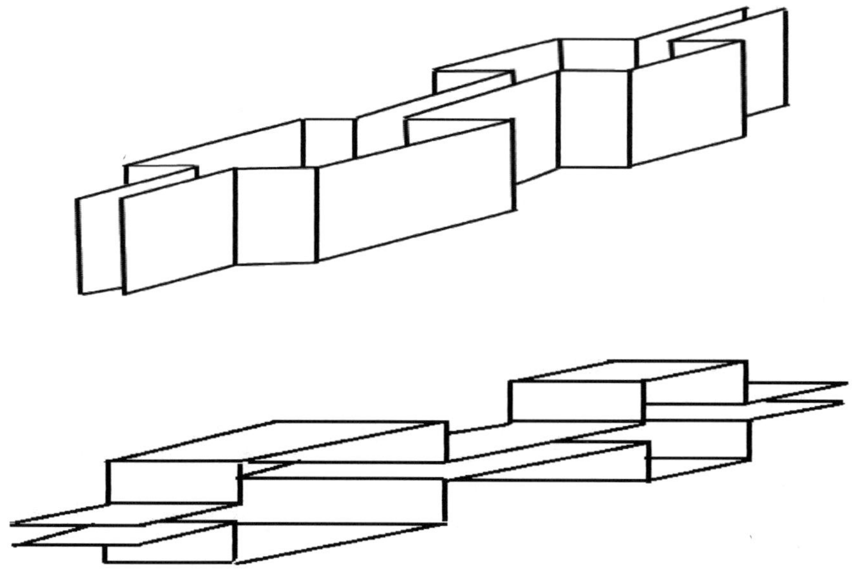

Figure 1. Orientation of slit ducts.

Taking into account the upper normalized frequency [19, 20] and the permissible degree of expansion the distance between the partitions is 4 mm at the sections of narrowing and 160 mm at the section of expansion.

2.2. *Model tests*

Model tests were carried out to determine [21, 22] the operational abilities of split silencers (figure 2).

The requirements of the similarity theory were fulfilled.

Acoustic characteristics of slit silencer were determined by the method of measuring in doubled reverberant chambers with removable elements:

- Mounting the removable elements of the duct connecting two reverberant chambers in one of which loudspeakers were placed.
- Measuring the 1/3-octave sound pressure levels in each reverberant chamber.
- Calculating the noise reduction in silencer model absence.
- Mounting the silencer model in place of the removable element.
- Measuring the 1/3-octave sound pressure levels in each reverberant chamber.
- Calculating the noise reduction in the silencer model presence.
- Calculating the noise reduction ensured by the device.

Aerodynamic characteristics of split silencer were determined by the method of measuring in pressure duct with the diaphragm:

- Setting the fan operation.
- Determining the airflow passed through the diaphragm.
- Measuring the total pressure on the model's inlet.
- Measuring the total pressure on the model's outlet.
- Calculating the pressure losses.

Spark arresting ability was tested in the stand with internal-combustion chamber:

- Mounting the silencer model at the stand.
- Initiating the ignition in the chamber.
- Visual fixing the breakthrough sparks.

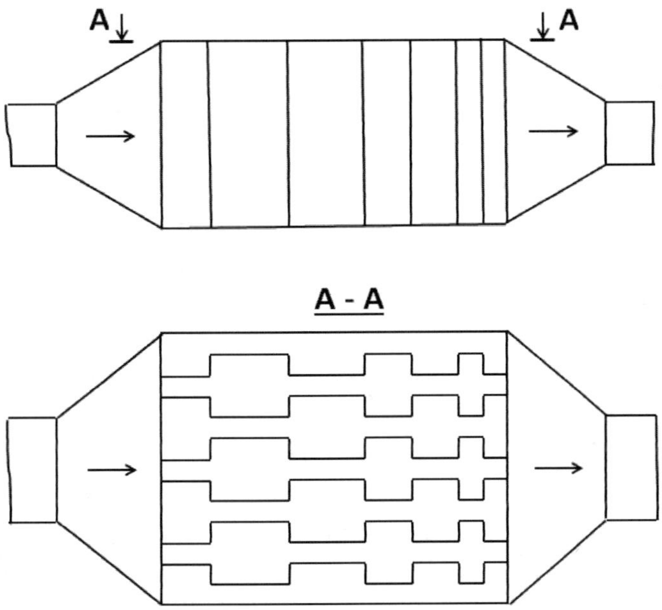

Figure 2. Slit silencer scheme.

Tests results show the following:

- Noise reduction was no less than 7.5 dB in the frequency range 125-4000 Hz.
- Maximum pressure losses in the slit silencer were 120 Pa.
- Sparks on the silencer's outlet do not pass.

2.3. Slit ducts section length

Based on the calculations of slit silencer variants it was found achieving the smooth broadband characteristic when the change indicator of length for slit ducts is $2^{-0.6}$.

The correction of the design is carried out on the example of the forming the slit silencers which acoustic and aerodynamic performances are close to standard plate silencers.

For plate silencers [23-25]:

- Plates with sound-absorbing material of 200 mm thickness are placed through 200 mm.
- Maximum pressure losses in the split silencer are 120 Pa.
- Device length is 1000 and 2000 mm.

Lengths of narrowing and expansion sections in formed silencers are shown in figures 3, 4.

The largest cross-sectional dimension of the ducts is taken to be 1000 mm.

The acoustic abilities of the slit silencers (figure 5, 6) indicate ensuring the effective noise reduction for a gas flue.

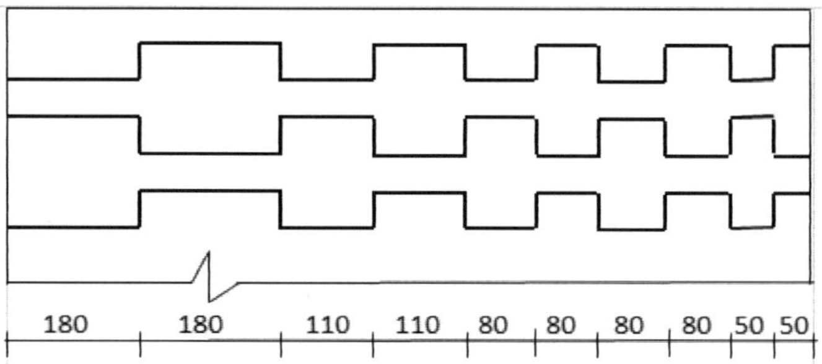

Figure 3. Sections of narrowing and expansion for slit silencer length of 1000 mm

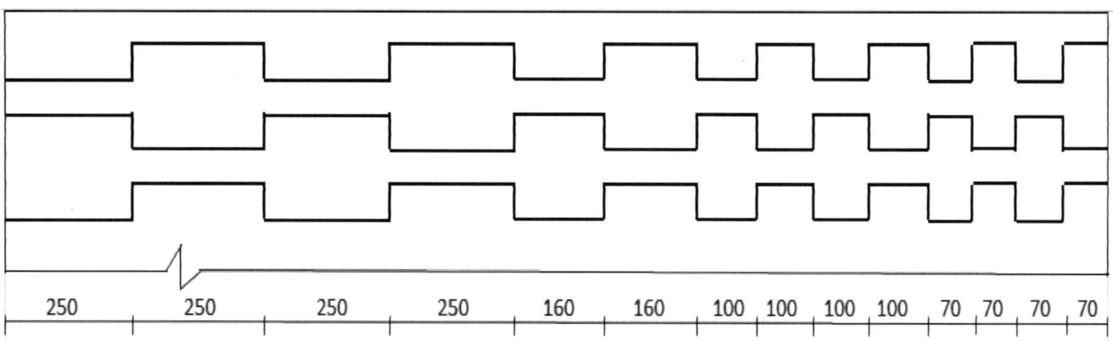

Figure 4. Sections of narrowing and expansion for slit silencer length of 2000 mm

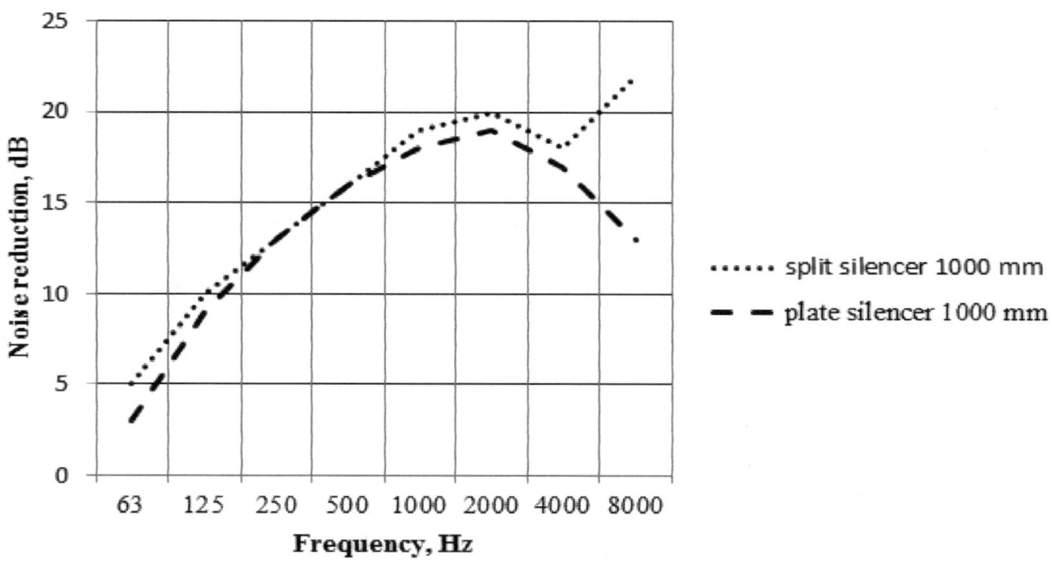

Figure 5. Acoustic characteristics for silencers of 1000 mm.

IOP Conf. Series: Materials Science and Engineering **828** (2020) 012019 doi:10.1088/1757-899X/828/1/012019

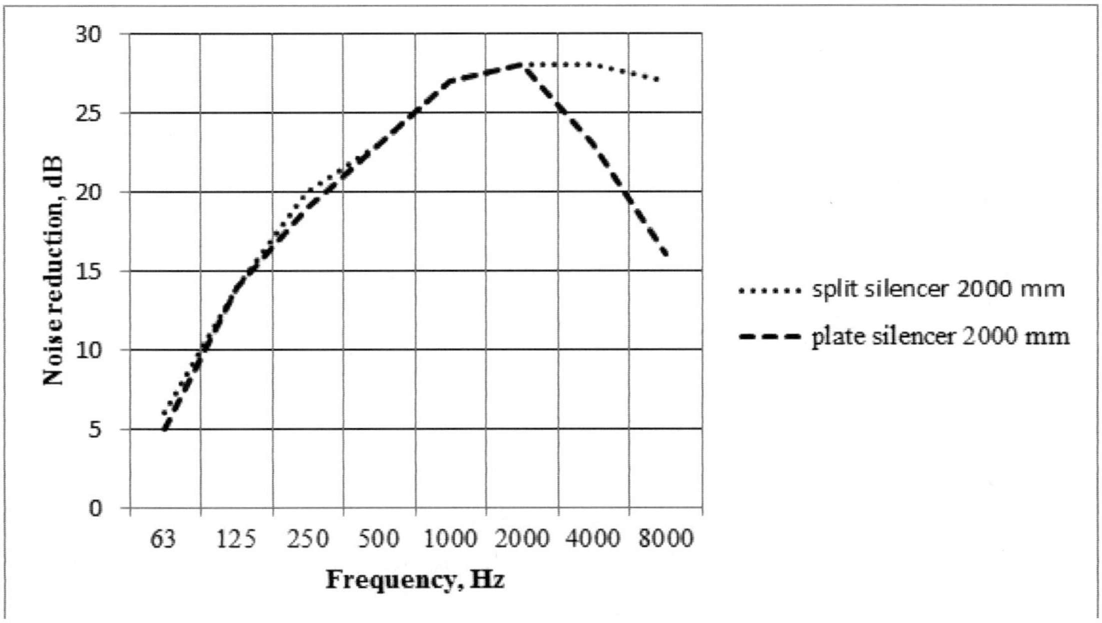

Figure 6. Acoustic characteristics for silencers of 2000 mm.

According to the calculations and experiments the successive placement silencer with vertical partitions and silencer with horizontal partitions increases the noise reduction twice (figure 7).

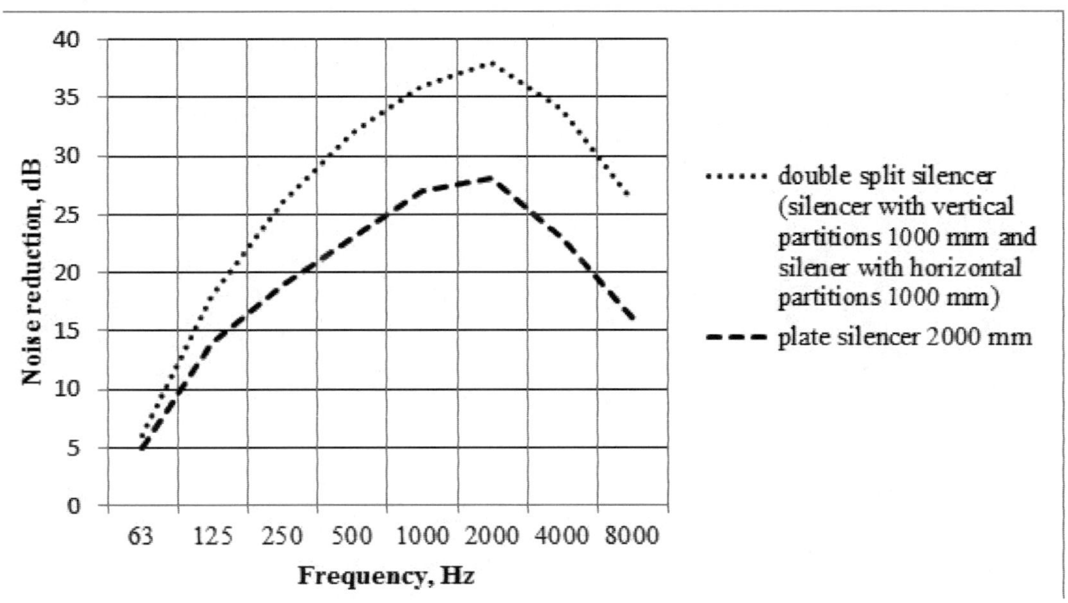

Figure 7. Acoustic characteristics for the double slit silencer.

3. Conclusion

Fire-fighting silencer formed by step partitions is recommended for systems with gas flues.

Identified geometry of the slit ducts creates conditions both for smooth broadband noise reduction with acceptable aerodynamic losses and for spark arresting.

IOP Conf. Series: Materials Science and Engineering **828** (2020) 012019 doi:10.1088/1757-899X/828/1/012019

References

[1] *ISO 14163:1998 Acoustics – Guidelines for noise control by silencers*

[2] *GOST R 53323-2009 Fire barriers and spark arresters. General technical requirements. Test method*

[3] Gusev V P 2009 *Academia. Architecture and Construction* **5** 184–192

[4] Rogova T N 2017 *Security Problems of the Russian Society* **1** 16–7

[5] Glav R, Regaund P L and Abom M 2004 *J. Sound Vib.* **273** 777–92

[6] Kar T and Munjal M I 2005 *J. Sound Vibr.* **285** 585–96

[7] Gerges S N Y, Jordan R, Thieme F A and Bento Coelho J L 2005 *J. Braz. Soc. Mech. Sci.* **27** 132–140

[8] Gusev V P and Leshko M Yu 2002 *AVOK* **4** 46–50

[9] Terebnev A V, Yemelyanov O N and Yablonik L R 2017 *Vesti gazovoy nauki* **5** 100–5

[10] Grudanov V Ya and Tkacheva L T 2017 *Bulletin BRU* **4** 17–28

[11] Ji Z L 2005 *J. Sound Vibr.* **283** 439-446

[12] Agafonova A A, Mironov S G and Tupov V V 2015 *Izvestiya vysshikh uchebnykh zavedeniy. Mashinostroenie* **9** 75–82

[13] Nikolaev V A, Bykov A I and Smirnov S G 2016 *Proc. 1st All-Rus. Conf. on Habitat acoustics* vol **9** (Moscow: BMSTU) pp. 127–134

[15] Plitsyna O V and Glinchikov D Y 2013 *Proc. Int. Congr. on Ecology and life protection of industrial-transport complexes* vol **6** (Togliatti: TSU) pp. 218–21

[16] Komkin A I and Vorobyeva L S 2013 *Izvestiya vysshikh uchebnykh zavedeniy. Mashinostroenie* **4** 50–9

[17] Agafonova A A and Komkin A I 2016 *Proc. 1st All-Rus. Conf. on Habitat acoustics* vol 9 (Moscow: BMSTU) pp. 14–23

[18] Gusev V P and Solodova M A 2010 *Academia. Architecture and Construction* **3** 211–8

[19] *SP 51.13330.2011 Sound protection*

[20] *SP 254.1325800.2016 Buildings and territories. Rules for designing of industrial sound protection*

[21] Plitsyna O V and Glinchikov D Y 2017 *Proc. Int. Congr. on Ecology and life protection of industrial-transport complexes* vol 6 (Samara: Samara scientific centre) pp. 144–7

[22] Gusev V P and Leshko M Yu 2002 *AVOK* **2** 75–6

[23] Gusev V P and Leshko M Yu 2006 *AVOK* **8** 34–40

[24] Gusev V P 2005 *AVOK* **4** 86–90

[25] Gusev V P, Ledenev V I and Leshko M Yu 2013 *Reference book on Calculation and design of noise suppression of ventilation systems, air conditioning and air heating* (Moscow: NIISF RAACS) pp. 20–80

Impact of artificial intelligence and industrial automation on territorial development: strategic guidelines

Natalia Polusmakova [1] **and Marina Glushchenko** [1]

[1] Volgograd State University, Institute of Management and Regional Economics, pr-t Universitetskij 100, Volgograd, 400062, Russia

E-mail: polusmakova@volsu.ru, malinat57@yandex.ru

Abstract. Automation and artificial intelligence become particularly relevant in the development of modern society that is focused on digitalization. The article considers the impact of automation and artificial intelligence on the sectoral development of the Russian economy and society taking into account their potentials and prospects. Keywords: artificial intelligence, automation, labor market, territorial development.

1. Introduction

The relevance of this topic directly depends on the existence of artificial intelligence and the development degree of production automation technologies in a single state and region.

The article is aimed at identification of strategic guidelines for artificial intelligence and automation development on the sectoral economies of countries as components of their territorial development.

To confirm the main goal, it is necessary to solve the following tasks:

1. Definition of states with the most favorable economic conditions for the creation of artificial intelligence;
2. Identification of professions which are most likely to be finally automated in the near future;
3. Analysis of the disadvantages and advantages of using artificial intelligence in production;
4. Research for the most effective resolution of the conflict between artificial intelligence and human competencies in the workplace.

2. Materials and methods (model)

The following methods were used to analyze the information:

1. Modeling method used to simulate the impact of automation on the labor market;
2. Analytical method compared the results of simulation and determined the optimal solutions to problems.

The innovation consists in the forecast of positive and negative impact of artificial intelligence and automation on industry types in the context of territorial development as factors of strategic stability of spaces. Hence, the need to study the current state of the market for automation and artificial intelligence, identify trends in their development and degree of impact, and determine the optimal strategic directions.

3. Results and discussion

The world community is becoming more high-tech and digital, forming the need for machines with elements of artificial intelligence. One main consequence of this process will be the increasing efficiency of the economy. In the United States production per worker doubled between 1995 and 2015 and continues to increase. There is a transition from separate "smart machines" to intelligent automation systems, for example, to connect all the equipment of the plant to a network managed by a single center or create a transport network of robotic vehicles transporting goods or people.

Table 1. Industrial robots in the USA and Europe (according to the International Federation of Robotics, IFR)

	1995	2000	2005	2010	2015
Robots per thousand workers in the United States	0.5	0.7	1.1	1.3	1.8
Robots per thousand workers in the Europe	0.4	0.5	1	1.4	1.6

It is expected that the robot market in the EU will develop and reach $3.7 billion by the end of 2018. According to 2016 statistics provided by World Robot Statistics, IFR, the number of industrial robots will increase by 2.6 million units in 2019. About 70% of industrial robots operate in metalworking, electrical and automotive industries. In 2015, due to the introduction of robotics, production in the EU increased in the electrical industry by 18%, in metallurgy - by 16%, and in the automotive industry - by 10%. The market for industrial robots is growing by 8% annually. The EU holds 32% of the global robotics market. Nevertheless, in the European Union it was decided to invest €700 million in innovation and robotic products by 2020. Statement was made at the World Economic Forum in Davos: "The proliferation of robots and robotics marked the IV industrial revolution" [1].

According to Reuters, in 2017, 9900 units of robotics were installed at enterprises in Central and Eastern Europe, it is by 28% more than in 2016. The International Federation of Robotics estimates that by 2020 robotics shipments to Eastern Europe will increase by 21%.

IFR indicates that Slovakia has the highest robotization rate in Eastern Europe (135 robots per 100,000 workers). In the Czech Republic - 101 robots, in Hungary - 57, in Poland - 32 per 100 thousand workers. Low robotization of Polish industry is due to the influx of cheap labor from Ukraine.

Interestingly, in Eastern Europe, robotization is often forced. The enterprises of these countries are increasing shortage of skilled workers. Skilled human resources go to Western Europe.

According to the IMF "Income Polarization in the United States" in 1970–2014 the share of US households with average incomes (50–150% of the median) decreased by 11% (from 58% to 47%) [2]. Consequently, the middle class is declining, the share of the poor and rich increases. This may be, inter alia, due to an increase in income for specialists working in the field of high technology and software, along with a lag in salaries among workers in the "old" industries.

Throughout the 20th century, automation, and then the introduction of artificial intelligence, helped to reduce the need for human labor. Job cuts in a number of industries, from 1 to 3 professions annually disappear. In catering and trade the automation and artificial intelligence threatens to halve the employment. Pessimistic forecasts indicate an increase in unemployment from 5-6% (natural level) to 15–20% in Russia and up to 30% in Europe.

In the West, the McKinsey Institute is thoroughly engaged in research into the effects of automation on the labor market. In the world, automation could put 1.2 billion people out of work whose total salary is $14.6 trillion. In this case, half of the cuts will fall on the four largest world economies: USA, China, Japan, India [3].

In Europe, the main job cuts will cover 5 countries: France, Germany, Italy, Spain and the UK [4]. In total, automation will replace 62 million people whose salary is $1.9 trillion a year.

Table 2. The expected impact of automation on the labor market of foreign countries [5]

Country	The automation potential (%)	Number of people affected by automation (million)
Japan	56	35.6
India	52	235.1

China	51	395.3
Russia	50	35.4
France	43	9.7
Germany	48	20.5
Spain	48	8.7
United Kingdom	43	11.9
USA	46	60.6
Italy	50	11.8
Canada	47	7.2

Significant changes in the labor market are expected in Russia

Table 3. Changes in the structure of jobs in the sectors of the Russian economy

Sector	The automation potential (%)	Number of people affected by automation (million)
Production	54	6.5
Retail	52	5.5
Administrative and government agencies	47	3.6
Education	20	1.3
Transport and logistics	75	4.3
Health and social support	41	2.2
Agriculture, forestry, hunting and fishing	64	3.3
Construction	64	3.1
Information	40	0.858
Finance and Insurance	43	0.92
Science and technology	36	0.673
Real estate transactions	37	0.462
Art, entertainment and recreation	33	0.373
Other services	47	0.479
Extractive industry	68	0.585
Hotel business and catering	67	0.57
Wholesale	54	0.428
Housing and Utilities:	52	0.313
Top management of organizations	65	0.014

Table 3 shows that the total number of jobs losses in the Russian economy is expected at about 35.5 million.

Using digital technologies in Russia over 6 years, it is possible to reduce the number of officials by one third and reduce government spending by 0.3% of GDP (Deputy Chairman of the Economic Council under the President of the Russian Federation Alexei Kudrin).

"Of today's seven jobs in the production of motors and other units, only one will remain," says Michael Brecht, Chairman of the Daimler team.

The report of the World Economic Forum (WEF) for 2016 predicts a reduction of 7.1 jobs by 2020 as a result of the introduction of new technologies. Mostly white-collar workers, office workers, will be fired. In exchange, only 2 million new jobs will appear in the field of information technology, architectural and engineering fields.

According to the UN report, the most significant reduction will affect developing countries - about ⅔ of the population will be unemployed. In developed countries, the first wave of reductions has already passed.

IOP Conf. Series: Materials Science and Engineering **828** (2020) 012020 doi:10.1088/1757-899X/828/1/012020

In 2016, Foxconn (an iPhone assembly company in China) installed 40,000 robots and laid off 60,000 people. The number of robots at the company's plants will increase by 20-30% annually, there will leave a minimum of employees responsible for production, logistics, testing and control.

The report of the International Labor Organization noted that in the textile and footwear industry of Southeast Asia more than 6 million workers out of 9.2 million can be fired as a result of automation.

Robots become cheaper than low-cost manual labour [6]. Automation improves speed and productivity. For this reason, Adidas intends to return footwear production to Germany (previously brought to factories in Asia).

In the European economy, automation has been growing for several decades, and yet the unemployment rate has not changed. What factors provide such sustainability?

First, the disappearance of jobs in some areas is offset by new industries and professions.

In the USA in 1900, the population employed in agriculture was 40%, in 2000 it was 2%; in industry in 1950 - 25% of the population, in 2010 - less than 10%. But, unlike the 20th century, when machines replaced the manual labour of low-skilled workers, now the threat of dismissal more affects highly paid specialists with higher education.

Secondly, automation cannot replace simple and important work - communication with clients (specifically with small businesses).

Thirdly, automation allows expanding production volumes and increasing the need for workers whose activities are not reduced as a result of the introduction of new technologies. For example, the appearance of ATMs could lead to the mass dismissal of bank employees. In reality, the ATMs saved money, due to which new bank branches were opened.

The introduction of new technologies does not always propose job cuts for all enterprises. A survey conducted in 2016 by ManpowerGroup agency revealed that only 12% of executives plan to reduce staff over the next 2 years due to the introduction of automation. The high cost of introducing new technologies, the lack of experience in working on new equipment stop many managers.

As a result, in 2011, McKinsey research showed that over 15 years the Internet has deprived 500,000 people of work, but created 1.2 million new jobs.

Often automation of production requires the creation of jobs. Amazon's online store plans to create 100,000 new jobs in the US over the next 18 months. The company requires both software developers and warehouse employees because new technologies increase turnover.

Statistics in Germany shows that automation and robotics in industry and transport, which have been accelerating sharply in recent years, does not lead to a decrease in the overall level of employment in the country. It is more likely opposite. Germany ranks third in the world in terms of robotization after South Korea and Japan, but employment in the industry is at a record high.

Already 40 percent of German enterprises complain about the shortage of qualified personnel. The problem of the lack of qualified personnel is especially acute in companies that provide IT services to industrial customers, as well as in machine-tool plants.

Over 350 thousand new jobs have been created in the German manufacturing industry over the past five years.

So, automation, robotization and digitalization, the introduction of digital technologies, on the one hand, lead to the destruction of certain jobs. However, they, on the other hand, invariably contributed to the creation of new jobs, and often - the emergence of new professions. Unmanned vehicles will leave many drivers and taxi drivers without earnings, but will overwhelm the work of the creators of software for self-driving vehicles. In factories, robots will replace classic conveyor workers, but someone will have to design, program and repair these mechanisms.

There was a polarization in the labor market: Automation affects the middle class, low-paying jobs are more profitable for people to do, and high-paying jobs cannot be done by robots yet. A new people activity type is machine learning. According to research by the McKinsey Institute, the global market for artificial intelligence services is growing at 25% annually, and by 2025 it will reach 125 billion euro. The number of employees in the information technology (IT) industry is constantly growing, in the West this is one of the few areas of activity where there are always vacancies.

According to The New York Times, in 2016, OpenAI, an artificial intelligence research organization founded in 2015 by Elon Musk and other investors, spent $7 million on wages for its 52 employees.

According to the report of Tencent Holdings, in the world there are from 200,000 to 300,000 specialists involved in the development or direct creation of artificial intelligence. But it is believed that only 22 thousand programmers can develop artificial intelligence systems.

The British laboratory DeepMind, owned by Google, spent $138 million on 400 employees, that is, an average of $345 thousand.

Google, Facebook, Apple, Amazon, Uber require specialists to work on face recognition programs, automatic natural language processing, digital assistants and autopilot cars.

The training of bachelors and masters in the field of artificial intelligence development is increasing and it is expected that in 3-4 years the shortage of specialists will be eliminated. In addition, Intel, Facebook and Google are introducing their internal training programs for AI professionals. The global spread of the Internet has made it possible to work remotely, as well as offshore outsourcing of software production when information technologies commissioned by companies are created by employees from another country.

There is a struggle between AI for specialists in AI, and the automotive industry, which creates drones, is competing with Silicon Valley. Companies also hire top academics, even professors from Stanford University.

Human activities will undergo change. Workers will often have to adapt to the machine, and not vice versa. Of course, there will be less routine monotonous work. But simultaneously, employees will need to switch to new tasks, learn flexible working methods. Thorough knowledge of AI machines operations will become a necessary requirement for specialists and management personnel. Understanding safety in working with "thinking machines" will become increasingly important.

4. Conclusion

New technologies have global importance, but the pace of their development and implementation varies significantly in different countries. In the field of artificial intelligence, the vast majority of innovations are created in the United States, Japan is the leader in robotics. Among European countries, Germany leads in new technologies, followed by France and the United Kingdom. In 2005, an average of 1 robot was involved in European factories per thousand workers, currently 1.6 robots. The market for industrial robots is growing by 8% annually. The EU holds 32% of the global robotics market.

Artificial intelligence and automation can reduce errors, improve quality and speed. As a result, in developed countries, labor productivity rises annually by 0.8-1.4%. If 1975 is taken as the starting point, then the total volume of production in developed countries has doubled.

At the current pace of development, by 2055, half of all work performed by people will be fully automated. In the 20th century mainly a simple routine work performed by unskilled workers has undergone automation. In the 21st century, new technologies allow the replacement of highly qualified specialists, such as engineers or office workers.

In the near future, robotization will affect the following activities:

- banking (computers independently assess the situation in the financial markets or the creditworthiness of the borrower);
- industry (many manufacturers of high-tech products plan to automate most of the technological processes);
- medicine (computerized systems for diagnosing diseases, robots for patient care, and even robotic surgeons have been introduced);
- transport (the transition from individual computerized functions, such as cruise control, to a car without a driver, such as a Daimler drone truck);
- trade and catering (in Europe there are fully automated stores without sellers and security, automated snack bars are being introduced);
- construction (projects have been developed for the automated construction of houses using a 3D printer).

Artificial intelligence continues to evolve. Voice and face recognition technologies, self-learning neural networks are introduced. These technologies will reduce a significant number of office workers.

There are areas of activity where artificial intelligence cannot replace humans. This is a job requiring complex decisions (for example, a politician or a judge), creative activity (architect, inventor, artist), work involving communication (teacher, doctor). The most advanced computer lags far behind a human in terms of flexibility of thinking, ability to solve non-standard issues.

Automation and implementation of artificial intelligence elements have a significant impact on the labour market of regions and countries. According to the forecast of the McKinsey Institute the world automation could lose 1.2 billion workers whose total salary is $14.6 trillion. In Europe, 62 million people may lose their jobs, mainly in France, Germany, Italy, Spain and the UK.

The above graph shows that the growth rate of median income, this level of income, that is shared by 50% of low-paid and 50% of high-paid workers, lags behind the growth rate of the gross domestic product in developed countries. Labour productivity is growing, but income is falling, and the number of jobs is declining.

Economic statistics shows that new technologies do not always increase wealth. In developed countries, the middle class is "eroded", the proportion of the poor and rich is increasing. The income of IT development specialists is growing significantly, while in developed countries the salary of skilled workers remained at the level of the 70s, taking into account inflation. Average income growth rates lag production growth rates. The unemployment rate is higher than in the 90s, while social programs mitigate the negative consequences.

In the coming years, it is expected further development of artificial intelligence and automation. Their impact on the labour market will increase. Regarding its nature, there are opposing points of view. Proponents of accelerated technological progress argue that new technologies will create more jobs than they eliminate. As an example, they point to the German industry experiencing a shortage of qualified personnel. Robotization reduced the number of workers at the assembly line but required a significant increase in software developers or robot maintenance specialists. Within a wide time frame, they consider the reduction in employment in agriculture from 50% at the beginning of the 20th century to 2% now. These workers have found a place in industry and services.

A more critical approach indicates that in recent decades, jobs for low-skilled workers who could undergo retraining and find a new job have been reduced. Now automation is becoming global. Jobs will be absent in catering, in offices. Job cuts will increase, and humanity will not have time to adapt to the new reality.

Even the owners of large companies are aware of the threat of uncontrolled development of automation because the unemployed will not be able to buy their products and make a profit.

To mitigate the effects of automation and robotization, a number of strategic solutions are proposed:

- development of social programs;
- unconditional basic income (any citizen of the country will receive it);
- partial nationalization of the economy (the economy will be controlled by artificial intelligence in the interests of the general population).

The analysis of the automation and implementation of the elements of artificial intelligence showed their high importance in terms of territorial development, the impact on the labor market and industry activities, taking into account potential opportunities. When developing a strategy and implementing relevant measures it is necessary to consider the prospects and trends of automation and robotization, the need for the use of artificial intelligence and digital technologies for the effective development of countries and regions.

References

[1] Watry G 2016 *The fourth industrial revolution and robot journalism.* Available at: http://www.rdmag.com/news/2016/01/fourth-industrialrevolution-and-robot-journalism (Date of reference 18.09.2018)

[2] Urunov A A and Rodina I B 2018 *Fundamental'nye issledovaniya* **1** 138-142. Available at:

http://fundamental-research.ru/ru/article/view?id=42064 (Date of reference 18.09.2018)

[3] *McKinsey Global Institute. Technology, jobs, and the future of work. briefing note prepared for the fortune vatican forum.* December 2016. Updated april 2017. Available at: https://www.mckinsey.com/~/media/mckinsey/featured%20insights/Digital%20Disruption/Harnessing%20automation%20for%20a%20future%20that%20works/MGI-A-future-that-works-Executive-summary.ashx (Date of reference 18.09.2018)

[4] Lizunkov V, Politsinskaya E, Malushko E, Kindaev A and Minin M 2018 *International journal of energy economics and policy* **8 (3)** 250-257.

[5] *Harnessing automation for a future that works*, ed James Manyika, Michael Chui, Mehdi Miremadi, Jacques Bughin, Katy George, Paul Willmott and Martin Dewhurst. Available at: https://www.mckinsey.com/featured-insights/digital-disruption/harnessing-automation-for-a-future-that-works

[6] Ergunova O T, Lizunkov V G, Malushko E Yu, Marchuk V I and Ignatenko A Yu 2017 *IOP Conference Series: Materials Science and Engineering* **177** 012046. DOI:10.1088/1757-899X/177/1/012046

IOP Conf. Series: Materials Science and Engineering **828** (2020) 012021 doi:10.1088/1757-899X/828/1/012021

GIS approaches to creating maps based on vegetation indices for forest management

Y Y Savchenko[1], O G Goleva[1], I A Korchagina[1], Y S Lobanova[1] and T S Borzhikov[1]

[1]Altai State University, 61 Lenina Street, 656049, Barnaul, Russia

E-mail: kolmakova@rb.asu.ru

Abstract. The article considers the main approaches to providing information support for improving forest management on the territory, taking into account the most up-to-date experience, modern approaches and implementing sustainable forestry management based on the data of remote sensing and geoinformational modelling. To identify the state of vegetation in the area under analysis, the authors of the article propose to introduce the definition of a normalized differential vegetation index. A characteristic feature of vegetation and its state is spectral reflectivity, which is characterized by substantial differences in reflecting radiation from different wavelengths. To obtain the necessary MODIS packages, we used the LAADS DAAC service supported by NASA. The data of this service are freely accessible, convenient for the user, and they are taken from a large number of satellites and their spectra. The optimal option for solving the tasks of compiling NDVI maps is using the geographic information system QGIS. Among the main advantages of this GIS we can consider user-friendliness and its multi-platform basis. For proper visualization of the NDVI map, a special discrete scale of values is used, the purpose of which is to differentiate between organic and non-organic nature, as well as determine the density of vegetation. The introduction of the proposed technology will significantly increase the mobility and effectiveness of forest management.

1. Introduction

Forests are renewable natural resources and play an important role in providing the environment suitable for human life. Forests also act as habitats of wild animals, they contain water resources and recreation areas. Forestry includes managing a wide range of natural resources. Forest management in the modern world is becoming increasingly complex.

In this regard, developing more advanced methods of processing spatial data on forest territories and proper presentation of the results prove to be extremely important. These methods include geoinformational technologies that allow recording any changes in the state of forests in time and space with high precision. Geoinformational technologies provide foresters with powerful tools for accounting, analysing and decision making. Developing GIS technologies, global positioning system (GPS) and remote sensing (RS) allows us to collect and analyze a large amount of spatial data.

In this article we look into the territory of the tape boron within the Ozerno-Kuznetsovsky forestry located in the Uglovsky district in the south-west of the Altai Territory. Forests on the territory of this forestry belong to the West Siberian subtaiga forest-steppe zone. The climate of the region is sharply continental. The species composition of trees is limited by a small amount of rainfall at high summer and low winter temperatures. Late spring and early autumn frosts with the vegetaion period of 150-160 days are typical of this zone. Sandy soil with deep location of groundwaters determines the dominance of pine plantations.

2. Research methodology

To identify the state of vegetation of the area under study, in his article we propose to introduce the definition of a normalized differential vegetation index. A characteristic feature of vegetation and its state is spectral reflectivity which is characterized by substantial differences in reflecting radiation from different wavelengths. NDVI makes it possible to see in what way a specific plant is capable to borb and reflect various wave spectra. This is the proper way to determine the state of the vegetation.

The main source of information is the data obtained from remote sensing which are widely used for accurate measurement and monitoring of vast forests [1, 2].

A GIS can be created to provide important, reliable spatial information about forest resources and can simplify forest planning and management, for example, accounting and updating forest inventories, reforestation, annual allowable cuts, biodiversity conservation, soil and water conservation [3].

Recent technological advances in the field of unmanned aerial vehicles (UAVs) and high-resolution satellite imagery with frequent time revisions have further expanded the possibilities of using information technologies [4, 5].

Thanks to satellite images, mankind can obtain rather numerous data on vital activity of plants in almost any part of the planet. Today, NDVI is the most common index for solving problems by using quantitative estimates of vegetation cover [6, 7, 8, 9, 10, 11].

NDVI provides an opportunity to see the difference in the intensity of reflected light in the red and infrared ranges, within a certain period of time.

NDVI is actively used to monitor soil moisture and saturation, evaporation and precipitation. In these cases, NDVI parameters are not always objective, because for measuring such data, one should take into account many features of the current location. The relevance of using NDVI is explained by its relative simplicity and speed of calculation, as well as the ability to work with satellite data of high resolution. In this regard, cartographers often use it for charting thematic and landscape maps.

Satellite images that allow you to work with the NDVI index are created by using various tools, one of which is the MODIS spectroradiometer. MODIS (Moderate Resolution Imaging Spectroradiometer) is a scanning image sensor designed for remote sensing of the planet. Currently, MODIS is located on two artificial Earth satellites «Terra» and «Aqua».

Due to the ratio of close spatial resolution and high temporal resolution, MODIS acts as an indispensable solution for daily monitoring of vegetation, nature landscapes, atmospheric dynamics and the earth cover as a whole.

To obtain the necessary MODIS packages, we are going to use the LAADS DAAC service which is supported by NASA and whose data are freely accessible, convenient for the user, and they are taken from a large number of satellites and their spectra. Most of the data obtained from LAADS DAAC are available in HDF format. Today, HDF4 is an obsolete version and is completely replaced by HDF5. The main difference between these versions is that HDF5 can contain a hierarchy of two types:

- Data sets that can be defined as sets of data and multidimensional arrays of objects of the same type;
- Groups that can be defined as certain containers for data sets and other groups.

Data in HDF5 are stored as a set of named object attributes that are contained in a special file system [12].

Primary packages with maps contain accurate information about 36 spectral channels. Each channel contains numerical data combined into a single database. Using the geographic information system, the process of constructing, calculating and visualizing the required index will be carried out. Storage of maps is provided for a sufficiently long amount of time.

For proper visualization of the NDVI map, a special discrete scale of values is used, the purpose of which is to differentiate between organic and non-organic nature, as well as determine the density of vegetation. As described earlier, a range of negative numbers tending to 0 are objects of non-organic nature. Positive numbers, within the range from 0 to 1, are objects of organic nature [13].

To display the NDVI index, a continuous gradient or discrete scale is used. More details about the belonging of values to types of objects are illustrated in table 1.

Table 1. Correspondence of object types to NDVI values.

Object type	NDVI value
Powerful, dense vegetation (forest)	0.7 – 1.0
Sparse vegetation (shrubs, pastures)	0.2 - 0.5
Open soil	0.025 - 0.2
Clouds	0
Snow, ice, sand, rocks	-0.1 - 0.1
Water	-0.42 - -0.33
Artificial materials (concrete, asphalt)	-0.5

After analyzing the table, we can conclude that values from -1 to 0 are assigned to inanimate objects and infrastructure: water, stones, sand, snow, etc. Wildlife objects, such as plants, are characterized by values ranging from 0 to 1.

The closer the value is to one, the denser vegetation prevails in the area. An example of a discrete NDVI scale is illustrated in figure 1.

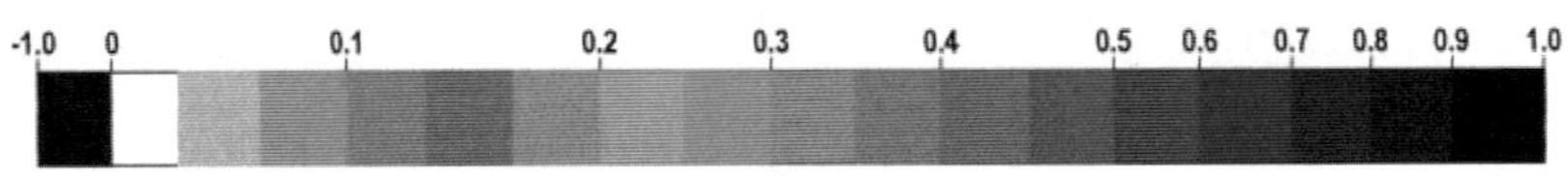

Figure 1. NDVI Discrete Scale

3. Results of research

The optimal solution for the implementation of the tasks of constructing NDVI maps was the use of the geographic information system QGIS. Among the main advantages of this GIS we can consider user-friendliness and its multi-platform basis.

In order for the map to acquire color shades, you must first calculate the NDVI index by the ratio of the two added spectral channels. The resulting NDVI map is shown in figure 2.

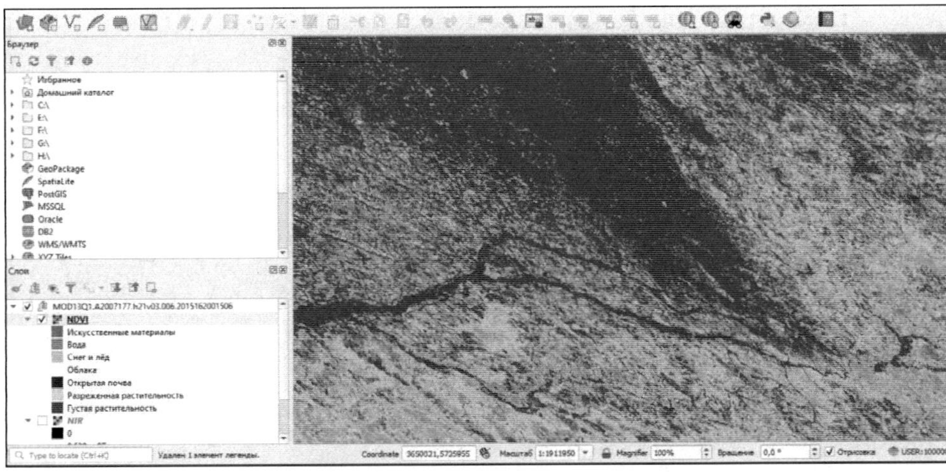

Figure 2. The resulting NDVI map

Next, we will create maps for 2008-2017.

All parts of the maps were grouped into subgroups, the names of which represent the year to which the image refers. The structure of these groups containing maps is shown in Figure 3.

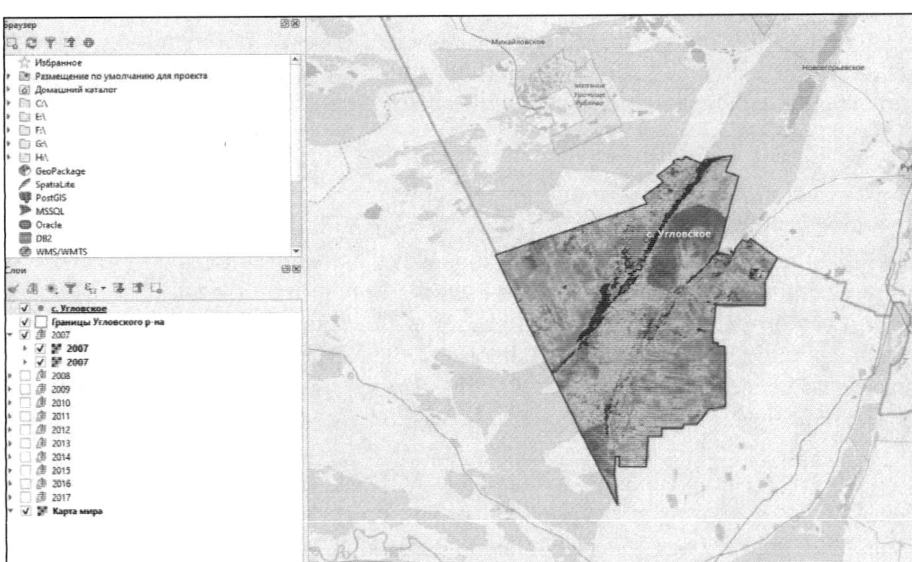

Figure 3. Data structure of groups containing maps

A chart was drawn up depicting the process of monitoring the state of the tape boron of the Uglovsky district of Altai Territory, with the introduction of NDVI technology. This model is shown in figure 4.

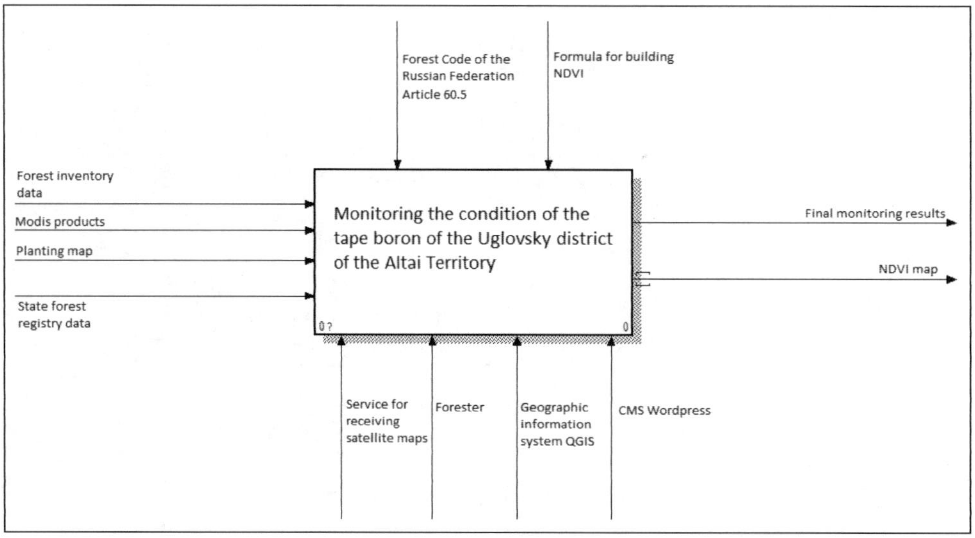

Figure 4. Context diagram of IDEF0 monitoring the status of the tape boron of the Uglovsky district of the Altai Territory

The final step is to provide the obtained data for free access. After confirming the received data by forest management specialists, the map is published on the site managed by the site control system Wordpress. Currently, the above-mentioned method is the only possible way of operating the necessary data and creating NDVI maps. Despite the simplicity of calculating the NDVI index, ready-made maps in the public domain, covering the entire planet for a certain period of time, do not exist so far.

4. Conclusions

The normalized vegetation index NDVI is capable of visualizing even the most inaccessible areas of the forest involving the minimum spending of financial and human resources.

The proposed method is universal and can be used not only to monitor the state of the vegetation of the tape boron. It can provide an opportunity to monitor the state of soils, precipitation, as well as the impact of climate on various natural objects. The use of the NDVI index becomes especially necessary when tracking planted vegetation: during the planting period (spring), and during harvesting (autumn). Thus, it is possible to evaluate problematic and arid areas of crops in which the NDVI value will be less than 0.5.

The introduction of the proposed technology will significantly increase the mobility and effectiveness of forest management. Publication of the processed images in free access will allow departmental bodies, organizations, legal entities and individuals to independently study and analyze published information, without any documentation and at no cost. This method makes it possible to simplify the process of obtaining complicated data from satellites, and thus facilitate their visualization and understanding, without requiring significant investments.

References

[1] Huete A 2012 *Geography Compass* **6** 513-32
[2] Pakhutchy V, Pakhutchaya L, and Shevelev D 2016 *Forestry Bulletin* **1** 17-23
[3] Yamashkin S A, Yamashkin A A, Zarubin O A, Tsibakov O V, Gurin V A and Kodulev A E 2019 *Int. J. of Civil Engineering and Technology* **10** 242-9
[4] Sagan V, Maimaitijiang M, Sidike P, Maimaitiyiming M, Erkbol H, Hartling S, Peterson K T, Peterson J, Burken J and Fritschi F 2019 *Int. Arch. Photogramm. Remote Sens. Spatial Inf. Sci.* **XLII-2/W13** 715-22

[5] Barbosa S G, Spletozer A G, Roque M P B, Ferreira N J A, Dias H C T, Ramos M P, Bonilla M A C, Ribeiro W S, Alcrńntara-de la Cruz R and Zanuncio J C 2018 *Scientific Reports* **8** 3959
[6] Martin H, Labbé S, Baldet P, Archaux F and Philippe G 2017 *iForest - Biogeosciences and Forestry*, **10** 215-19
[7] Jovanović M M and Milanović M M 2015 *Pol. J. Environ. Stud.* **24** 529-35
[8] Meera Gandhi G, Parthiban S, Nagaraj Thummalu and Christy A 2015 *Procedia Computer Science* **57** 1199-210
[9] Ahmadi H and Nusrath A 2012 *J. of geography and geology* **2** 58-67
[10] Doumit J and Kiselev E 2017 *11 Edition of the international congress Geo-Tunis* **5** 11-21
[11] Gong Z, Kawamura K, Ishikawa N, Goto M, Wulan T, Alateng D, Yin T and Ito Y 2015 *Solid Eart* **6** 1185–94
[12] Schedlbauer M 2010 *The Art of Business Process Modeling: The Business Analyst's Guide to Process Modeling with UML & BPMN* (Scotts Valley: CreateSpace) p 120
[13] Peltier T R 2016 *Information Security Fundamentals* (Boca Raton: Auerbach Publications) p 440

IOP Conf. Series: Materials Science and Engineering **828** (2020) 012022 doi:10.1088/1757-899X/828/1/012022

Ecological state of surface waters of rivers in industrial cities

S V Senotrusova[1], V G Svinukhov[1,2], V V Zhoglitcheva[1,2] and O L Kurgaeva[1,2]

[1] Plekhanov Russian University of Economics
Stremyanny lane 36, Moscow, 117997, Russian Federation
[2]Lomonosov Moscow State University,GSP-1, Leninskie Gory, Moscow, 119991, Russian Federation

E-mail: inmar2000@mail.ru

Abstract. The paper analyzes the chemical composition of the surface waters of a river flowing through an industrial city. Samples were taken in the Ussuri River, flowing through the city of Lesozavodsk. Four factors were derived that describe 78% of the variance of the variables. The first factor is associated with elements such as oxygen, BOD5 (Biological Oxygen Demand in 5 days), NH4, Fe, as well as phenols. Oil products and SSAS (Synthetic Surface-Active Substance) stood out in the second factor. The third factor is related to COD (Chemical Oxygen Demand), and NO3. Suspension and copper stood out as the fourth factor.

1. Introduction

The application of statistical methods to the analysis of results of studies of the chemical composition of surface waters is of particular interest. Therefore, the factor analysis method [1-2], which is widely used to identify sources of emission of tracer elements was applied as a reliable method of analysis. Since there is a correlation between the concentrations of pollutants in discharges and their content in river streams, the factor analysis method applied to the data on the chemical composition of surface waters makes it possible to distinguish groups of elements coming from different sources.

The method allows us to represent each variable (in this case, concentration) as a linear combination of independent factors. As the first step of the calculations, the matrix of pair correlation coefficients between the variables is produced. As the next step, the eigenvalues and eigenvectors are determined from the correlation matrix. Normalized vectors are factors that influence the variability of features. Factors are ranked according to the distribution of the total variance of the data. The first several factors usually describe more than 80% of the variance, which is sufficient for the interpretation of the original variables. Factors and coefficients of factor mapping (i.e. factor loads) are not uniquely determined and have several equivalent solutions. Factors Fj can be changed by orthogonal transformation (the varimax method is used in the work) with the corresponding conversion of factor loads Aij. The transformation is carried out in such a way that the resulting coordinate axes satisfy the criteria of a simple factorial structure [1]. According to the Horn rule [1], factors are selected for eigenvalues 1.

2. Materials, methods and organization of research

To properrly discuss the results of the study, we should first take into account some features of the elements migration in the hydrosphere. Chemical composition of river waterin cities is mainly influenced by the following factors: 1) the release of conditionally pure industrial waste into the waterways after treatment; 2) the influx of storm water polluted by natural dumps of waste and raw materials, soils, concentrating industrial emissions; 3) the emission of sewage by treatment facilities. Indeed, in recent years it has been established that for most of the chemical elements, the most important source is storm runoff, with which up to 85% Zn, 70% Pb, 50-70% Cd, Cu, Mn, about 40% Mo and Ni fall into the waterways [3]. The composition of storm runoff, which is the most important technogenic supplier of chemical elements to water courses, is formed by washing away soils, washing away dust from asphalt surfaces, erosion of buildings and structures, and washing out landfills. Atmospheric precipitation has a significant impact on the pollution of stormwater. According to [4], the percentage of chemical elements of aerial origin in the surface rainwater of urban areas can be from 35 to 54% for Mn, Zn, Pb and 90% for Cu. For Sn and Cr, the most significant supply is with sewage (about 50%). The input of chemical elements with conditionally pure industrial effluents into watercourses is about 5% for Cu, Cd, and Ni. Among man-made sources, the contribution to the pollution of watercourses is increasing in the following series: industrial wastewater of conditionally clean water - sewage after treatment facilities (if available) - storm surface runoff.

For research, the city of Lesozavodsk (Primorsky Territory) was chosen, in which about 30 thousand people live. The main sources of anthropogenic emissions and discharges are industrial enterprises, heat power enterprises (boiler houses), and motor vehicles. In total, the city has about 20 enterprises with significant emissions and discharges of harmful substances. The main pollutants emitted into the atmosphere: solids, sulfur dioxide, carbon monoxide, nitrogen oxides, hydrocarbons, phenols. Emissions and discharges contain a number of specific pollutants: yeast dust, furfural, methanol, organic acids, and sulfuric acid.

The research material was the data of annual observations at a network of stations of the Primorsky Administration for Hydrometeorology and Environmental Monitoring. Samples were taken in the Ussuri River, flowing through the city. The analysis involved data on the average annual concentrations of such substances as: 1) suspended solids, 2) oxygen content, 3) BOD5, 4) NH4, 5) NO3, 6) Fe, 7) Cu, 8) oil products, 9) SAS, 10) phenols; 11) COD.

3. Results and their consideration

The results of the correlation analysis are given in table 1, and the results of factor analysis are given in table 2.

Table 1. Matrix of paired correlation coefficients (samples taken in the city)

№	1	2	3	4	5	6	7	8	9	10	11
1	1	0,03	0,31	-0,06	-0,14	-0,13	0,51	0,04	0,10	-0,25	-0,26
2		1	0,61	-0,35	0,00	-0,49	0,25	-0,11	-0,14	-0,37	-0,08
3			1	-0,40	-0,47	-0,43	0,47	0,36	0,22	-0,50	0,00
4				1	0,00	0,35	-0,31	0,34	0,25	0,47	-0,13
5					1	-0,05	-0,38	-0,30	-0,21	0,11	-0,41
6						1	-0,21	0,22	0,12	0,69	0,12
7							1	-0,10	-0,16	-0,43	-0,38
8								1	0,88	0,08	-0,12
9									1	0,08	-0,17
10										1	0,10
11											1

Hereinafter, the variable number means: 1) suspended solids, 2) oxygen content, 3) BOD5, 4) NH4, 5) NO3, 6) Fe, 7) Cu, 8) oil products, 9) SSAS, 10) phenols, 11) COD.

Significant values with a confidence of 95% are values of 0.47. This is due to the fact that the significance of the correlation coefficients and factor loads is determined by the number of observations [5].

Table 2 presents the matrix of factor loads calculated for water samples taken in the city. From the data in Table 2, it can be seen that, as a result of statistical data processing, four factors were identified that describe 78% of the total variance of the variables.

Table 2. Factor load matrix (samples taken in the city)

Element	A1	A2	A3	A4
Suspension	0,05	0,07	-0,05	-0,81
O2	0,79	-0,01	-0,02	0,01
BOD5	0,71	0,38	0,34	-0,32
NH4	-0,62	0,36	-0,18	0,08
NO3	-0,02	-0,25	-0,85	0,32
Fe	-0,82	0,09	0,17	0,04
Cu	0,28	-0,12	0,09	-0,01
Petroleum Products	-0,08	0,97	0,09	-0,01
SSAS	-0,07	0,93	-0,01	0,01
Phenols	-0,76	0,05	0,00	0,28
COD	-0,04	-0,21	0,80	0,45
Total Variance, %	31	52	66	78

The first factor, describing 31% of the variance of the variables, significantly positively loads such elements as oxygen, BOD5 and negatively: NH4, Fe, phenols. This distribution of ingredients is quite acceptable since it indicates the effect of pollution by these components on the oxygen content in water. The first factor can be interpreted as an anthropogenic component of discharges of industrial enterprises. The second factor, which describes 21% of the variance of the variables, was oil products and surfactants. Surfactants can mainly enter watercourses with sewage and the release of petroleum products into this factor indicates one source of these pollutants entering the city's watercourses. Falling into the third factor, which describes 14% of the variance of variables, COD with a significant load and NO3 with a negative load is completely logical since the content of NO3 in water should be inversely related to the COD content. In the fourth factor, which describes 12% of the dispersion, two elements stand out: suspension and copper. Suspended matter should mainly flow into streams with storm runoff. Thus, copper can enter watercourses most likely as a flush from the underlying surface of the city.

4. Conclusion

Thus, the first factor can be interpreted as an anthropogenic component of discharges of industrial enterprises. The second factor can be defined as domestic sewage. The third factor characterizes the concentration of oxygen in water. The fourth factor can be defined as flushing from the underlying surface.

References

[1] Iberla K. *Factor Analysis* 1967 (Moscow: Mir)
[2] Lawley D and Maxwell A 1967 *Factor Analysis as a Statistical Method* (Moscow: Mir)
[3] Bondareva D G, Surits O V, Khristoforova N K, Elinskaya E O and Revutskaya I L 2017 *Problems of Regional Ecology* **6** 88-93
[4] Shulkin V M, Chernova E N, Khristoforova N K and Kozhenkova S I. 2014 *Geoecology. Engineering geology, hydrogeology, geocryology* **6** 483-94
[5] Brooks K and Carusers N 1963 *Application of Statistical Methods in Meteorology* (Leningrad:

Gidrometeoizdat)
[6] Surits O V, Khristoforova N K, Kopylov P V and Bondareva D G 2014 *Public Health and Habitat* **4 (253)** 24-6
[7] Klinskaya E O and Khristoforova N K 2013 *Problems of Regional Ecology* **6** 232-6
[8] Khristoforova N K and Salomai M S 2006 *Investigated in Russia* **9** 1380-6
[9] Kondratiev I I, Svinukhov V S and Svinukhov G V 1994 *Meteorological, Geochemical and Medical Aspects of the Natural Environment of Spassk-Far Primorsky Territory* (Vladivostok: FENU) pp. 21-3

IOP Conf. Series: Materials Science and Engineering **828** (2020) 012023 doi:10.1088/1757-899X/828/1/012023

Import substitution as encouragement method of industrial production of Volgograd region

Inna M Shor[1], Dmitriy M Shor[1], Dildarahon A Shelestova[1], Victoria V Batmanova[1] and Lyubov I Galamyan[2]

[1]Volgograd State University, 400062, Prospekt Universitetskiy, 100, Volgograd, Russia
[2]Russian Presidential Academy of National Economy and Public Administration (RANEPA), 119571, Prospect Vernadskogo, 82, Moscow, Russia

E-mail: ShorIM@volsu.ru, ShorDM@volsu.ru, ShelestovaDA@volsu.ru, batmanovavv@volsu.ru, galamyan@mail.ru

Abstract. The import substitution development which contributes to the encouragement of national industrial production under foreign sanctions is capable of providing the sustainability of enterprises in Russian regions. Herewith the measures which are taken today for the import substitution encouragement do not allow fully displaying the potential of its stimulating constituent. The goal of the research is the development of measures for import substitution growth as a method of import production in Volgograd region. The system approach, statistical analysis and dialectical method were used during the research. Volgograd region was taken as the example and the modern situation in import subsitution was analyzed. The SWOT analysis of industrial enterprises was conducted what allowed presenting the measures aimed at import subsitution growth as a method of encouragement of regional industrial enterprises. The authors made the conclusion about the necessity of an optimal combination of import substitution and export orientation which was based on the creation and development of enterprises which are focused on the world practice and which are profitable in global competition. This allows eliminating a number of drawbacks of import substitution which deal with the problem of a limited volume of the domestic market, construction of new industrial facilities, optimization of production processes, reduction of the monopolism, illegal economy and corruption. Key words: import replacement, efficiency, industrial production, encouragement, measures, enterprise, Volgograd region

1. Introduction

Under modern conditions the import substitution is one of strategic directions of the public policy of the Russian Federation and its regions which is meant to provide a sustainable industrial production. In this case we can speak about the reduction or cessation of imports of definite products by means of manufacturing of the same or analogous production (import-substituting products). Herewith according to the President's Address to Parliament dated 04.12.2014 [1], import substitution is a long term priority which is expected to work for the establishment of a big number of new industrial enterprises without regard to external conditions.

The import substitution is the state program of the Russian Federation [2], the monitoring of which since 2015 is carried out by a special governmental commission [3] which is directed to the creation of conditions for the national production of goods in the industries where a high degree of import dependence remains.

During the crisis the Ministry of Economic Development of the Russian Federation analyzed the structure of consumption and the inventory coverage of the imported equipment and goods which are necessary for the provision of stable employment and national economy development. The Ministry also studied the existing industrial, technical and technological restrictions of the corresponding industries.

In the result the critical types of production were found and the priority industries for the development of import substitution were identified, i.e.:

- components and equipment for food industry;
- heavy, power, oil and gas engineering, electrochemical and cable industry;
- machine tool industry, shipbuilding equipment;
- chemical and petrochemical industry;
- pharmaceutical and medical industry;
- production of conventional weapons, civil aviation, engines, components and equipment for transport engineering [4].

In the result the intensification of work of executive authorities for the development of import-substituting technologies and plants lead to the situation when today we can speak about certain success in this field which is mainly manifested in such industries as:

- chemical industry;
- vehicle manufacturing industry;
- medical industry;
- oil and gas production and others.

It is important to note that the development of a set of measures for the acceleration of the process of import substitution is a determining vector of modern government industrial policy of the Russian Federation and its regions. Herewith an important role here has the provision of an optimal correlation between the import substitution as a result of sanctions' overcoming and the diversification growth of national economy on the one hand and large scale industrial policy which reflects long term interests of the development of Russia as a country which on the other hand pretends to have the leader's ambitions in the world.

Under these circumstances the regions of the Russian Federation, taking into account industrial and technological, institutional, social, economic and other features of regional industry and also the potential of their development identify a system of import substitution priorities at the implementation of industrial policy which do not contradict but complement the similar positions at the federal level. One of such regions, which is actively involved in this process is the Volgograd region. The peculirarity of its industrial development is a large share of processing industries (more than 85%), including oil refining, chemical, metallurgical and machine-building industries [5].

The regional authorities have approved and they are implementing the Concept of import substitution in the Volgograd region for 2015 – 2020 and a plan of actions encouraging the import substitution in the Volgograd region [6]. Also, in order to ensure the transparence of import substitution measures an advisory council on import substitution was established [7].

The increase of import-substituting products is one of the strategic goals of industrial development of the Volgograd region. The priority sectors for import substitution are machine-building, chemical, metallurgical industries, military-industrial and agricultural complexes, as well as light industry. In 2018, 58 enterprises in the region produced 80 types of import-substituting products for an amount exceeding 57 billion rubles. The output growth was 9.3%. At the same time, in 2017, 53 enterprises were involved in the import substitution and they produced 75 types of import-substituting products [8]. It should be noted that the growth of import substitution became possible due to the implementation of large investment projects and the planned modernization of industrial complexes. However, in some cases, the industrial enterprises in the region face with a lack of capital, technology, labor, fixed assets, what in turn requires the removal of all these restrictions to the maximum extent and the creation of the necessary conditions for the growth of production volumes.

IOP Conf. Series: Materials Science and Engineering **828** (2020) 012023 doi:10.1088/1757-899X/828/1/012023

Thus, today the need for the development of a set of measures which improve the existing mechanism of import substitution which would contribute to sustainable development of national industrial production becomes relevant.

A significant contribution to the study of the mechanism of import substitution and the features of its implementation was made by Plotnikov V.A., Vertakova Yu.V. [9], Polovinkin V.N., Fomichev A.B. [10], Yakovlev G.I. [11] and by some other authors.

However, the analysis of the degree of development of this subject area indicates the absence of a unified approach to the formation and implementation of the import substitution mechanism in modern conditions of industrial enterprises' work. This does not contribute fully to their sustainable development and, therefore, the realization of the production potential of the country and its regions. The relevance of the problem under analysis and the insufficient level of its study determined the goal (development of recommendations for the efficiency growth of import substitution as a method of encouragement of industrial production in Volgograd region) and the tasks of the current study (to analyze the statistical data of modern examples of import substitution in Volgograd region and to present the proposals of the efficiency increase of import substitution as a method of encouragement of the activity of mentioned enterprises). Taking into account the goal and tasks the structure of this work was determined. The results of the study can be used by public authorities at the implementation of the import substitution policy in industrial enterprises.

2. Materials and Methods (Model)

The methodological base of the study is presented by theoretical and applied scientific papers of modern authors devoted to the problems of the implementation of import substitution mechanism; information published in periodicals and on the Internet. During the research a systemic approach, method of statistical analysis and the dialectical method of study by means of the application of general scientific methods and techniques (observation, synthesis and comparison methods, grouping technique, generalization of theoretical and practical material) were used.

In order to determine the directions of efficiency growth of import substitution, we used the SWOT analysis, which allowed us to identify the strengths and weaknesses, opportunities and threats for industrial enterprises.

3. Results and Discussion

3.1. The analysis of statistical data characterizing the results of import substitution in the Volgograd region [12] allowed us to draw the following conclusions:

- in 2018 the region was in the top ten of the leading constituent entities of the Russian Federation in the production of import-substituting products;
- from 2015 to 2017 the regional enterprises shipped import-substituting products at the amount of 128 billion rubles. At the same time, if in 2015 the share of these products in the total volume of shipped products by enterprises of the manufacturing industry was 4,8%, then in 2017 it was 7,3%. In 2018, 58 enterprises of the region produced 80 types of import-substituting products at the amount of more than 57 billion rubles, while in 2017, 75 types of these products were produced by 53 enterprises participating in the import substitution (Table 1).

At the same time, in 2018 the volume of production of heating radiators doubled. The volume of production of pipes for oil and gas production grew by 75,8%, the production of polyester fiber increased by 32,1%. The production of ceramic tiles increased by 4,7%, the production of cotton fabrics grew by 7,9%, and the production of silicon carbide increased by 9,2%.

The positive dynamics of import substitution indicators is facilitated by the implementation of investment projects and the modernization of industrial sites. Over the past five years, 50 new facilities were established in the region, 163 billion rubles were allocated for the development of existing firms and for the creation of new enterprises. The implementation of projects with investment funds and state support gives a positive result. According to the results of 2018, the total volume of industrial production in the region for the first time exceeded a trillion rubles.

Table 1. Volume of import-substituting products shipped by enterprises of Volgograd region in 2017

Industry	Number of enterprises involved in import substitution	Volume of shipped products obtained in the process of import substitution, billion rubles
Chemical industry	16	27,3
Machine-building industry	16	7,9
Metallurgy	7	11,6
Shipbuilding, timber industry complex, medical industry, children's products industry	7	0,014
Light industry	5	2,28
Manufacture of building materials	2	2,98

The chemical industry of the region continues to lead in the volume of shipped import-substituting products. According to the results of nine months of 2018, 16 chemical enterprises of the region produced import-substituting products (tires for agricultural machinery, polyester fibers, plastic products, insulating polymer composites, hydroxide and magnesium oxide, petroleum-polymer aromatic resins and more) at the amount of 23,6 billion rubles.

One of the key tasks of the industrial development of the region is the import substitution of equipment.

An important role in the solution of this problem has the "Volgogradneftemash" company which fills the domestic market for gas, oil and petrochemical industries with technological gas equipment. The "Kamyshin experimental plant" produces installations for the bottom discharge of oil and oil products, railroad equipment. The plant "Neftegazmash" produces tube block furnaces, pistons, blow out prevention equipment and spare parts. The steelplant "Akhtuba" produces metal structures.

According to the results of 2018, the largest growth in the production of import-substituting products in comparison with 2017 showed the following enterprises. The "RUSAL Volgograd Aluminum Plant" produced (calcined petroleum coke, aluminum powders and pastes, components for the manufacture of car engines and more). The firm "Zika" produced cement-based fast-hardening compounds and cement and polymer-based anchoring compounds. The "Forte-Prom" produced heating radiators. The firm "Sealed Air Kaustik" produced multilayer barrier coextrusion film made from polyamide, multilayer polymer packaging with printing for food, medicines, household chemicals and personal care products. The "Raduga Zink List" produced galvanized and color coated rolled product. The "Volzhsky pipe plant" produced stainless seamless pipes, pipes for gas stations, pipes for oil and gas. The "EPK Volzhsky" produced large-sized, thin-walled special ball bearings, bearings of high-loaded units, roller wheel bearing. The "Stekloplast" company produced fittings for special environments.

3.2. Without humiliating the significance of the achieved success in the import substitution program at the industrial enterprises of the Volgograd Region, it should be borne in mind that the largest part of these enterprises significantly depends on foreign supplies.

In the manufacture of import-substituting products, the imported materials, components, technologies, machines, software is used to a large extent. Accordingly the tasks of efficiency growth of import substitution should be solved, not only by means of the expansion of the volume of products, but also by means of the use of domestic capabilities.

It should also be mentioned that there is a staff problem that the industrial enterprises of the region face at import substitution. In this case, it is spoken about the migration of skilled labor to enterprises with more favorable remuneration. Another problem with the personnel is the absence of the necessary qualifications in the labor market.

At the same time, it is possible to train highly qualified and stable personnel in the industrial enterprises of the region only with a high salary, employment benefits, professional training, education and advanced vocational training.

It is worth adding to the mentioned above that, when the import substitution is implemented, the industrial enterprises are characterized by a large volume of accounts receivable. In some cases, the generated portfolio of orders consists of contracts with the following payment for shipped products or with minimal advance financing. According to a number of contracts, under which the targeted credit financing is attracted for the purchase of raw materials needed for the implementation of orders, then the contractual delivery periods are increasing without extending the terms of credit contracts. At the same time, only if the work is carried out constantly in order to accelerate the turnover of receivables, it is possible to release the additional cash and to reduce the need of industrial enterprises in borrowed funds.

3.3. To determine the directions of efficiency growth of import substitution as a method of encouragement of industrial enterprises in the Volgograd region, we will use one of the research methods or the SWOT analysis, which allows you to identify the strengths and weaknesses, opportunities and threats for these enterprises in the region (Table 2).

In our SWOT analysis, we emphasized a sufficient number of weaknesses and threats what indicates the importance of taking a number of measures by executive authorities for their reduction (elimination) in order to stimulate the development of industrial enterprises in the region by means of import substitution.

Table 2. SWOT-analysis of industrial enterprises of Volgograd region

Strengths	Weaknesses
Diversity of industrial production. In the structure of industrial production, the largest share has the basic branches of manufacturing, the products of which have a significant competitive market value. Growth prospects for oil production and gas production. Favorable geographical position. River and rail links.	High wear and low technical sophistication of fixed assets. Inconsistency of the quantity and quality of labor potential with the needs of industrial development. A high proportion of the energy component in production. High-interest rates on borrowed funds.
Opportunities	**Threats**
Sufficient volume and diversity of mineral resources. Growth of educational level of labor resources. Increase of scientific and technical potential.	Natural decline and drain of the working-age population. Dependence of products on imported raw materials and equipment. Tariffs in the energy and transport sectors, fluctuations in world prices. Poor infrastructure development.

The experience of industrial enterprises of the Volgograd region shows the importance of tax incentives for the development of production of import-substituting products. First of all, we are talking about a direct investment incentive. A part of the tax base for income tax is exempt from the

IOP Conf. Series: Materials Science and Engineering **828** (2020) 012023 doi:10.1088/1757-899X/828/1/012023

taxation if it is invested in the expansion or restoration of fixed assets for industrial purposes in fixed directions. The reduction of the tax burden on income tax makes it possible to accumulate funds for the modernization of production capacities and it allows increasing the budget revenues from the growth of turnover taxes (for example, VAT). Currently, the tax policy is focused on the implementation of the fiscal function, while the potential of the stimulating function, creating the conditions for investment in the development of import-substituting production remains practically unused.

In general, in our opinion, the import substitution should be combined with the export orientation, based on the creation and development of industries focused to the world market and profitable at global competition since the export orientation can eliminate a number of disadvantages of import substitution. They are as follows:

- Narrowness of domestic market is eliminated;
- New production facilities are created taking into account the comparative advantages regard of the international division of labor;
- Production processes are improved under the influence of competition in foreign and domestic markets;
- The monopolism scale, the informal sector and corruption reduce.

It seems that the import substitution at an optimal combination with export orientation should be implemented in three areas:

- modernization of production ensuring the growth of output of imported industrial products, analogues of which are not produced in Russia in sufficient quantities;
- development of imported industrial products during a short period of time, but its import substitution is economically profitable or possible due to some objective reasons;
- search for ways to substitute indirectly the imported industrial products that are not manufactured in Russia, since the import substitution is economically unprofitable or impossible due to objective reasons.

The implementation of these measures implies the solution of a number of key tasks:

- analysis of the region's commercial industrial turnover in recent years, identification of the most critical imported products in the context of specific industries of the regional economy;
- study of the production, scientific and technical potential of industrial enterprises in the region that produce goods which would compete with imported goods. They can be also good substitutes. The assessment of a load of production capacities and depreciation of fixed assets;
- development of a list of equipment and technologies necessary for industrial enterprises in the region and the assessment of the economic feasibility of their production in the region [13];
- development of technical specifications for the production of equipment subject to import substitution by means of an open discussion of these documents with the consumers of products, i.e. industrial enterprises of the region;
- identification of the list of products, the needs of which the region can satisfy by means of the increase of the own production using the existing resources and by means of the government support of industrial enterprises.

4. Conclusion

4.1. The analysis of statistical data on the implementation of the import substitution program in Volgograd region allowed to make the following conclusions. The Volgograd region is among the ten leading federal entities of the Russian Federation in the production of import-substituting products. In 2018 58 enterprises of the region produced 80 types of import-substituting products at the amount of more than 57 billion rubles, while in 2017 only 75 types of these products were produced by 53 enterprises participating in import substitution. The implementation of investment projects and the modernization of industrial sites contribute to the positive dynamics of import substitution indicators. According to the volume of shipped import-substituting products, the regional chemical complex continues to lead among the enterprises, taking part in import substitution.

4.2. Without humiliating the significance of positive results of the implementation of import substitution in industrial enterprises of Volgograd region, it is necessary to mention that there are problems of legal, resource, personnel, financial and other nature that impede the full disclosure of the potential of the encouraging component of import substitution in the development of the region's industry.

4.3. Having used the tools of the SWOT analysis, we came to the conclusion that it is necessary to combine optimally the import substitution with the export orientation at industrial enterprises of Volgograd region what should be implemented in three areas. The first direction is the modernization of production in order to ensure an increase in the output of imported industrial products, analogues of which are not produced in Russia in sufficient quantities. The second direction consists in the development in a short time of production of imported industrial products that are not produced in Russia, but their import substitution is economically profitable or possible due to objective reasons. The third area of activities is the search for ways of an indirect replacement of imported industrial products that are not produced in Russia since its import substitution is economically unprofitable or impossible due to objective reasons. The implementation of these measures providing an efficiency increase of import substitution involves the solution of a number of key tasks: analysis of the region's commercial industrial turnover in recent years, identification of the most critical imported products in specific sectors of the regional economy; study of the production, scientific and technical potential of industrial enterprises in the region that produce commodities that compete with imports or substituting goods; assessment of the utilization level of their production capacities and depreciation of fixed assets.

References

[1] Address of the President of the Russian Federation to the Federal Assembly of 04.04.2014. 2014 *Rossiiskaya gazeta* **278**
[2] Decree of the Government of the Russian Federation dated April 15, 2014 No. 328 "On approval of the state program of the Russian Federation "Development of industry and increasing its competitiveness" 2014 *Corpus of Legislative Acts of the Russian Federation*, **18 (IV)** Art. 2173.
[3] *Decree of the Government of the Russian Federation, dated August 4, 2015 No. 785 "On the Government Commission on Import Substitution".* URL: http://www.pravo.gov.ru.
[4] Semenov M V 2017 *Import substitution Industry: accounting and taxation* **4** 52 – 60.
[5] *Federal Service for State Statistics in Volgograd region*. URL: volgastat.gks.ru.
[6] Resolution of the Administration of Volgograd Region dated September 14, 2015 No. 527-p "On Approval of the Concept of Import Substitution in Volgograd Region for 2015–2020 and the Action Plan for the Promotion of Import Substitution in Volgograd Region for 2016–2020" 2015 *Volgogradskaya pravda* **153**
[7] Resolution of the Governor of the Volgograd Region dated December 30, 2015 No. 1157 "On the formation of an expert council on import substitution under the Governor of Volgograd Region" 2016 *Volgogradskaya pravda* **11**
[8] The enterprises of the Volgograd region produced import-substituting products which cost more than 57 billion rubles 2019. URL: volgograd.ru.
[9] Import substitution: theoretical foundations and prospects for implementation 2014 *Economics and Management* **11** 38–41.
[10] Polovinkin V N and Fomichev A B 2018. *Expert Union* **12** 30–3
[11] Yakovlev G I 2015 *Bulletin of Samara State University of Economics* **5** 59–64
[12] *Committee for Industry and Trade of Volgograd Region* 2019. URL: promtorg.volgograd.ru.
[13] *Import substitution: foreign experience and lessons for Russia* 2019. URL: research-journal.org.

IOP Conf. Series: Materials Science and Engineering **828** (2020) 012024 doi:10.1088/1757-899X/828/1/012024

Human gait modeling method

S V Sivolobov[1], A V Khoperskov[1] and V V Bumagin[1]

[1]Volgograd State University, 100 Prospect Universitetsky, Volgograd, 400062, Russia

E-mail: sivolobov@volsu.ru

Abstract. The article deals with problem of human gait modelling. Human gait analysis is used for various human identification tasks and for various purposes in medicine. Lagrange equations of the second kind in many research are used to model gait. We use a model of an anthropomorphic mechanism of five links to describe human movements. The mathematical model is implemented in the form of software that uses input data on the kinematics of a real gait. An important result of the study is the creation of an algorithm for calculating model parameters that best reproduce the gait of a particular person. Our approach is based on the application of a genetic algorithm to minimize a new special objective function, which depends on discrete generalized coordinates that depend on time. The dynamic picture significantly improves the quality of the model, which is able to reproduce the subtle features of the human gait.

1. Introduction

Gait is an important biometric indicator of a person along with a face photograph, voice or handwriting [1]. It can be used to identify a person [2, 3], for diagnostic purposes in medicine [3]. The practice of clinical studies of gait shows promising prospects for a sample of children with developmental disorders with cerebral palsy and myelomeningocele. DeLuca and others [4] examined 91 patients who had a recommendation for surgery from experienced doctors. Gait analysis was the reason for reviewing surgical recommendations for 52 percent of patients, which reduced the cost of surgery and patients avoided unnecessary procedures [5]. Gait modeling is of considerable interest for identifying people, as well as for studying gait pathologies [6, 7]. Gait modeling is used in the construction of stably moving anthropomorphic systems with two legs, when creating prostheses and orthoses of the lower extremities [8], medical exoskeletons for the rehabilitation of disabled people, patients with paraplegia [9].

The one-limb support and two-limb support phases of movement alternate in the person walking process. During a one-limb support movement, the support leg is on the surface and the second leg is in the process of being transferred. During the two-limb support phase, both legs rest on the surface [10]. The first phase is more than 76 percent of the step duration and it is being actively studied in both experimental and theoretical approaches [5, 11, 12].

The work [12] describes the modeling of a person's gait by analytically calculating the trajectories of the lower extremities in a single support phase using the Lagrange equations of the second kind, which are fitted with real trajectories using the MPC controller in MATLAB / Simulink. We offer an alternative way to calculate the trajectories of a particular gait by varying the parameters of the model using dynamic data.

2. Mathematical model

The Lagrange equations for a system of a planar anthropomorphic mechanism of five pivotally connected massive elements are the basis of our model [10]. All parts of the mechanism can only move in the sagittal plane (Figure 1).

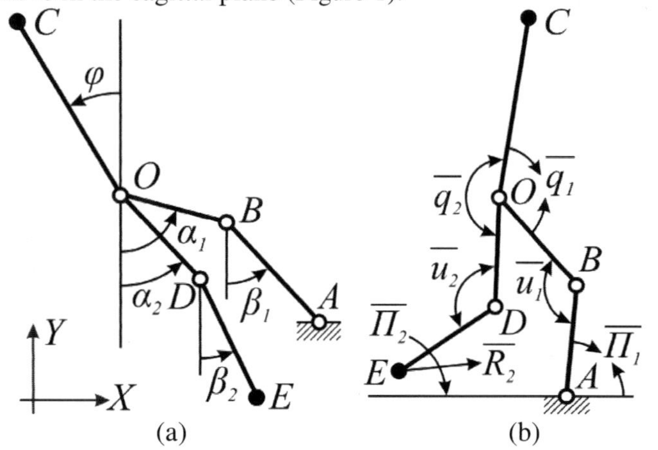

Figure 1. Kinematic scheme of a five-link anthropomorphic model (a – angles between rods, b – non-conservative forces acting on the mechanism).

The OC element denotes the body, and the identical (identical) two-link (OBA and ODE) are legs.

The distance between the hip joints of the legs in the frontal plane is small, and therefore we will consider a flat spatial structure. The links OB and OD are called the hips, BA and DE are the lower legs, and O is the hip joint. All joints are assumed to be ideal, since friction in the joints of a person is small [11]. We study only the single-limb support phase of movement, therefore, the OBA support leg is pivotally connected to the surface at point A, and the double-link system designates a moving leg that is not connected to the surface. Feet and hands are not considered in this model. The research work [10] describes a seven-link mechanism, which additionally contains two feet, which are considered massless.

The five generalized coordinates are angles φ, α_1 ,α_2 ,β_1 ,β_2. They denote the angles that form the body, thighs and lower legs with a vertical line, respectively. Non-conservative forces acting on the mechanism are shown in Figure 1b. We distinguish moments of forces in the knee joints $\bar{u}_1$ and $\bar{u}_2$, between the body and hips $\bar{q}_1$ and $\bar{q}_2$. The vector $\bar{R}_2$ denotes the external force applied to the end of the leg E (R_{2x} and R_{2y} are its horizontal and vertical components), and $\overline{\Pi_1}$ and $\overline{\Pi_2}$ are the moments of external forces applied to the legs.

The equations of motion in matrix form for such a five-link are as follows [10]:

$$B(z)\ddot{z} + gA|\sin z_i| + D(z)\left|\dot{z}_i^2\right| = C(z)w\,, \tag{1}$$

where g is the gravity,

$$z = |z_i| = \begin{vmatrix} \varphi \\ \alpha_1 \\ \alpha_2 \\ \beta_1 \\ \beta_2 \end{vmatrix}, \quad |\sin z_i| = \begin{vmatrix} \sin\varphi \\ \sin\alpha_1 \\ \sin\alpha_2 \\ \sin\beta_1 \\ \sin\beta_2 \end{vmatrix}, \quad \left|\dot{z}_i^2\right| = \begin{vmatrix} \dot{\varphi}^2 \\ \dot{\alpha}_1^2 \\ \dot{\alpha}_2^2 \\ \dot{\beta}_1^2 \\ \dot{\beta}_2^2 \end{vmatrix}, \quad w = \begin{vmatrix} \overline{u_1} \\ \overline{u_2} \\ \overline{q_1} \\ \overline{q_2} \\ \overline{\Pi_1} \\ \overline{\Pi_2} \\ R_{2x} \\ R_{2y} \end{vmatrix}. \tag{2}$$

The eight-dimensional column w is the vector of forces and moments of forces. The matrices B(z), A, D(z), C(z) have the form:

$$B(z) = \begin{vmatrix} J & L_aK_r\,cos(\varphi-\alpha_1) & 0 & L_bK_r\,cos(\varphi-\beta_1) & 0 \\ L_aK_r\,cos(\varphi-\alpha_1) & J_a-2L_aK_a+L_a^2M & -L_aK_a\,cos(\alpha_1-\alpha_2) & \begin{matrix}(J_{ab}-L_aK_b-L_bK_a+\\ +L_aL_bM)\,cos(\alpha_1-\beta_1)\end{matrix} & -L_aK_b\,cos(\alpha_1-\beta_1) \\ 0 & -L_aK_a\,cos(\alpha_1-\alpha_2) & J_a & -L_bK_a\,cos(\alpha_2-\beta_1) & J_{ab}\,cos(\alpha_2-\beta_2) \\ L_bK_r\,cos(\varphi-\beta_1) & \begin{matrix}(J_{ab}-L_aK_b-L_bK_a+\\ L_aL_bM)\,cos(\alpha_1-\beta_1)\end{matrix} & -L_bK_a\,cos(\alpha_2-\beta_1) & J_b-2L_bK_b+L_b^2M & -L_bK_b\,cos(\beta_1-\beta_2) \\ 0 & -L_aK_b\,cos(\alpha_1-\beta_2) & J_{ab}\,cos(\alpha_2-\beta_2) & -L_bK_b\,cos(\beta_1-\beta_2) & J_b \end{vmatrix}, \quad (3)$$

$$A = \begin{vmatrix} -K_r & 0 & 0 & 0 & 0 \\ 0 & K_a-L_aM & 0 & 0 & 0 \\ 0 & 0 & K_a & 0 & 0 \\ 0 & 0 & 0 & K_b-L_bM & 0 \\ 0 & 0 & 0 & 0 & K_b \end{vmatrix}, \quad (4)$$

$$D(z) = \begin{vmatrix} 0 & L_aK_r\,sin(\varphi-\alpha_1) & 0 & L_bK_r\,sin(\varphi-\alpha_1) & 0 \\ -L_aK_r\,sin(\varphi-\alpha_1) & 0 & -L_aK_a\,sin(\alpha_1-\alpha_2) & \begin{matrix}(J_{ab}-L_aK_b-L_bK_a+\\ +L_aL_bM)\,sin(\alpha_1-\beta_1)\end{matrix} & -L_aK_b\,sin(\alpha_1-\beta_2) \\ 0 & L_aK_a\,sin(\alpha_1-\alpha_2) & 0 & -L_bK_a\,sin(\alpha_2-\beta_1) & J_{ab}\,sin(\alpha_2-\beta_2) \\ -L_bK_r\,sin(\varphi-\alpha_1) & \begin{matrix}-(J_{ab}-L_aK_r-L_bK_a\\ +L_aL_bM\,sin(\alpha_1-\beta_1)\end{matrix} & L_bK_a\,sin(\alpha_2-\beta_1) & 0 & -L_bK_b\,sin(\beta_1-\beta_2) \\ 0 & L_aK_b\,sin(\alpha_1-\beta_2) & -J_{ab}\,sin(\alpha_2-\beta_2) & L_bK_b\,sin(\beta_1-\beta_2) & 0 \end{vmatrix} \quad (5)$$

$$C(z) = \begin{vmatrix} 0 & 0 & -1 & -1 & 0 & 0 & 0 & 0 \\ -1 & 0 & 1 & 0 & 0 & 0 & -L_a\cos\alpha_1 & -L_a\sin\alpha_1 \\ 0 & -1 & 0 & 1 & 0 & 0 & L_a\cos\alpha_2 & L_a\sin\alpha_2 \\ 1 & 0 & 0 & 0 & -1 & 0 & -L_b\cos\beta_1 & -L_b\sin\beta_1 \\ 0 & 1 & 0 & 0 & 0 & -1 & L_b\cos\beta_2 & L_b\sin\beta_2 \end{vmatrix}, \quad (6)$$

where $B(z)$ is the symmetric and positive definite kinetic energy matrix, A is the diagonal matrix of potential energy, $D(z)$ is the matrix of Christoffel symbols of the first kind for the matrix $B(z)$, L_a and L_b are the lengths hips and shanks, respectively, J is the moment of inertia of the body, relative to the point O, $K_r = m_k r$, (m_k is the body mass), r is the distance from the hip joint O to the center of mass of the body, J_b is the moment of inertia of the shank relative to the knee joint (B for the lower leg BA), $K_b = m_b b$, b is the distance from the knee joint to the center of mass of the shank, J_a^o is the moment of inertia of the hip relative to point O. The mass of the entire five link M is calculated by summing the mass of the body and the double mass of the same legs: $M = m_k + 2m_a + 2m_b$, where m_a and m_b are the masses of the thigh and shank, respectively. The parameters K_a, J_a and J_{ab} are calculated by the formulas:

$$K_a = m_a a + m_b L_a, \quad J_a = J_a^o + m_b L_a^2, \quad J_{ab} = K_b L_a, \quad (7)$$

where a is the distance from the hip joint to the center of mass of the hip.

The control has an impulsive pattern at the beginning ($t = 0$) and the end of the step ($t = T$), and the mechanism moves along the ballistic trajectory in the time interval $0 < t < T$ in the absence of any control actions ($w(t) = 0$). The simulated mechanical system passes from the state $z(0)$ to $z(T)$, in which the moving leg is on the surface. Model of impulse control of an anthropomorphic mechanism motion was proposed in [10] and experimentally confirmed in [8]. Thus, the solution $z(t)$ satisfies a homogeneous system in the time interval $0 < t < T$:

$$B(z)\ddot{z} + gA|sin\,z_i| + D(z)\left|\dot{z}_i^2\right| = 0. \quad (8)$$

Equation (8) resolved with respect to the second derivatives has the form:

$$\ddot{z} = -gB^{-1}(z)A|sin\,z_i| - B^{-1}(z)D(z)\left|\dot{z}_i^2\right|. \quad (9)$$

To solve the boundary value problem, which involves finding the initial angular velocity vector $\dot{z}(0)$, in which the solution (8) starting from the initial state $z(0)$, $\dot{z}(0)$ passes through $z(T)$, it was proposed in [10] to minimize the residual function:

$$I = \sum_{i=1}^{5} k_i\big(\Delta z_i(T)\big)^2, \quad (10)$$

where $\Delta z_i(T)$ is the difference between the given and calculated values of the generalized coordinate, and k_i are some weight coefficients. The resulting trajectories of motion with this solution of equations (8) are very different from the real trajectories obtained from data on the movement of a person. To approximate the calculated trajectories to the real ones, a scheme in MATLAB / Simulink using the MPC controller was proposed in [12]. The authors estimated the error of the calculated trajectories within 5 percent. This article further proposes a method for constructing solutions of equations (8), in which the calculated trajectories are closer to real ones.

3. The method of calculating trajectories

We used the Biped file for the 3Ds Max program from open sources as a source of real motion paths of the anthropomorphic mechanism using motion capture technology. The control values of the generalized coordinates $\hat{z}(t)$ are calculated on the basis of 17 frames of human movement in the single-limb support phase. In contrast to the previously described method, it is proposed to use the following function as a residual function that must be minimized to find a solution to equations (8):

$$I = \sum_{t=0}^{T} \sum_{i=1}^{5} \left(z_i(t) - \hat{z}_i(t)\right)^2 \quad (11)$$

Minimization of function (10) ensures that only the initial and final positions of the mechanism match, while minimization (11) also ensures passage through points $\hat{z}(t)$. In the process of minimization, it is necessary to repeatedly solve the Cauchy problem. To solve the Cauchy problem, the fourth-order Runge-Kutta method with a constant integration step was used, as proposed in [10].

In the beginning, it was supposed to calculate the trajectory closest to the real one by finding only the suitable $\dot{z}(0)$. Since the biometric parameters of the person in the frames used were unknown, the values from [10, p. 86]

$$M = 75\,kg,\ m_a = 8.6\,kg,\ m_b = 4.6\,kg,\ J = 11.3\,kg \cdot m^2,\ J_a^o = 0.535\,kg \cdot m^2,$$
$$J_b = 1.02\,kg \cdot m^2,\ r = 0.386\,m,\ a = 0.18\,m,\ b = 0.324\,m,\ L_a = 0.41\,m,\ L_b = 0.497\,m. \quad (12)$$

The initial value of $z(0)$ is equal to $\hat{z}(0)$. To minimize, we used the *fminunc* function from the MATLAB package, for which it is necessary to specify an optimized function and a vector of initial values. The initial values of $\dot{z}(0)$ for optimization were calculated as follows:

$$\dot{z}_{i0}(0) = \frac{\hat{z}_i(dt) - \hat{z}_i(0)}{dt} \quad (13)$$

where dt is the time step. The result of optimization is the value $\dot{z}(0)$, minimizing the value of function (11):

$$\dot{z}(0) = |0.5136 \quad -4.4374 \quad 2.3095 \quad -0.0857 \quad -0.7415|^T \quad (14)$$

However, the residual I with such a value of $\dot{z}(0)$ is 2.4298, which is a sufficiently large value and the trajectories of the mechanism are very different from real, which is clearly visible in the kinogram of the five-link system in Figure 2.

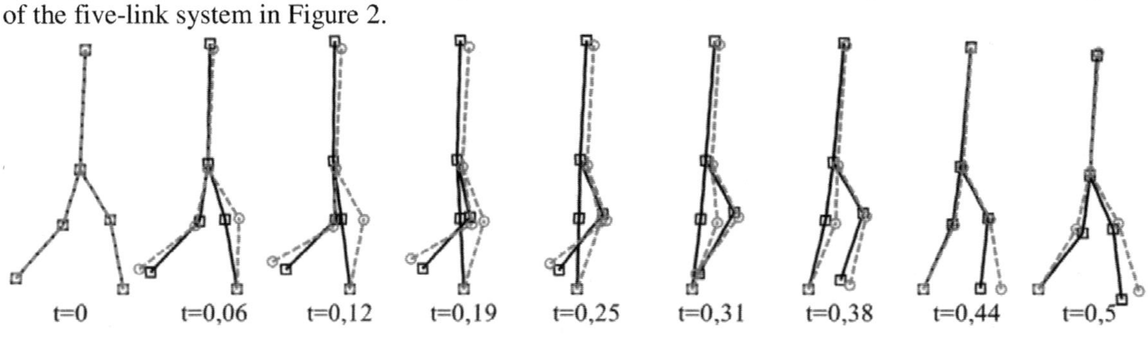

Figure 2. Comparisons of the real (gray dashed lines) and modelled (solid black lines) five-link system motion kinograms.

To obtain a more accurate solution, optimization was carried out using a larger number of parameters. First of all, we varied the values of parameters that are difficult to determine from the video recording (masses of limbs, moments of inertia, location of centers of mass). In addition to $\dot{z}(0)$, the values of J, J_a^o, J_b, M, m_a, m_b, r were selected for optimization. Optimization results with a different number of parameters to be optimized are shown in Table 1. In addition to the fminunc function, the fminsearch function from the MATLAB package was also used for optimization. A dash in the table indicates an error while optimizing this function.

Table 1. Optimization results

№	Optimized parameters	Optimization result (residual value)	
		fminunc	**fminsearch**
1	$\dot{z}(0)$	2.4298	3.3181
2	$\dot{z}(0)$, J	2.4296	2.4413
3	$\dot{z}(0)$, J, J_a^o	1.8036	2.0484
4	$\dot{z}(0)$, J, J_a^o, J_b	0.5576	1.0271
5	$\dot{z}(0)$, J, J_a^o, J_b, M	0.3218	0.9379
6	$\dot{z}(0)$, J, J_a^o, J_b, M, m_a	0.2826	0.3565
7	$\dot{z}(0)$, J, J_a^o, J_b, M, m_a, m_b	0.2826	1.0098
8	$\dot{z}(0)$, J, J_a^o, M, m_a	0.6807	1.4156
9	$\dot{z}(0)$, J, M, m_a	—	0.5912
10	$\dot{z}(0)$, J, J_a^o, J_b, M, m_a, r	—	8.4122
11	$\dot{z}(0)$, J, J_a^o, J_b, M, m_a, m_b, r	—	14.5106

Options 7 - 11 show that an increase in the number of parameters often leads to an increase in the discrepancy due to the fact that the objective function has several local minima in conditions of a large number of variable parameters. To solve this problem, global optimization methods are more suitable, and we used a genetic algorithm to optimize for 14 parameters: $\dot{z}(0) = |\varphi, \alpha_1, \alpha_2, \beta_1, \beta_2|^T$, J, J_a^o, J_b, M, m_a, m_b, r, a, b. Optimization was carried out in the MATLAB R2010b package with standard settings for the genetic algorithm (the size of the population is 50 individuals). To increase the accuracy and reduce the working time of the genetic algorithm, the hybrid method was used: in the beginning, a genetic algorithm was used to approach the global minimum, and after reaching a certain small residual value (after the genetic algorithm was completed), a minimum was searched using the fmincon function. As a result of the genetic algorithm, the following values of the model parameters are obtained:

$$\dot{z}(0) = |-0.04755 \quad -1.72029 \quad 2.13511 \quad -3.14159 \quad -3.14159|^T,\ M = 28.251\,kg,$$
$$m_a = 0.100\,kg,\ m_b = 9.996\,kg,\ J = 250\,kg \cdot m^2,\ J_a^o = 25.765\,kg \cdot m^2,$$
$$J_b = 2.062\,kg \cdot m^2,\ r = 0.100\,m,\ a = 0.100\,m,\ b = 0.497\,m \qquad (15)$$

Values unchanged during optimization of the parameters L_a and L_b were taken as in (12). The discrepancy calculated by formula (11) is $I = 0.3917$ under the above calculation conditions. The result of optimization using the fmincon function following:

$$\dot{z}(0) = |-0.0484 \quad -1.4885 \quad 2.0153 \quad -3.6294 \quad -3.9709|^T,\ M = 8.734\,kg,$$
$$m_a = 1.122\,kg,\ m_b = 2.910\,kg,\ J = 250\,kg \cdot m^2,\ J_a^o = 50\,kg \cdot m^2,$$
$$J_b = 0.513\,\text{кг} \cdot m^2,\ r = 0.717\,m,\ a = 0.010\,m,\ b = 0.492\,m. \qquad (16)$$

The discrepancy decreased and amounted to $I = 0.2844$ for the above parameters. As can be seen in (15) and (16), some parameters have unrealistic values, however, precisely at such values, it was possible to obtain trajectories close to real ones. In addition, it may be worthwhile to set restrictions on the parameter values during optimization, however, at the same time, it may not be possible to achieve such small residual values. Figures 3a – 3d show a comparison of the graphs of the dynamics of the angles φ, α_1, α_2, β_1, β_2 obtained for a real person and by modeling.

Figure 4 shows the time dependences of the coordinates of the knee of the supporting leg (Figs. 4a, 4b) and the knee of the moving leg (Figs. 4c, 4d) to compare our results with the data [12]. The solid lines are the simulation results, the dashed lines are the experimental coordinates of the knee joints.

IOP Conf. Series: Materials Science and Engineering **828** (2020) 012024 doi:10.1088/1757-899X/828/1/012024

The coordinates of the joints were calculated according to the formulas published in [10, p. 88]. Figure 5 shows a visual comparison of the kinogram of a modelled and real human gait [13].

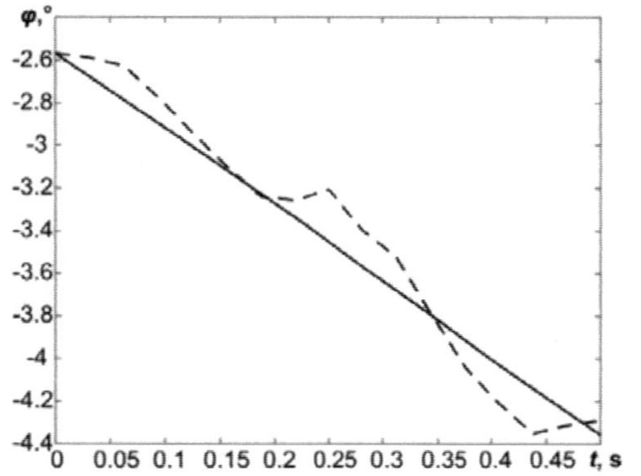

(a)

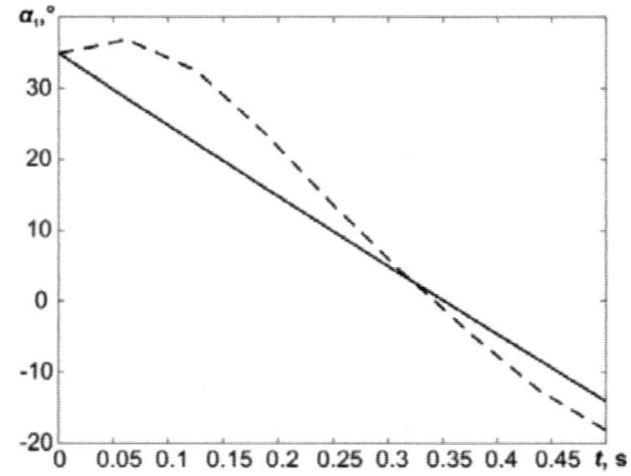

(b)

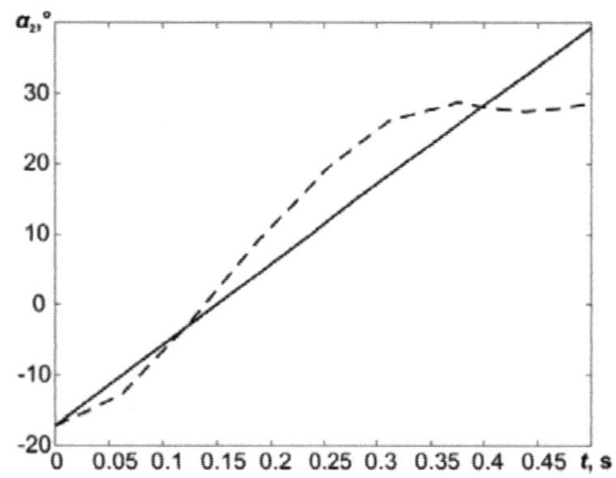

(c)

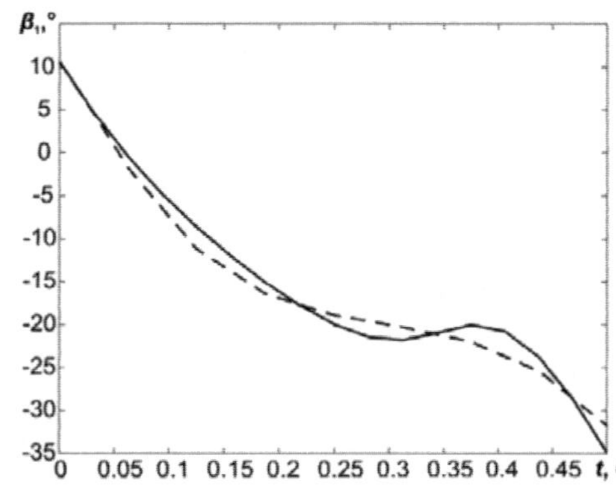

(d)

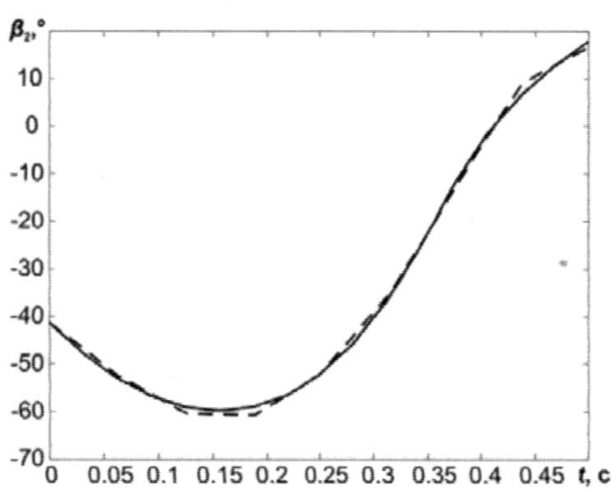

(e)

Figure 3. Comparison of the dynamics in angles for a real person (dash lines) and results of modeling (solid lines)

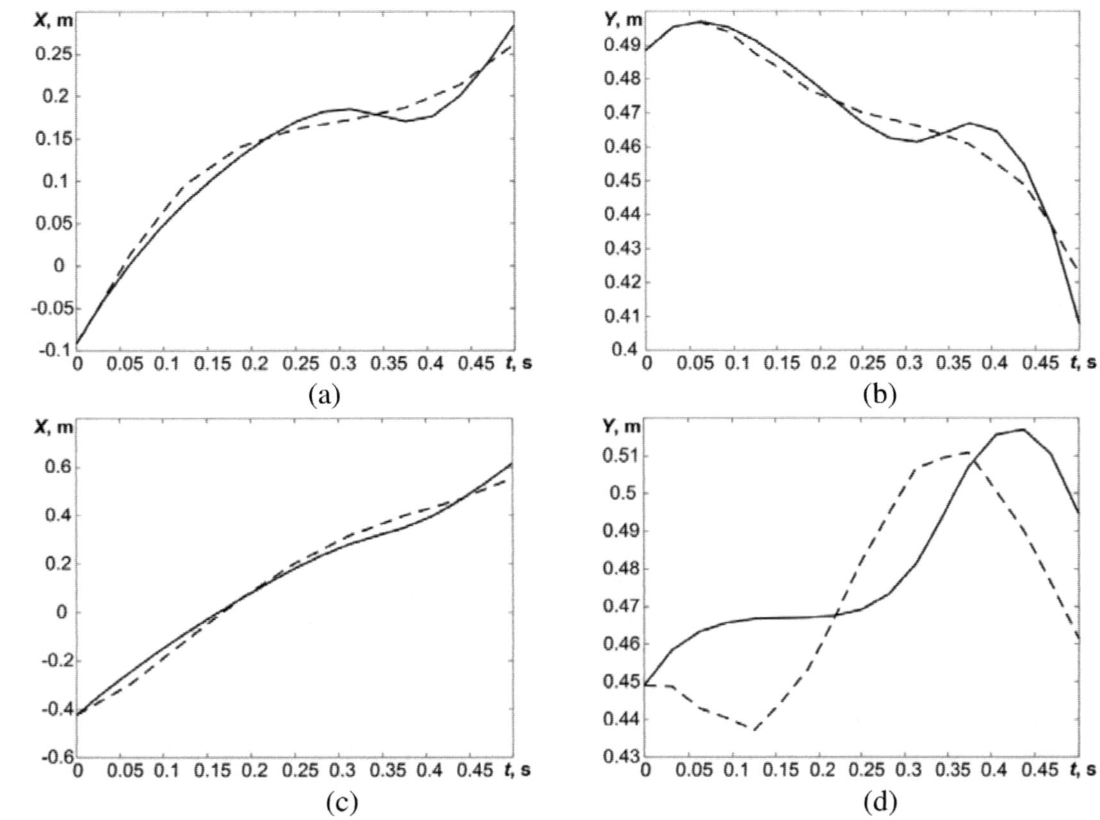

Figure 4. Time dependences of the coordinates of the knee of the supporting (top) and moving (bottom) legs; solid lines are the result of modeling, dashed lines are the coordinates of the movement of a real person

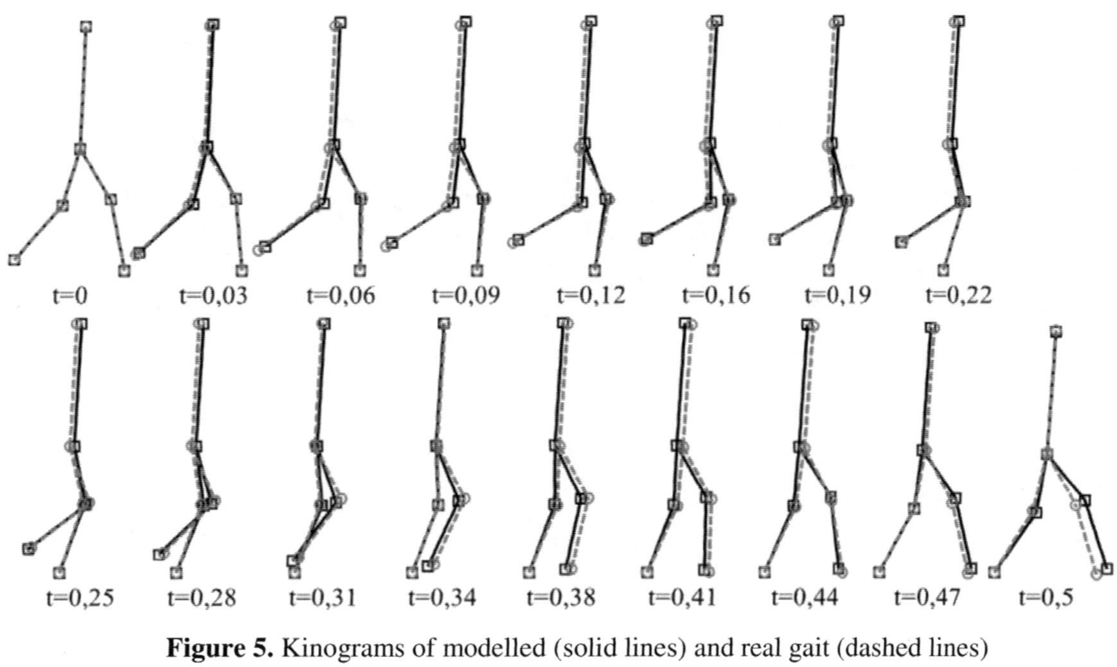

Figure 5. Kinograms of modelled (solid lines) and real gait (dashed lines)

4. Conclusion

This article is aimed at improving methods for modeling the trajectories of the five-link mechanism close to the gait of a real person by finding the appropriate model parameters using the genetic algorithm to solve the optimization problem. The simulation results show that the mathematical model of the five-link described in [14], [15] can quite accurately describe the gait of a real person. Modeling errors are caused by several factors: the use of a flat model instead of a spatial model, feet and hands are not taken into account in the motion model. In addition, finding the optimal solution requires the use of complex algorithms due to a large number of arguments of the objective function.

Acknowledgment
This work was supported by the Ministry of Science and Higher Education of the Russian Federation when creating software for the numerical simulation (government task No. 2.852.2017/4.6).

References

[1] Bulgakov V G and Bumagin V V 2011 *Forensic examination* **3**
[2] Sokolova A I and Konushin A S 2019 *Proceedings of the Institute for System Programming of the RAS* **31(1)**
[3] Elif Surer and Alper Kose 2011 *Computer Analysis of Human Behavior* p 412
[4] DeLuca P A, Davis R B and Ounpuu S 1997 *Journal of Pediatric Orthopaedics* **17(5)**
[5] Chambers H G and Sutherland David H 2002 *The Journal of the American Academy of Orthopaedic Surgeons* **10(3)**
[6] Hicks J L, Schwartz M H and Delp S L 2009 *The Identification and Treatment of Gait Problems in Cerebral Palsy* p 660
[7] Martinez F, Cifuentes C and Romero E 2013 *Journal of NeuroEngineering and Rehabilitation* **10(73)**
[8] Chigarev A V and Borisov A V 2010 *Russian Journal of Biomechanics* **15(1)**
[9] Ivanov A V and Formalskiy A M 2015 *Journal of Computer and Systems Sciences International* **2**
[10] Formalskiy A M 1982 *Movement of anthropomorphic mechanisms* (Moscow : Science) p 368
[11] Kolesnikova G P and Formalskiy A M 2014 About one way to model a human's gait *Engineering Journal: Sciense and Innovation* **1**
[12] Ovchinnikov I A, Kovalenko P P and Minh Vu 2016 Human gait modeling in MatLab/Simulink. *Izvestiâ vysših učebnyh zavedenij* **58(8)**
[13] Ergunova O T , Lizunkov V G, Malushko E Yu, Marchuk V I, Ignatenko A Yu 2017 *OP Conference Series: Materials Science and Engineering* **177(1)** 012046. DOI: 10.1088/1757-899X/177/1/012046
[14] Shamne A N 2019 *IOP Conf. Ser.: Mater. Sci. Eng.* **483** 012099 DOI: https://doi.org/10.1088/1757-899X/483/1/012099
[15] Lizunkov V, Politsinskaya E, Malushko E, Kindaev A, and Minin M 2018 *International journal of energy economics and policy* **8 (3)** 250-257

Features of prosecutorial supervision over the implementation of legislation on specially protected natural territories in modern Russia

Natalya Solovyeva [1], Vladimir Shinkaruk [1], Pavel Fantrov [1] and Davlatali Kakhhorov [1]

[1] Volgograd State University, 100, Prosp. Universitetsky, Volgograd, 400062, Russia

E-mail: solovieva_na@volsu.ru

Abstract. One of the ways to protect natural territories is the obligation authority to carry out prosecutorial supervision over the implementation of environmental legislation which directly establishes the obligation of individual entities to exercise this type of control. However, this obligation is not duly enforced. It creates a situation where on the one hand, there are positive goals of the state in the field of environmental protection, on the other hand, the positive desire of the nature user to comply with the law, and it leads to a negative result. The issue to enforce the law in the field of environmental control has remained relevant for a long time, and each stage of reforming environmental legislation increase its importance. The article describes the subject, object, tasks and purpose of prosecutorial supervision over the implementation of laws in the field of environmental safety and environmental management. So, the subject of prosecutorial supervision is the activity of prosecutors aimed at regulating public relations in implementing the law. The article classifies the subjects of prosecutorial supervision over the implementation of legislation on the specially protected natural territories as supervised and non-supervised. The purpose of prosecutorial supervision over the implementation of legislation on specially protected natural territories is to ensure the rule of law in the field of environmental safety and environmental management of specially protected natural territories. The authors divided the tasks of prosecutorial supervision in this area into general, special and private. The paper assesses the problems influencing the effectiveness of prosecutorial supervision over the implementation of legislation on specially protected natural territories, which include: the imperfection of the legislative framework; inefficiency in management, control and cadastral registration of such territories; contradictions arising from the interaction and coordination of bodies that manage, protect and control specially protected natural territories with prosecution authorities. The authors offer recommendations aimed at improving the efficiency of prosecutorial supervision over the implementation of legislation on specially protected natural territories.

1. Introduction

The environmental prosecutor's office of the Russian Federation has a major focus on supervisory activity overseeing the implementation of laws on environmental protection and nature management. In this supervision area, the environmental prosecutor's offices in Russia are implementing the provisions enshrined in Article 42 of the Constitution of the Russian Federation on everyone's right to a favourable environment, reliable information on the state of the environment and compensation for damage caused by violations of environmental laws.

The research topic has scientific interest since specially protected natural territories is of great importance and consist of a scientific, historical, cultural, aesthetic, recreational and health-improving character. In Russia, it detected everywhere criminal facts of non-compliance with the regime of specially protected natural territories. Over the past 2.5 years, prosecutors have banned access to more than a thousand sites containing information on the sale of Red Book animals. Over the same period, it identified 30 thousand environmental violations including those of illegal logging, poaching, unauthorized seizure of land and environmental pollution. To compensate for the damage to the unique natural complexes, the courts received almost 5 thousand lawsuits for 132 million rubles. According to the materials of the prosecutors, 180 criminal cases were instituted. Failure to comply with the regime of specially protected natural territories established by law may lead to negative consequences both for the ecology as a whole and for the life of people and the economy. The need to comply with the statutory regime of specially protected natural territories is determined by the significant damage to the unique climate of such territories, their flora and fauna can be caused. The area of specially protected natural territories occupies 18% of the Russian territory, and violations of legislation in the field of specially protected land will result in the risk for the entire ecology of the Russian Federation.

2. Materials and methods

The methodological basis of the study was a set of scientific methods used in the process of identifying features of prosecutorial supervision over the implementation of Russian legislation on specially protected natural territories. This article provided the following scientific methods:

- general scientific: from the abstract to the concrete, analysis, synthesis, induction, deduction, modelling;
- systematic method, in assessing prosecutorial supervision over the implementation of legislation on specially protected natural territories, which is an integrated system that includes a combination of the subject, object, tasks and goals of supervision over the implementation of laws related to environmental safety and environmental management;
- institutional, method focused on the study of institutions overseeing prosecutorial supervision over the implementation of legislation on specially protected natural territories;
- sociological, method assessing the effectiveness of prosecutorial oversight of implementing legislation on specially protected natural territories based on evidence.

3. Results and discussion

The centralized system of prosecution authorities of the Russian Federation performs prosecutorial supervision over the implementation of legislation as the main function. General provisions on prosecutorial supervision to enforce the law are enshrined in the third section of the Federal Law of the Russian Federation "On the Prosecutor's Office of the Russian Federation" and are the rule of law at the stage of criminal proceedings. The Order of the Prosecutor General of the Russian Federation "On the organization of prosecutorial supervision over the implementation of laws, the observance of human and citizen freedoms" provide the implementation of the environmental everybody's rights established by the second chapter of the Constitution of the Russian Federation. Following the invariable foundations of the constitutional system, human and civil rights are recognized as the highest value and are protected by the state.

According to A.A. Amirdzhanyan, prosecutor's supervision is the Prosecutor's General of the Russian Federation and its subordinate prosecutors` activity on behalf of the state aimed at ensuring accurate and uniform implementation of all laws of the Russian Federation throughout the state" [1, pp. 139]. Environmental prosecutor's offices carry out prosecutorial supervision over the implementation of legislation on specially protected territories and ensure legality in the field of environmental safety and environmental management.

Article 21 of the Third chapter of the Federal Law of the Russian Federation "On the Prosecutor's Office of the Russian Federation" enshrines the general provisions on prosecutorial supervision over the execution of laws which include a ban on supervision by other state bodies, except for the prosecutor's office; conducting an audit in accordance with the information received on the legal violations. If this information cannot be confirmed or disproved, this provision is a distinguishing feature of prosecutorial supervision from the control and supervision carried out by other public authorities.

The totality of the subject, object, tasks and goals of supervision over the implementation of laws in the field of environmental safety and environmental management constitute the content of prosecutorial supervision.

An important component of prosecutorial oversight is the subject. The subject of activity always shows the direction of activity, that is, the direction of the prosecution authorities' activity of the Russian Federation in prosecutorial supervision over the implementation of legislation on specially protected natural territories [2, pp. 48]. The subject of activity exists regardless of the will and consciousness of people. Consequently, the subject of prosecutorial supervision is the activity of prosecutors aimed at regulating public relations in implementing the law.

Implementation of legislation in the field of environmental safety of the Russian Federation is of two types:

- the application, it is the activity of mainly authorized officials in the publication of regulatory legal acts and the implementation of significant legal actions;
- compliance, it is the behaviour of officials, heads of commercial and non-profit organizations and citizens, which is lawful and does not contradict the legislation of the Russian Federation.

The subjects of prosecutorial supervision over the implementation of legislation on specially protected natural territories may include: The Ministry of Natural Resources and Ecology of the Russian Federation, the Federal Service for State Registration, Cadastre and Cartography, the authorities of the constituent entities of the Russian Federation and local authorities that manage, control, supervise and coordinate the activities of specially protected natural territories.

The Government of the Russian Federation is not supervised subject. Therefore, when adopting a normative legal act that contradicts the legislation of the Russian Federation, the prosecutor can only inform the President of the Russian Federation who is also not supervised.

The goal is always activities intended to achieve the result, in this case, it is the activities of the prosecutor's office in the implementation of prosecutorial supervision. We have identified the following goals of prosecutorial supervision over the implementation of legislation on specially protected natural territories:

- ensuring the rule of law in the field of environmental safety and environmental management of specially protected natural territories;
- ensuring the unity of legislation in the field of environmental safety, i.e. all subordinate regulatory legal acts must not contradict superior regulatory legal acts or be brought into conformity with higher ones;
- strengthening the rule of law in the field of environmental safety and protection of land with the special regime, which includes two elements: all subjects of prosecutorial supervision must act on current legislation and all newly created regulatory legal acts should not violate the unity of the legal system;
- protecting everybody's rights and freedoms as well as the legal interests of society and the state; regarding the rights of citizens, prosecutorial supervision is focused on implementing everybody's rights and freedoms established by article 42 of the Constitution of the Russian Federation.

The content of the prosecutor's supervision over the implementation of legislation on specially protected natural territories includes its tasks. The tasks of prosecutorial supervision in this area include general, special and particular tasks. The general tasks are applied to all types of prosecutorial supervision and are regulated by the Constitution of the Russian Federation and the Federal Law "On the Prosecutor's Office of the Russian Federation" and other regulatory legal acts, and consist in the supervision of compliance with the Constitution of the Russian Federation and the implementation of laws in force throughout the Russian Federation by all entities without exception.

Particular tasks are tasks that the prosecutor solves during the implementation of specific activities in decision-making. They are numerous and diverse comparing with general and special tasks.

Special tasks are tasks solved in the implementation of certain types of prosecutorial supervision. In the field of environmental safety and environmental management, it set relevant tasks regarding specially protected natural territories. V.N. Shelestyukov and V.V. Erin distinguish the following tasks:

- identification of legal violations that regulates the organization, protection and use of specially protected natural territories;
- establishment of the causes and conditions conducive to these offences;
- elimination of violations and revealing the circumstances of offences in the field of environmental safety and environmental management;
- adaptation of measures by the prosecutor: measures to compensate for damage caused to health or property as a result of the legal violation governing specially protected natural territories; measures related to the accountability of persons committed violations;
- prevention of environmental offences [3, pp. 126].

In modern Russia, control bodies do not always take timely and adequate measures to eliminate violations and hold perpetrators accountable. Therefore, the task of the prosecution authorities is to increase the efficiency of the bodies exercising control over specially protected natural territories, as unacceptable reduced control, lack of proper assessment and response to non-compliance with the law [4, pp. 51]. This task is necessary to achieve the goal of promptly detecting violations of legislation on specially protected natural territories. One task of increasing the efficiency of prosecutorial supervision over the implementation of legislation on specially protected natural territories is the systematic coordination of law enforcement agencies in combating environmental crime for productive interaction not only with state authorities but also with non-profit associations engaged in environmental protection environment [5, pp. 128]. Correctly assigned tasks contribute to the actual elimination of identified legal violations on specially protected natural territories.

Prosecutorial supervision over the implementation of legislation on specially protected natural territories is subject to the information about the offence, and when data cannot be confirmed or refuted, the prosecutor or his deputy decides to conduct an audit and notifies the head or authorized representative of the organization or body that will be subjected to verification no later than the day it begins. It is possible to conduct prosecutorial inspections of compliance with the laws of specially protected natural territories by one's initiative, the frequency established by the Prosecutor General's Office for certain types of inspections and based on an interested party's complaint. The prosecutor's decision to conduct such an inspection must indicate the subject, purpose and grounds of the audit. If the prosecutor's check found information that indicates other legal violations by the body or organization requiring for actions then the prosecutor or his deputy makes a motivated decision to conduct a prosecutor's check with the extension of the check subject [6, pp. 116].

The prosecutor's check lasts for thirty calendar days, this period can be extended by another thirty calendar days with a reasoned decision of the prosecutor or his deputy. The subsequent extension of the prosecutor's inspection requires the decision of the Prosecutor General of the Russian Federation. If the check object is an organization or body with branches in different entities of the Russian Federation, then the term is established relative to such a branch. Also, the prosecutor or his deputy may repeatedly suspend this period following the implementation of actions established by law. The legislation does not limit the number of suspensions of prosecutorial checks but limits its duration. The total inspection period may not exceed 6 months, and this period may be extended by the Prosecutor General of the Russian Federation.

The powers of the prosecutor to supervise the implementation of legislation on specially protected natural territories should include the identification of violations contributing to circumstances and the identification of perpetrators. When conducting a prosecutor's check of a legal violation on specially protected natural territories, the prosecutor initially identifies responsible for this area. Specially protected natural territories can be under federal, regional and local jurisdiction. This classification facilitates the identification of the body to inspect. Upon receiving reports of the offence, the following bodies are subject to prosecutorial verification: relating to specially protected natural territories of federal significance, this is the Ministry of Natural Resources and Ecology of the Russian Federation managing in these territories, as well as the Federal Service for State Registration of Cadastre and Cartography monitoring and recording these territories; territories of regional significance - state authorities of the constituent entities of the Russian Federation and the territorial body of the Federal Service for State Registration of Cadastre and Cartography, territories of local importance - local authorities.

During a prosecutor's audit of the activities of the Ministry of Natural Resources and Ecology, the Federal Service for State Registration of Cadastre and Cartography and its territorial departments, state authorities of the constituent entities of the Russian Federation, local authorities, authorities managing and monitoring specially protected natural territories, the prosecutor and his deputy need to check: compliance with the normative legal acts adopted by them, regulating relations in the sphere of organization and functioning of specially protected natural territories, with federal legislation; the procedure for conducting cadastral registration of such territories; provisions approved by these bodies on specially protected natural territories for their compliance with applicable law; creation of security zones; implementation of zoning these territories; land transfer procedure; definition of forest boundaries in these territories; authorization for the construction and reconstruction of capital objects, reconstruction and commissioning of objects located within the border of the indicated territories; accounting for the number of the animals and plants and methods for their protection [7, pp. 647].

Federal Service for Supervision of Natural Resources by the order of Ministry of Natural Resources and Ecology of Russian Federation No. 400 dated June 30, 2004, exercises control and supervision in the field of environmental management as well as within its competence in the field of environmental protection.

In the Federal Service for Supervision of Natural Resources, the prosecutor must establish and verify: identification of legal violations to specially protected natural territories by authorized bodies; the legality of the measures taken by authorized persons to prevent legal violations in this area; calculation of damage caused to specially protected natural territories owing to legal violation in this area; compliance with administrative law; transfer of materials to law enforcement authorities if there are grounds for violation, misconduct and acts; control over the issuance of orders for legal violation and the establishment of respective fines.

Prosecutorial supervision over the implementation of legislation on specially protected natural territories in modern Russia has the following acts of the prosecutor's response to identified violations: the prosecutor's protest; prosecutor's submission; prosecutor's resolution; warning for the inadmissibility of legal violation [8, pp. 38].

The prosecutor's protest consists in the fact that the prosecutor or his deputy protest the body, organization or official issued the normative legal act contradicting the higher laws on specially protected natural territories. Also, the prosecutor may protest higher authority, organization or official. The protest must be considered no later than 10 days following the date of its admission to the organization or official. If the prosecutor protests the regulatory legal act of the legislative body of the constituent entity of the Russian Federation or local government, then it appears at the next scheduled meeting of this body, the results of this meeting or the results of the protest's consideration are immediately reported to the prosecutor who brought the protest. Under Russian law, the prosecutor may set a shorter period for consideration of the protest and also withdraw the protest.

The submission of the prosecutor is an official document on elimination of a legal violation submitted by the prosecutor or his deputy to the body or official who is authorized to eliminate this violation in the field of specially protected natural territories or the non-compliance of regulatory acts with the higher law on these territories, and is subject to immediate consideration. The submission of eliminating the legal violation of the special regime, the legal status of the land, natural resources and other real estate objects, protected zones of specially protected natural territories requires taking measures to eliminate the corresponding violations within one month. The results of the measures taken must be submitted in writing to the prosecutor who brought such a submission. A decision to take measures eliminating a violation or non-compliance with legislation by a collegian body requires the notification of prosecutor about the meeting day. If the decision of the Russian Government does not comply with the environmental human rights and other provisions of the rights established by the Constitution of the Russian Federation and other laws of the Russian Federation, the Prosecutor General shall inform the President about this.

High efficiency is essential in the implementation of prosecutorial supervision over the implementation of legislation on specially protected natural territories. So, the environmental prosecutor's offices annually identify more than three thousand legal violations in these territories by state authorities in the federal, regional and local governments, and administrations that manage, control, supervise and organize this area, as well as citizens and legal entities having land plots located within the boundaries of specially protected natural territories. In total, the prosecution authorities identify annually more than 280 thousand legal violations in the field of environmental protection and nature management. Prosecutors have a range of powers in supervising the implementation of legislation on specially protected natural territories and must ensure legality in this area. Nevertheless, in modern Russia, there are many problems in the field of prosecutorial supervision over the implementation of legislation on specially protected natural territories.

One main problem affecting the effectiveness of prosecutorial supervision is the contradiction of the Federal Law of the Russian Federation "On Specially Protected Natural Territories" to the provisions of other legislative acts regulating these relations in the sphere of organization, protection and use of such territories [9, pp. 236]. In practice, these contradictions or issues not fully resolved lead to inconsistency and arbitrary interpretation of certain provisions and cause difficulties for the prosecutor and his deputy to hold prosecution of persons violating the implementation of the legislation. So, due to the lack of normative consolidation of the issue related to the abolition of specially protected natural territories, the removal of the legal status of such land plots, prosecutors and their deputies in several constituent entities of the Russian Federation revealed the following violations: reduction and change of borders, reduction of the area or even the abolition of such territories in violation of the legal requirements.

Another problem is the inefficiency in the management, control and cadastral registration of such territories: violation of the cadastral record for such territories, lack of data on the quantity and their registration and borders. Ineffective legal regulation of these issues leads to unauthorized seizure of land, the construction of buildings and structures on them. A similar situation for the illegal use of reserved lands leads to the impossibility of collecting damages caused to natural objects and holding perpetrators accountable. In prosecutorial supervision, the prosecutor or his deputy must carefully check for the identification of the offences: cadastral registration and check their borders according to the data established by law, cadastral plans and regulations established by the passport of territories.

The existing problems in the field of prosecutorial supervision over the implementation of legislation on specially protected natural territories became the prerequisite for the creation of a department for the supervision of the implementation of laws on environmental protection and nature management in the structure of High Office for Oversight and Execution of Federal Law in 2019. The new Directorate of the Prosecutor General's Office solves the tasks of strengthening prosecutorial oversight over the implementation of environmental laws and the observance of citizens' rights to a favourable environment. Besides, his tasks include the whole range of measures to combat illegal logging, water and air pollution, offences in the extraction and use of wildlife, aquatic biological resources and minerals. An equally important task of the new body is the organization of prosecutorial supervision in the field of preserving the unique natural complexes of the Russian Arctic, the Baikal natural territory, the Far East, and the Volga basin.

A feature of prosecutorial supervision over the implementation of legislation on specially protected natural territories is state oversight bodies' inspections in cooperation with prosecution authorities in this area [10, pp. 128]. The plans for such inspections require approval by the environmental prosecution authorities, and unscheduled inspections require agreement with the prosecution authorities at the location of the inspected object. The interaction and coordination of the bodies that manage, protect and control specially protected natural territories with the prosecution authorities are an effective means to ensure the legality of the adopted regulatory legal acts and to comply with the statutory regime of these territories.

One main feature of prosecutorial supervision over the implementation of legislation on specially protected natural territories in modern Russia is the presence of a wide range of legal acts regulating these relations. So, in the prosecutorial supervision over the implementation of legislation on specially protected natural territories, prosecutors need to check not only the compliance of regulatory legal acts with the Constitution of the Russian Federation and the main legal acts regulating relations in the field of such territories but also to identify the conformity of administrative regulations and the availability of documents certifying such a territory, i.e. passport, cadastral plan established the boundaries and regimes of a specific territory. This problem was also pointed out by the President of the Russian Federation V.V. Putin in a message to the Federal Assembly on February 2019, suggesting to terminate all outdated regulatory acts in the sphere of control and supervision on January 1, 2021, leaving only modern and updated documents. V.V. Putin also noted that in the field of prosecutorial supervision, it is necessary to restore order, adopt a new law on the prosecutor's office, and to review the duties of the prosecutor. In our opinion, the change in regulatory legal acts in the field of prosecutorial supervision over the implementation of legislation on specially protected natural territories requires a systematic approach.

A feature is also the analysis and improvement of the practice of prosecutorial supervision over the implementation of legislation on specially protected natural territories [11, pp. 5]. Such an analysis should consider current state tasks in the field of protection of specially protected natural territories, legislative changes and complex environmental regulations both in modern Russia and around the world. It is necessary to use the accumulated prosecutorial supervision and the totality of such potential.

To ensure the effectiveness of prosecutorial supervision over the implementation of legislation on specially protected natural territories, we offer some recommendations: to identify the illegality of normative legal acts regardless of information flow prosecutors should use the right to participate in meetings of legislative, executive bodies of state power and local authorities, for periodic inspections; systematically analyze prosecutorial and law enforcement practices in the field of environmental safety and environmental management; for each legal violation in specially protected natural territories to hold perpetrators accountable.

4. Conclusion

Prosecutorial supervision over the implementation of legislation on specially protected natural territories is one highest priority activity of prosecution authorities of the Russian Federation [12]. At the present stage of development, there is the most effective mechanism for prosecutorial supervision over the legal enforcement on the protection of nature and its resources, and the prosecution authorities play an important role in the field of environmental safety and environmental management. In modern Russia, there is a network of environmental prosecutor's offices that prevent and suppress offences in the field of environmental safety and environmental management. Given the Russian environmental situation for each violation in the field of environmental safety and environmental management, prosecutors and their deputies essentially raise the question of holding the perpetrators accountable and compensation for material damage caused by the environmental violation. Indeed, non-compliance with legislation on specially protected natural territories may entail the destruction of natural complexes and the loss of the unique habitat and recreational value of the territories. For each offence in specially protected natural territories, the prosecutors and their deputies issue many

IOP Conf. Series: Materials Science and Engineering **828** (2020) 012025 doi:10.1088/1757-899X/828/1/012025

response action reports and monitor its execution. Upon the prosecutor's inspection, the case can be referred to the court, and the court must make a motivation decision. The prosecution authorities must monitor not only the received information received but the media because legal violations on specially protected territories attract public attention. Establishing the obligation of interaction between prosecution authorities and the media will be an appropriate source of information on legal violations. At present, prosecutorial supervision does not respond effectively to reports of legal violations on specially protected natural territories. Thus, it is necessary to create the possibility of the prosecutor and his deputy to respond to such offences independently without the received information.

In prosecutorial supervision over the implementation of legislation on specially protected natural territories, prosecutors and their assistants face problems that aggravate the effectiveness of this function. For example, issues related to the management, protection and use of natural resources in specially protected natural territories are regulated by many regulatory legal acts, many of which have unresolved issues or contradict other regulatory legal acts. The problem of prosecutorial supervision over the implementation of legislation on specially protected natural territories in modern Russia lies in the fact that the legislation is not fully implemented by state authorities, bodies of constituent entities of the Russian Federation and local governments that manage, control, supervise and coordinate activities in specially protected natural territories, as well as commercial and non-profit organizations, legal entities and individuals with land plots within the given territories, that is, in many supervised facilities.

References

[1] Amirjanian A A 2019 *State Service and Personnel: Law* **1** 139-140 DOI: 10.24411/2312-0444-2019-10035
[2] Hanyukova O Yu 2016 *Domestic Jurisprudence: Law* **9** 47-50
[3] Shelestukov V N and Erin V V 2016 *Black Holes in the Russian Legislation* **2** 125-7
[4] Kolmykova M A and Solovieva K O 2018 *Step into Science* **2** 49-53
[5] Gulyaikhin V N and Fantrov P P 2016 *Science Journal of VolSU. History. Area Studies. International Relations* **3** 122-31
[6] Kozhin A V 2011 *Bulletin of Vyatka State University* **4** 112-9
[7] Boltanova E S 2018 *Bulletin of St. Petersburg University: Law* **4** 640-52
[8] Berdinsky S V 2016 *Legality* **4** 38-9
[9] Bizhanova K A 2018 *Actual Problems of Russian Law* **9** 232-42
[10] Klyueva K I 2018 *Almanac of the Young Researcher* **5** 126-30
[11] Palamarchuk A V 2016 *Legality* **8** 3-9
[12] Ergunova O, Lizunkov V, Blaginin, V, Politsinskaya E, and Shaykina O 2018 *International Journal of Energy Economics and Policy* **8(1)** 122-127

Significance of specially protected nature areas in protection of resource potential of urbanized regions (the case of Volgograd region)

G I Starokozheva[1], I V Mitrofanova[1,2] and V V Batmanova[1]

[1]Volgograd State University, 400062, Prospekt Universitetskiy, 100, Volgograd, Russia
[2] Federal Research Centre the Southern Scientific Centre of the Russian Academy of Sciences (SSC RAS), 344006, Prospekt Chekhova, 41, Rostov-on-Don, Russia

E-mail: stargala@volsu.ru

Abstract. Strengthening of the functional role of cities in the development of regions is always associated with a change in local geographical characteristics connected with the joining of the areas beyond the fixed boundaries of urbanized territories. As a result of this process, the areas occupied by natural ecosystems are reduced, new resources are involved into the technogenic use, and the natural potential of the neighboring territories is exhausted and depleted. That is why such a method of resources' protection as conservation of nature areas is of a particular importance for the regions with a high level of urbanization, for example, Volgograd region. The main idea of the article determined the logic and methods of research. The cartographic analysis made it possible to distinguish specially protected nature areas of the region according to their location in respect of urbanized administrative entities. On the basis of the statistical processing of the data found in the reports of the authorized regional authority, the current state and dynamics of development of specially protected nature areas are found according to categories, level of subordination, number, areas, types of resources' diversity. Particular attention is paid to the degree of vulnerability and to the threats of the impact of human activity on nature and resource potential of specially protected nature areas when the cities are spatially planned. Attention is also paid to the ways of protection from negative effects. The study showed that the limitation of anthropogenic processes within the boundaries of specially protected nature areas makes it possible to preserve thousands of biological resources from destruction, protect hundreds of water sources from pollution, and provide a compliance of "friendly" human habitat to industrial and technological requirements. As a result, the article proves high importance of specially protected nature areas as a way of conservation of the natural potential of Volgograd region. It also determines the relevant tasks of security activities and methods for its effectiveness growth.
Key words: specially protected nature areas, urbanized territories, anthropogenic effect, resource potential, nature protection, efficiency, Volgograd region

1. Introduction

The methodology of territorial nature protection in Russia has been formed for a hundred of years [1], beginning from the establishment of the first sable reserve in January 1917 in Buryatia to the present day, when the area of protected areas occupies more than thirteen percent of the territory of the Russian Federation and they are today a heterogeneous mixture of areas of various levels of

IOP Conf. Series: Materials Science and Engineering **828** (2020) 012026 doi:10.1088/1757-899X/828/1/012026

administrative subordination.

In the opinion of A. A. Tishkov, the asymmetry in the distribution of specially protected nature areas and the frequency of their establishment were significantly affected by the resource component. The dynamics and direction of spatial influence of global commodity markets, changes in human motivation concerning consumption and use of resource potential are reflected in the geographical location of specially protected nature areas determining their specific place in the territorial structure of the country's regions. A. A. Tishkov carried out the analysis of the programs for the organization of specially protected nature areas in recent decades and found their relatively low efficiency [2].

One of the reasons explaining this conclusion is violation of usual hierarchy in the interaction between the federal government and regional executive authorities resulting in the decentralization concerning the arrangement of specially protected nature areas, the conservation and their resource potential use. At the same time the shift of managerial functions towards the constituent entities of the Federation took place, which made the regional authorities solve independently the contradictions in the system "resources – their use – challenges". The solution of the problem is complicated by the fact that the specially protected nature areas within the boundaries of the largest part of the regions of the Russian Federation present a collection of dissipated conservation areas of various types. The specially protected nature areas of regional level of administration have the largest share among all specially protected nature territories in Russia. Their total area exceeds one half of the total area of all specially protected nature areas in Russia. Hence, in each region, the "destiny" of the most part of specially protected nature areas depends on the subjective ideas of authorized persons about the real value of the resource potential of specially protected nature areas and very often the variants for the conservation of natural resources are less profitable than the options for their use in economically viable industries.

The problem of choice is especially urgent in the regions with a high level of urbanization. The strengthening of the function of cities in the development of such regions is accompanied by a change in local geographic characteristics [3] through the use and joining of suburban areas. As a result, the natural state of the territories involved into the urbanization process is changing, the areas occupied by natural ecosystems are reduced, new resources are exploited, the region's resource potential is exhausted and depleted.

In the view of the foregoing, the main purpose of the article is to determine special importance of the protection of nature areas as a way of protection of the resource potential of urbanized regions.

Volgograd oblast was chosen as the object of the study as a region with a high level of urbanization, in which the ratio of the area of specially protected nature areas to the region's surface does not reach the average all-Russian level (10%).

The study is based on the idea of the correlation of settlement patterns in the territories of the constituent entities of the Russian Federation and the world experience in the design of specially protected nature areas, when the total surface of protected areas should be at least 20% of the total area of the territory which is being designed.

The data about the administrative and territorial structure, size and share of the urban population and economic structure were obtained from the official electronic sources of executive authorities and the Federal Service for State Statistics in Volgograd region. The analysis of location, ratio of the number and area of the specially protected nature areas of the region was carried out according to the territorial planning schemes of the region, municipal areas, general lay-out of rural settlements. The cadastral documents of the specially protected nature areas of the region, reports of authorized executive authorities [4] in the field of ecology and use of natural resources of different years became the source of the information on the availability and the state of natural resources.

2. Results and Discussion

Since late 1970s of the twentieth century, Volgograd region has been one of the regions with a high level of urbanization. A favorable geographical location, rich resource potential ensured an active industrial development of the region and the concentration of the population in urban settlements next to industrial facilities. Between 1979 and 2016, the share of the urban population of the Volgograd region grew by five percent from 71.5 to 76.66 percent. The administrative and territorial units of the region include six cities of regional significance (among them is the city with one million of inhabitants, i.e. Volgograd), thirteen cities of regional subordinance and seventeen industrial townships, which are centers of economic activity, concentration of industrial production, consumption of resources, goods, services, exchange and distribution. The statistical data show that in the cities the economic growth is 20-30% faster than in the rest of the region. In the region about 70% of the revenues of the regional budget and gross regional product are formed in large cities.

Like any process, the urbanization, along with the pros, has its cons. The spatial planning of cities and the development of infrastructure which depend on industrial and technological imperatives quickly led to the destruction or reduction of qualitative characteristics of the human environment by reducing the area of natural ecosystems. The depletion of natural biological diversity, displacement of native flora species by alien components in the process of the emergence of anthropogenic corridors of agglomeration and artificial gardening of cities, the loss of certain structural elements of plant and animal communities, the appearance of urban landscapes (wastelands, landfills, solid waste landfills and abandoned fields) are observed. In addition, over several decades, Volgograd region has been one of the twenty regions with a high level of pollution of atmospheric air, soil and surface watercourses. All the factors mentioned above negatively affect the assimilation potential of the territory of the region and, as a result, the living conditions of the population and their health condition. That is why the protection of nature areas as a way of protection of the natural resource potential is of particular importance for the region with a high level of urbanization.

The issues of effective spatial organization of protected areas are widely discussed in the scientific community. So, in his studies of A.A. Chibilev [5] reveals the integrated nature of the natural resources of protected areas, their social and economic importance, problems of identification, location and concentration of the basic frames of ecological networks as a basis for spatial planning in steppe regions. A group of Perm scientists under the aegis of S.A. Buzmakova [6] determined the numerical efficiency of the development of the network of specially protected nature areas of the Perm region and analyzed the concept of "geographical diversity". An obvious achievement of many years of the work of the authors is a comprehensive methodology for the calculation of an optimal share of the territory of protected areas in order to ensure the protection of various types of valuable resources. The scientific paper by O. A. Klimanova, E. Yu. Kolbovskiy, O. A. Illarionova [7], in which the specially protected nature areas, as an object of ecosystem services play a key role and a promising place in the construction of ecological frames of the largest cities of Russia is especially relevant. The deep theoretical research does not reduce the seriousness of practical problems in the areas of spatial distribution of specially protected nature areas and their targeted use. In other words when speaking about the level of concentration of specially protected nature areas in urbanized regions, one should take into account the already existing ratio of transformed areas and nature areas, individual climatic features of the region under analysis, the specificity of anthropogenic impact and a qualitative assessment of the nature areas under protection.

2.1. Analysis of the modern state of specially protected nature areas of Volgograd region

By the end of 2018, 58 specially protected nature areas of three levels of administrative subordinance are situated in Volgograd Region (Figure 1). Among them, the share of the number of specially protected nature areas of federal subordinance is 8.62%, the share of areas referring to local administration level is 1.72%, and percentage of territories belonging to the regional level of administration is 89.66%. The total area of all specially protected nature areas is 999.4 thousand hectares, which amounted to 8.85% of the area of the Volgograd region (11290 thousand hectares). At the same time, the share of areas of federal specially protected nature areas is 0.21% (2078.4 ha), the share of regional areas is 99.74% (996850.1 ha) and the percentage of local areas is 0.05% (508.6 ha) [8].

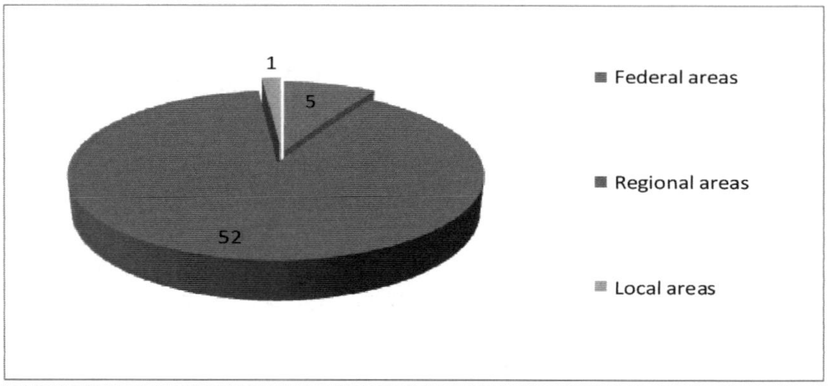

Figure 1. The number of specially protected nature areas according to the level of subordinance in Volgograd region in 2018

Taking into account the predominant number of regional specifically protected nature areas, we limit our efforts by the analysis only of their structure and location. The total number of specially protected nature areas of regional subordination in 2018 amounted to 52, the total area is 996.85 thousand hectares. The nature parks are the largest among the specially protected nature areas. The weight of the areas of seven nature parks amounted to 71.42%; the share of the areas of eight nature reserves is 25.64%; the most numerous categories are especially valuable territories (18 units) and natural monuments (18 units) which have shares of 2.66 and 0.27%, respectively; the area of the only natural landscape does not exceed 0.01% of the total area of regional specially protected nature areas.

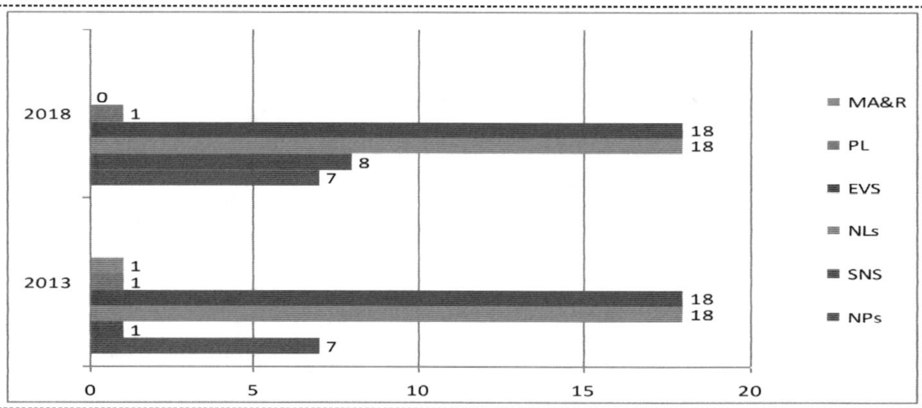

Figure 2. Ratio of the number of specially protected nature area of Volgograd region according to categories in 2013 and 2018 are as follows: MA&R – Medical areas and resorts; PL – Protected landscapes; EVS – Especially valuable territories; NLs – Nature Landmarks; SNS – State Nature Sanctuaries; NPs – Nature parks

The difference that can be seen when comparing the structure of regional specially protected nature areas by categories, number, size of areas in 2018 with the same indicators of 2013 (Figure 2, 3) arose as a result of the work of the state environmental control services and the harmonization of regional regulatory legal acts with the requirements of federal law. The main transformations took place due to some measures: exclusion from the borders of the nature parks "Ust Mdeveditskiy", "Volgo and Akhtuba flood plain", "Eltonskiy" of several settlements (town of Serafimovich, Krasnoslobodsk, Elton settlement); replacement of ten residential plots within the "Ust-Medveditskiy" nature park with three well-preserved plots that were not previously part of the park; temporary exclusion and then inclusion into the list of specially protected nature areas of seven hunting reserves after the addition of species; exclusion from the regional areas of the health and recreation area (deposit of mineral waters

“Gornaya Polyana”) from regional specially protected nature areas.

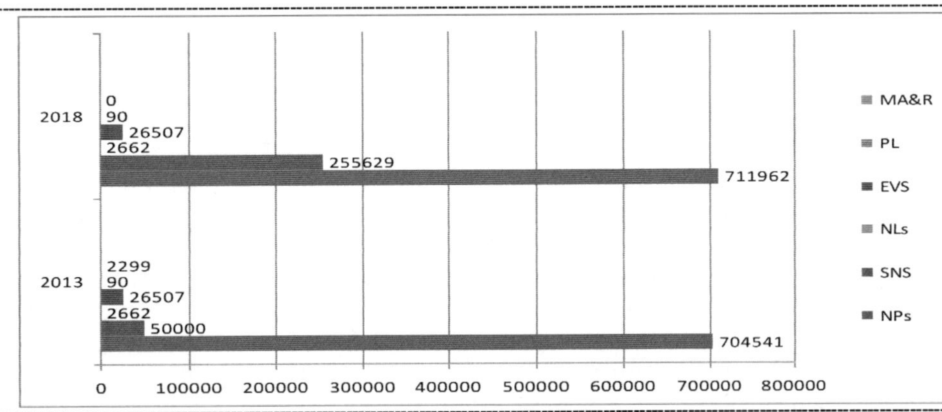

Figure 3. Ratio of the area of specially protected nature area of Volgograd region according to categories in 2013 and 2018 are as follows: MA&R – Medical areas and resorts; PL – Protected landscapes; EVS – Especially valuable territories; NLs – Nature Landmarks; SNS – State Nature Sanctuaries; NPs – Nature parks

As for the relative position of specially protected nature areas in space, the cartographic analysis showed their scattered, uneven distribution in twenty-five municipal districts and two cities of oblast subordination. The borders which are adjacent with urban areas were identified at the following sites: “Volga-Akhtuba floodplain” nature parks (Volgograd, Volzhskiy, Krasnoslobodsk, Leninsk, the towns of Srednyaya Akhtuba and Svetly Yar), nature park “Donskoy” (the town of Ilovlya), Ust - Medvedetskiy” (town of Serafimovich); “Razdorsky hunting reserve” (Mikhailovka); natural monuments “Chernichkin Garden” (Uryupinsk), “Iris” (Kalach-on-Don). The especially valuable territories are the “Green Ring” of the city of Volgograd (industrial township Gorodishche). The total area of these specially protected nature areas made up 363.84 thousand hectares, the total area of specially protected nature areas located next to rural settlements (villages, towns, villages, farms) was 632.88 thousand hectares. As practice shows, the location has a significant impact on the degree of exploitation of the nature resource potential of specially protected nature areas.

2.1.1. Resource potential of nature parks: threats and factors of negative impact.

The organization of regional nature parks in the early 1900s is explained by the need for legal protection of unique natural, historical and cultural resources, which were subject to such an impact which by that time was so serious and its consequences could not be ignored any longer. The resources under protection in the territories of nature parks are systematized in Table 1. The table is compiled according to the cadastral documents and annual reports of budgetary institutions which were created in each park for operational management.

The information mentioned above is not full, as it is subject to regular corrections in accordance with the results of scientific studies of the Volgograd scientists [9] and the systematic monitoring of the status of areas taken under protection. That’s why the values of most indicators are approximate. But it is already obvious that the nature parks have a high resource potential, however, the conservation status does not guarantee their full protection against the anthropogenic pressure. The threats and negative impact factors are associated with the location of nature parks, the specifics and intensity of the economic development of neighboring territories.

So, on the territory of the nature park “Volgo-Akhtubinskaya floodplain”, the level of negative impact on individual elements of the ecosystem and floodplain as a whole is evaluated on a scale from moderate to significant and critical. The reduction in area and productivity of forests, meadows, and wetlands is affected by the transformation of the hydro regime after the construction of the Volga

Hydroelectric Power Station [10], intensive agricultural development, forest cutting, road construction, hydraulic structures, mass unorganized tourism and recreation. The floodplain as a whole suffers from an increase in traffic load on the bridge across the river Volga, air pollution, soil, emissions from mobile sources, biological pollution during the operation of the storage ponds of Big Liman, fires, burns, loss of floodplains of fishery significance by many reservoirs. The poaching leads to a decrease in the population of certain types of biological resources. The high level of anthropogenic influence is explained by the border location of the Volga-Akhtuba floodplain with the territory of the Volgograd-Volga agglomeration where about 40% of the territory is broken by anthropogenic landscapes. Some scientists [11] consider the Volga-Akhtuba floodplain as an object of evolution of a technogenic and natural system, the vital activity of which depends on human control.

The nature park "Ust-Medveditskiy" is situated in the central part of the Serafimovich district along the Don and Medveditsa rivers and a significant part of the park is occupied by forests. The main sources of air pollution are: road transport, agricultural machinery, heating boilers, service plants and facilities and civil buildings. An unsatisfactory state of water resources is determined by the discharge into the water of untreated sewage from settlements, production sites and personal subsidiary plots, pollution of the banks of rivers, especially in places of spontaneous public recreation, and solid household waste.

Table 1. Resources of nature parks of the Volgograd region

Type of resource	Nature parks						
	Volgo-Akhtubinskaya floodplain	Donskoy	Eltonskiy	Nizhnekhoperskiy (Lower Khoper)	Tsimlyanskie peski	Scherbakovskiy	Ust-Medveditskiy
Types of biological resources	Number of biological resources						
Birds	≈230	≈100	≈85	≈164	≈140	≈182	≈130
Mammals	≈30	≈80	≈32	≈30	≈50	≈50	≈50
Insects	≈1400	≈300	≈3000	≈500	≈600	≈1800	≈900
Fishes	≈52	≈50	≈12	≈52	≈50	≈50	≈25
Plants	≈800	≈700	≈562	≈1500	≈400	≈600	≈638
Reptiles	≈5	≈10	≈8	≈30	≈10	≈11	≈8
Other	≈8	≈74	≈3	≈34	≈4	≈4	≈4
Number of types of biological resources included into the Red Book of Endangered Species of the Volgograd region							
Flora	19	37	24	88	19	46	28
Fauna	62	43	47	37	35	57	35
Types of hydrological resources							
Rivers and brooks, number/length, kilometers	3/165	4/92	7/95	22/1253	1/62	12/46	3/133
Lakes, number/surface, km²	2827/104	9/32.3	1/17480.3	958/48.3	4/53	12/4.58	2/0.3
Shallow rivers, number/surface, km²	178/53	-	-	-	-	-	-
Ponds, number/surface, ha	-	2/2.7	5/145	28/1.8	-		-
Water reservoirs, number/surface, ha	-	-	-	-	1/2700	-	-
Marshes, number/surface, km²	2/2.06	-	-	1/10.2	-	-	-
Springs, number/capture, litres per second	-	7/0.3-1	6/0.1-1.6	20/0.8-1.6	-	13/0.1-3.7	4/0.3-0.3
Forest resources, surface, hectares							
Forests, windbreaks, shrubs	41852.9	9608.7	89.2	69808.3	23529.9	10424.2	23403.6

Particular environmental importance was assigned to Nature Park "Donskoy" due to its location in the area with a low population and weak economic development. The landscapes and natural ecosystems of the park are well preserved and serve as a biological reserve for the Red Book of Endangered Species of flora and fauna.

Relative purity of water bodies is observed. The threats of a negative impact are associated with unorganized activities of residents of several rural settlements and rural areas including the Ilovlya settlement for cattle grazing, waste disposal in residential and recreational areas, harvesting and deforestation, the use of technologies that reduce the soil fertility.

The list of threats mentioned above is also typical of other nature parks which take into account the types of threats that are typical of specific kinds of resources which are not reflected in Table 1. For example, in the "Tsimlyanskie Peski" nature park, the greatest threat is the violation of hunting and fishing rules, as the numerous reservoirs of the park serve as the habitats for waterfowls and spawning grounds for valuable fish species. The Nature Park "Nizhnehoperskiy" has deposits of sample and rare soils, so an unauthorized plowing of land by local residents threatens with the loss of unique elements of the ecosystem. The environmental objective of the "Shcherbakovskiy" nature park is to prevent the degradation of rare nature complexes representing a combination of various landforms with plant groups. The anthropogenic threats are associated with a violation of the sanitary condition in places of recreation of the population, in highway rest stops, viewing platforms, fishing and bivouac places. The destruction also occurs under the influence of climatic factors. The features of the Nature Park "Elton" are its location in the semi-desert zone, contrasting landscapes which have no analogues within the region (desert steppes, grass-wormwood semi-deserts, salt-dome landscapes, sors and silted estuaries). The lake Elton has unique deposits of mud and salt brine that are unique in composition and balneological properties. This park also has river systems belonging to Elton basin. The negative factors of influence in this nature park are land reclamation systems, arable land, soil cover deflation, dirt and unauthorized collection of healing mud from the Lake Elton; unauthorized development of clay and sand, steppe fires; spring and autumn hunting, poaching; uncontrolled recreation.

According to the set of target functions, the nature parks are the most integrated natural-territorial entities and play a sensible environment stabilizing role. The natural resources located within their borders are limited in civilian circulation; change of the purpose of land under nature parks is prohibited; lands under valuable nature territories are not subject to privatization.

2.1.2. Target functions of specially protected nature areas of small and medium size.

According to the size of the occupied areas, natural monuments, protected landscapes and especially valuable territories can be attributed to small specially protected nature areas of Volgograd region, their total area is about 25259 hectares, the total number of small and medium sized areas is 37 units. According to the specification, the natural monuments are divided into integrated areas (protection of artificially created forest massifs and places of high concentration of species included into the Red Book of Endangered Species); geological areas (protection of geological exposure with outcrops of Neogene and Paleogene rocks, landscapes with species of Cretaceous flora, objects of quartz sandstone with imprints of ancient flora); paleontological areas (protection of places of accumulation of the remains of ancient animals); biological and botanical areas (protection of the habitats of species included into the Red Book of Endangered Species, nature massifs of rare and endangered tree species, communities of floodplain meadows and forests, areas of virgin steppes).

Especially valuable territories and protected landscapes provide the natural living conditions for animals and plant communities in systems of lakes, floodplain forests and meadows, estuaries, in natural and artificially created forests, steppe, sand massifs, isolated terrain features and oak forests.

Seven from all state nature reserves have a hunting specialization and are intended for the conservation, reproduction and rational use of hunting resources. The faunal area "Drophiny" was created in order to preserve bustards, their habitat and migration routes.

The largest part of small and medium-sized specially protected nature areas is situated within the boundaries of municipal areas with the agricultural specialization. The main threats for them are illegal fishing, hunting, cattle grazing, building of bonfires, plowing, deforestation and hay-mowing. The anthropogenic influence of ecosystem communities in these territories can be qualified as weak or moderate. The hunting reserve "Razdorsky", the natural monuments "Chernichkin Garden" and "Iris Garden", the "Green Ring" area of Volgograd, which are located near urban settlements, in addition to the listed threats, are used intensively for recreational purposes.

3. Conclusion

A characteristic feature of the economic development of the Volgograd region is a high level of urbanization and anthropogenic changes in the environment under the influence of agro-, urban- and technogenic imperatives. The methodology of territorial protection in such conditions is of particular importance if it helps to prevent threats, reduce the negative impact, protect and restore the natural potential of the region. On the basis of this statement, the presented study uses a certain sequence of research.

The analysis of the current state of the system of specially protected nature areas of Volgograd region revealed the presence of objects of various types of subordinance and categories. The size and structure of specially protected nature areas over the past five years remain stable. The main elements are the specially protected nature areas of regional importance, among them the largest ones are the nature parks. The nature parks operate in the form of budgetary institutions as economic entities. In relation to natural complexes, flora and fauna objects, landscapes, the management of nature parks carry out the following stages of work: "accounting-protection-monitoring-assessment-conservation-restoration-supervision". In the absence of specially protected nature areas of a higher administration level in the region, the tasks of scientific research, environmental education, and the provision of conditions for regulated recreation were given to nature parks [12]. As the most integral natural and territorial formations with multifunctional targets, they have a high environmental and stabilizing value.

Within the boundaries of small (natural monuments, protected landscapes, especially valuable territories) and medium sized (nature reserves) specially protected nature areas, the analysis showed a rich species diversity of valuable natural resources. However, an uneven distribution, territorial fragmentation, remoteness from the center for the protection and monitoring of these elements of the systems of specially protected nature areas enhance their vulnerability to threats and negative impact factors.

The following measures are relevant for the increase of importance of specially protected nature areas in the protection of the natural resource potential of Volgograd region. They are: toughening of the regimes of individual zones of nature parks, creation of buffer and transit zones and other elements of environmental protection having compensatory and sanitizing properties at their periphery provides the corridors which ensure the environmental links to specially protected nature areas of various levels of administration and subordinance. It will be possible to implement the suggested measures without the damage to the economic development of the region only if an integrated scientifically justified management of all parts of the system is ensured.

Acknowledgement

The article is prepared with the financial support of RFFR grant № 18-010-00637 A "Improvement of management of specially protected nature areas in the regions of Russia (exemplified by Volgograd region)".

References

[1] Tishkov A A 2017 *Bulletin of the Russian Academy of Sciences. Geographical series* **1** 8–19
[2] Skorobogatov A S and Skorobogatov A S 2017 *Economic issues* **1** 81–102.
[3] Tishkov A A 2017 *Geography and Natural Resources* **3** 13–21
[4] *Report "On the state of the environment of Volgograd region in 2018"* 2019 (Izhevsk: «Print-2» Publ) p. 330

[5] Brylev V A and Ovcharova A Yu 2016 *Geomorphology* **2** 18–26
[6] Brylev V A, Streltsova E N and Aristov A V 2001 Geomorphology **3** 87–93
[7] Sagalaev V A , Bochkin V D 2002 *Botanical journal* **87(11)** 115–22
[8] Chibilev (Jr.) A A and Chibilev A A 2019 *South of Russia: ecology, development* **14(1)** 117–25
[9] Zaitsev A A, Voronov G A and Buzmakov S A 2012 *Russian Journal of Earth Sciences* **12** 139–46
[10] Vishnyakov N V and Zelenskaya O Yu 2018 *South of Russia: ecology, development* **13(4)** 119–28.
[11] Klimanova O A, Kolbovskiy E Yu and Illarionova O A 2018 *Bulletin of St. Petersburg University Earth sciences* **63(2)** 127–46
[12] Mitrofanova I V, Starokozheva G I, Shkarupa E A, Batmanova V V, Mitrofanova I A and Tlisov A B 2016 *Revista Galega de Economia* **25(3)** 57–68.

Mathematical model for assessing environmental risk

I V Sukhorukova and N A Chistyakova

Department of Higher Mathematics, Plekhanov Russian University of Economics, Stremyanny lane, 36, Moscow, 117997, Russia

E-mail: suhorukovaira@yandex.ru

Abstract. The article considers the urgent problem of assessing environmental risks. The problem of environmental protection is now becoming global. This is primarily due to the growth of industrial production and large-scale mining. To quantify damage, the concept of environmental insurance is introduced. This means that generally accepted insurance risks are directly applied in the field of environmental protection. In addition to the general formula for a comprehensive risk assessment to predict harmful environmental impacts, it would be natural to take into account the results of these impacts over time. In this study, the authors proposed more detailed mathematical models of risk factors. We got a calculated and analytical expression in which the probability of an insured event by risk factors varies with time. Examples of models for predicting harmful effects in the environment are given. In solving this problem, probabilistic mathematical models were used. The proposed methods and tools allow taking into account possible environmental risk factors at all stages of decision-making. This will allow in practice to avoid the occurrence of adverse economic consequences in the implementation of economic activity. This contributes to the prospects of implementing business strategies for integrating environmental requirements into their activities. Keywords - environmental insurance, environment, mathematical model, economic damage, integrated risk assessment

1. Introduction

The problem of environmental protection is currently becoming increasingly widespread and becoming global. This is primarily due to the constant influence of energy and the growth of industrial production, large-scale extraction of natural resources. The tasks of environmental protection and the problems of environmental risks over the past decade are among the priorities for the leadership of the Russian Federation. At the state level, programs are being adopted to preserve the country's environmental safety. The Federal Law of the Russian Federation “On Environmental Protection” provides for a whole range of measures for economic regulation in the field of environmental protection. Based on the accumulated knowledge and experience of both Russian scientists and foreign researchers [1-4], the use of environmental insurance and environmental audit is proposed as the basis for these measures. Their joint use is aimed at the development of economic regulation in the field of environmental protection. When deciding on environmental insurance, well-known insurance risks should be directly applied in the field of environmental protection. Environmental risks are understood as risks caused by the negative impact of a person on the environment. Of particular danger are man-made disasters, as a result of which significant damage is caused not only to nature, but also to people, their health and property.

2. Subject, objectives and method of research

When deciding on environmental insurance, well-known insurance risks should be directly applied in the field of environmental protection. Environmental risks are understood as risks caused by the negative impact of a person on the environment. Of particular danger are man-made disasters, as a result of which significant damage is caused not only to nature, but also to people, their health and property. The fundamental principles of environmental management that are currently in force are based on administrative management methods, such as issuing licenses, issuing permits for emissions, discharges, waste disposal, determining the maximum amount of resource withdrawal. At the same time, they also contain economic, market mechanisms for environmental management [5-8] Payments for damage caused by pollution were calculated, payments for the development of natural resources were made, and compensation for damage was calculated. Accordingly, environmental insurance and the environmental audit procedure are additional progressive sources of economic regulation of environmental activities. Calculation of environmental insurance tariffs allows you to develop a mechanism for accounting for liability for damage caused as a result of accidental environmental pollution. Promotes the use of reserves of insurance companies to develop preventive measures for the implementation of technological and environmental measures that reduce environmental risk. This fact is beneficial to all participants in the insurance process: insurance company, company, third parties and the environment.

It should be noted that the existing system of eliminating the consequences of pollution prompts the local leadership, farmers, land users to the most intensive development of these territories. This is due to the fact that the allocated budget funds are mainly spent on the restoration of facilities located in high-risk areas, which leads to a further increase in damage [9-12]. In this situation, to reduce the amount of damage, it is necessary to use not only traditional measures. The adoption at the state level of a policy of regulating the economic use of these territories is required. Economic methods of regulation should be combined with a complex of administrative and legal restrictions. Environmental audits and environmental insurance are aimed at implementing measures stipulated by the state program of the Russian Federation "Environmental Protection" for 2012-2020. The proposed methodology will ensure the rights of the state as the owner of natural resources to maintain an acceptable quality of the environment. Promotes conservation and restoration of the natural environment. Ensure rational use and the necessary level of reproduction of natural resources. The joint mechanism of the environmental audit procedure and environmental insurance helps prevent the negative impact of economic and other activities on the environment and helps to eliminate its consequences.

It is known that in the Russian Federation as of 2017, the largest volume of generation of production and consumption waste was accounted for by the type of economic activity "mining" - 93% of the total amount of generated waste. This is due to the fact that the extraction and enrichment of minerals generates the largest amount of waste, the basis of which is overburden. Therefore, when issuing licenses for the operation of hazardous production facilities, including agricultural enterprises on contaminated lands, it is necessary to present a contract of insurance of civil liability risk for causing harm during the operation of a hazardous production facility. In turn, the contract of insurance of civil liability risk for causing harm during the operation of this facility is concluded on the basis of an environmental audit conclusion regarding identification of environmentally hazardous production facilities, as well as a conclusion on environmental risk assessment.

The object of insurance is the environmental risk of civil liability, expressed in the presentation to the policyholder of property claims by individuals and legal entities for compensation for damage from pollution of land, water or air in the territory of the specific insurance contract. Tariff rates for insurance payments may be set as a percentage of the annual turnover of the enterprise. It should be noted that the terms of environmental liability insurance provide for the establishment of limit amounts for insurance compensation payments (liability limits) and the insurer's own participation in the payment of losses. Insurance compensation includes compensation for damage caused by damage or loss of property; the amount of losses associated with the deterioration of living conditions and the environment; expenses for cleaning up the contaminated area and bringing it into a state that complies with the standards.

An environmental audit in the environmental management system should be carried out when developing a plan of preventive measures while reducing environmental risks and assessing damage in the event of an insured event. Environmental auditing requires an answer to questions about assessing the likelihood of environmental risk and negative effects, the principles for determining the size of losses that occur in the national economy, and methods for representative ranking of enterprises according to their environmental hazard.

3. Results and discussion

The most important issue, the solution of which is relevant to date, is the form of environmental insurance. To effectively take into account environmental risks, a compulsory insurance procedure for all potentially hazardous industries is required. At the level of adoption of federal laws, for the first time, this mandatory form of environmental insurance is mentioned in Article 18 of the Federal Law "On Environmental Protection". In the mandatory form of environmental insurance, it is also necessary to take into account the features of liability insurance in terms of compensation for damage caused. Since when concluding an environmental insurance contract, it is possible that there is an injured party in addition to two parties to the process: the insurer and the insured. The mandatory form of the insurance contract guarantees the assistance to this third injured party in the event of an insurance event. Due to the mandatory insurance procedure, insurance rates will be calculated to compensate for the damage caused. When calculating compensation for damage, it is also possible to take into account the different contributions of each individual risk of the potential hazard. When solving the problem with the form of environmental insurance, one can rely on the experience of foreign countries, in particular European countries. Companies carrying out activities with possible environmental consequences receive a license for a specific type of activity related to potential health and property risks of third parties. The way to ensure financial guarantees for licenses obtained is of various types. Most often, in practice, insurance coverage is provided by the reserves of the insurance company itself, concluding an insurance contract. Also, a fixed amount on the Company's deposit account may act as a financial guarantee. In certain cases, the solvency of the Company is confirmed by the bank by providing a targeted loan. The priority remains the mandatory conclusion of an insurance contract. It is a mandatory form that serves as a fundamental condition for the effective operation of the Company. The insurance contract should be concluded only with leading insurance companies guaranteeing insurance coverage in the event of an insurance event. In the mandatory form of concluding an insurance contract, it is necessary at the legislative level to fix the total compensation paid to the injured party. It should also be borne in mind that damage can be caused not only to individuals but also to legal entities. The compensation received upon the occurrence of an insurance event takes into account the level of damage to property or its total loss. Financial coverage allows you to compensate for damage caused to air, water, forests, soil. It is also necessary to provide financial resources to eliminate the consequences of pollution. These are quite significant expenses for the budget of our country. So far, there is clearly no certainty in the financial support of the environmental insurance fund. Previously, the state insurance company Gosstrakh functioned in our country. All mandatory types of insurance, functioning at that time, were financed from the budget of the Gosstrakh. After the collapse of the USSR, the company Gosstrakh ceased to exist. Therefore, in modern Russia, the formation of an environmental fund is one of the most important tasks. To assess the value of this fund, it is necessary to take into account insurance payments from enterprises, organizations and institutions. When calculating environmental insurance tariffs, the state will receive certain negative socio-economic consequences. Mandatory payments of enterprises will affect the increase in the value of their products. This will cause discontent among many people. There are certain risks associated with filling the state environmental insurance fund. To solve the problem of assessing the necessary funds, it is necessary to calculate the optimal value of insurance coverage. Using mathematical methods allows us to assess the risk of environmental management. For this purpose, the amount of damage is first calculated. associated with adverse economic, social, environmental and economic reasons. The integrated risk [13] assessment currently used is determined by the following formula:

$$R = \sum_i p_i \cdot D_i, \tag{1}$$

where p_i - probability of occurrence of the i - th risk factor;

D_i - damage from the implementation of the i-th risk factor, rub.

The risk of environmental management from the perspective of insurance can be considered in several sections:

• the risk of the occurrence of a specific phenomenon that is spontaneous in nature, such as floods and floods, etc .;

• the risk in the operation of a particular environmental management facility when unforeseen events are correlated with a specific environmental management facility;

• the risk associated with the likelihood of damage or death of the natural resource object (for example, the risk of the death of valuable ecological systems, loss of soil fertility, etc.).

In addition to the general formula for integrated risk assessment for predicting harmful environmental impacts, it would be natural to take such impacts into account in dynamics. To do this, we can propose more detailed mathematical models of risk factors and use formula (1), in which the probability of occurrence of an insured event by risk factors varies over time. For different risk factors, different probabilistic models of exposure may be natural. Before constructing specific models, we consider [14] the random variable T - the time before the occurrence of the insurance event by the risk factor and the concept of the survival function that is natural for such a probability model

$$S(t) = P(T > t), \quad t > 0$$

In addition, since the insured event is associated with a point in time, which is characteristic of the fact that before that everything worked in a normal mode, and at that moment an emergency situation arose, it is natural to introduce a function, which we will call a risk intensity function, which depends on the time x,

$$r_x = \lim_{\varepsilon \to 0+} \frac{1}{\varepsilon} P(T < x + \varepsilon | T > x), \, x > 0$$

It is easy to obtain a connection between the introduced probabilistic characteristics. On the one hand, we express a risk intensity function across survival function :

$$\begin{aligned} r_x &= \lim_{\varepsilon \to 0+} \frac{1}{\varepsilon} P(T < x + \varepsilon | T > x) = \lim_{\varepsilon \to 0+} \frac{1}{\varepsilon} \frac{P(x < T < x + \varepsilon)}{P(T > x)} = \\ &= \lim_{\varepsilon \to 0+} \frac{1}{\varepsilon} \frac{S(x) - S(x + \varepsilon)}{S(x)} = \frac{-S'(x)}{S(x)}, \, x > 0 \end{aligned} \tag{2}$$

On the other hand, solving a differential equation (2) for survival function and given that at zero it is equal to unity, we obtain the expression survival function across risk intensity function

$$d \ln S(x) = -r_x \quad \Rightarrow \quad \ln S(x)\Big|_0^t = -\int_0^t r_x dx \quad \Rightarrow \tag{3}$$

$$\Rightarrow \ln\frac{S(t)}{S(0)} = -\int_0^t r_x dx \quad \Rightarrow S(t) = e^{-\int_0^t r_x dx}, t > 0$$

The upper bounds for x and t in equalities (2) and (3) are not indicated here, since for specific statements of the problems they can be either finite or conditionally infinite. Here are some examples of models for predicting harmful effects in the environment.

The simplest example is when the risk factor acts with constant intensity r. Then from (3) we obtain the forecast for the survival function

$$S(t) = e^{-\int_0^t rdx} = e^{-rt}, t > 0$$

Another example is when the intensity of risk increases over the interval of influence of the factor $(0, \Delta)$. If it grows by law $r_x = \frac{1}{\Delta - x},\ 0 < x < \Delta$, then the survival function is equal to $S(t) = e^{-\int_0^t \frac{1}{\Delta - x}dx} = \frac{\Delta - t}{\Delta} = 1 - \frac{t}{\Delta},\quad 0 < t < \Delta$.

A rather natural model can be an example of risk intensity when equipment wear leads to an exponential increase in risk intensity. In this case

$$r_x = ce^{ax}, x > 0, a > 0, c > 0,$$

$$S(t) = e^{-\int_0^t r_x dx} = e^{-\int_0^t ce^{ax}dx} = e^{-\frac{c}{a}\left(e^{at}-1\right)}, t > 0$$

Constants are selected based on statistical observations when monitoring the influence of a harmful factor.

Having obtained a survival function for a specific environmental risk model, it is possible to predict the probability of the occurrence of an insured event at any time interval.

4. Conclusions

The proposed mathematical model for assessing environmental risk and calculating the amount of damage caused to the environment due to pollution as a result of anthropogenic activities determines the procedure and procedure for applying environmental audit to assess the amount of harm. The authors of this scientific study have proposed a formula for integrated risk assessment for predicting the dynamics of harmful effects on the environment. More detailed mathematical models of risk factors were developed and the probability of an insured event on risk factors changing over time was calculated. The calculated economic damage in accordance with the calculations can be used to determine the insured amount for civil liability insurance of enterprises and organizations - sources of an increased environmental hazard.

References

[1] Krstić M, Stajković N, Lazić S and Bokonjić D 2017 *Vojnosanitetski pregled* **74 (4)** 341-348
[2] Broderick J W, Lavoie D R and Perel A J 2000 *Environmental quality management* **10 (1)** 3-9
[3] Freeman P K and Kunreuther H C 1997 *Managing environmental risk through insurance* (Boston: Kluwer) p. 45-46.
[4] Kunreuther H C 2001 *The Geneva Papers on Risk and Insurance* **26 (2)** 276-295

[5] Sukhorukova I V and N A Chistyakova 2018 *Jornal of Reviews on Global Economics* **7** 146-151, DOI: 10.6000/1929-7092.2018.07.12.
[6] Sukhorukova I V and Chistyakova N A 2018 *Astra salvensis* **6** 847-857
[7] Wernick N 2002 Pollution engineering **34 (9)** 16
[8] Sparrevik M and Breedveldy G 2010 *Integrated environmental assessment and management* **6(2)** 240-248
[9] Ermekbayeva B J, Bimendieva L A and Tovma N A 2016 *International journal of innovative technologies in economy* **1 (3)** 29-33
[10] Tanaka Y 2003 Chemosphere **53(4)** 421-425
[11] Sukhorukova I V and Chistyakova N A 2019 *Jornal of reviews on global economics* **8** 269-274
[12] Popov VA 2018 *Jornal of reviews on global economics* **7** 453-456
[13] Motkin G A 1996 *Basics of environmental insurance* (Moscow:: Nauka) 192 p.
[14] Sukhorukova I V and Chistyakova N A 2018 *Journal Regional Science Inquiry* **X (2)** 195-203

Development of automation and virtualization instruments for the housing and utility services system

Dmitry Andreevich Toporov

Rostov State Economic University (RINH), 69, Bolshaya Sadovaya Street, Rostov-on-Don, 34006, Russian Federation

E-mail: toporovda@yandex.ru

Abstract. The rationale of the research topic is determined by the importance of public administration instruments for the housing and utility services sector in the country. In this article, we considered theoretical and practical aspects of modern electronic technology used to improve the public administration of H&U and the principal trends in their development. We analyzed the "Smart H&U" system: its content, targets, value, advantages, along with the current practice of implementing this project and its results. We detailed the concept of the state information system for H&U (SIS H&U) and studied the legal foundations of the creation and operation of SIS H&U. We analyzed the importance of the state information system for ensuring the openness, accessibility and transparency of activities in the housing and utility services sector. The author listed the prospects and advantages of this system.

1. Introduction

In modern urbanized society, housing and utility services (H&U) is one of the most important sectors of the social tertiary area. Among the numerous problems requiring attention in H&U, special attention should be given to the quality of public utilities, as well as transparency, reliability and accessibility of information about the housing facilities and residents, energy consumption, accrued fees for utility services, the current condition of H&U facilities and networks [1]. We believe these problems can be addressed successfully only with the use of advanced electronic (information) technologies.

2. Materials and methods

The object of the study is public relations in the public administration of H&U, the interaction of management organizations and citizens.

The subject of the study is the legal instruments regulating the public administration of H&U, scientific works, Internet articles dedicated to the study topic.

The methodological basis of the study includes the following methods: dialectical, historical, systemic, logical, comparative-legal, formal-legal, etc.

3. Results

Within the national project "Housing and Urban Environment" and national program "Digital Economy", the government is implementing the "Smart City" project, the purpose of which is to

IOP Conf. Series: Materials Science and Engineering **828** (2020) 012028 doi:10.1088/1757-899X/828/1/012028

digitally transform and automate the processes of resource management and services provision, along with the comprehensive improvement of the efficiency of urban infrastructure [2].

In March 2019, Russia Ministry of Construction approved the standards for "Smart Cities". They are to be implemented within five years in all towns and cities with the population of 100 thousand and more. The "Smart City" concept focuses on an ordinary citizen, with their goals and requirements, as the center point of all processes [3].

Today, the level of urbanization is constantly growing, cities take up about 70% of the global economy [4]. These growing numbers make control of the processes running in cities more difficult. "Smart cities" provide for the integrated application of IoT solutions in H&U and transport infrastructure in order to ensure the safety of city residents and other areas.

"Smart H&U" is a part of the "Smart City" concept, consisting of 6 subsections:

1. Introduction of intelligent accounting systems for utility resources;
2. Reducing energy consumption;
3. Implementation of automated control of response to consumers' requests and accident response;
4. Introduction of a digital management model for public utility facilities;
5. Introduction of the tool for holding a general meeting of premises owners in multi-apartment buildings by electronic vote.
6. Introduction of automatic systems for monitoring the condition of buildings, including noise level, temperature, serviceability of lifts, fire safety systems and gas equipment.

"Smart City" innovative technologies for the housing and utility services sector are aimed at improving the efficiency of utility and energy infrastructure, reducing the costs of resource-supplying organizations using concession, energy service and service models [6].

All the ideas of the digital transformation of H&U can be divided into two groups. First, those changing the system itself. Among the priorities of the "Smart H&U" concept, the Construction Ministry wants to introduce the modern technologies which will literally keep a "medical history" of every building. The idea is that one can track online all information about the technical condition of the premises, its engineering networks, all previous problems and their solutions.

"Smart H&U" is a special system, the main objective of which is full control over the automatic operation of housing and utility services. "Smart H&U" is able to have full control over the equipment operation, take readings from all meters at the right time, ensure efficient work of H&U employees and facilities, prevent accidents, etc.

The main goals of "Smart H&U" include [7]:

- reasonable use of energy and water resources;
- automation of the meter reading process with the subsequent data transfer to the respective authorities;
- facilitating the work of H&U employees;
- minimization of errors caused by the human factor;
- control over the quality of utility services through a system of special services and personal accounts;
- improving the efficiency of instrument control.

"Smart H&U" should reduce the monthly utilitiy fee for tenants. This direction is being actively studied and developed. A number of studies are conducted in order to launch trial versions of the system in different cities of the Russian Federation. Today, such large companies as Vega, Matrix, Betar, etc. are engaged in the development and production of devices to support the "Smart H&U" system in the country.

"Smart H&U" has 3 main levels:

- Local: location of meters - houses, apartments.
- The level of readings collection: smart meters use wireless (and wired) technologies for data transfer: Wi-Fi, PLC, NB-IoT, LPWAN, LoRaWAN, etc.

IOP Conf. Series: Materials Science and Engineering **828** (2020) 012028 doi:10.1088/1757-899X/828/1/012028

- The level of information processing and analysis: the user can view the final results of calculations on their personal account page upon authorization.

Through the system of personal account pages (cabinets), users can track all information in the form of tables or graphs, which greatly facilitates the perception of digital data, allows predicting the consumption of water, electricity and gas in the future, notifies about changes in utility fees, notifies of emergencies and events.

In Russia today there are several large organizations developing the dispatching of energy resources. All dispatch companies have their individual approach to organizing a business structure. Some companies produce and implement their own devices, which means that their software can only be used by a limited number of devices manufactured by them [8].

Other organizations, such as NEKTA, operate with a universal system, focusing on the development of targeted software systems [9]. These complexes support the work of most existing metering devices, while the number of supported equipment is increasing on a daily basis. Some companies have a narrow-focused specialization (one type of energy resource), while others are engaged in the integration of software for various devices (meters, sensors).

In 2017, the comprehensive “Smart H&U” solution was successfully implemented by Megafon in Innopolis residential buildings in Kazan. This is the first project in Russia to implement Internet-based tools in H&U based on NB-IoT technology. The “Smart H&U” solution combines an automated system for collecting and transmitting data from utility meters, a convenient web application for users, an extensive library of drivers for various models of digital metering devices, and numerous tools for integration with external systems. The solution allows metering and controlling the consumption of power, heat, gas and water.

In addition, the entire range of fire and security sensors, such as leakage, movement, break-in, etc. can be connected to the system. Using the appropriate meter models, MegaFon’s solution also allows you to program and manage metering devices remotely.

The solution transmits data using NB-IoT technology. It is characterized by small amounts of data exchange, the ability to use autonomous power and flexibly manage the power consumption of devices. Up to 80000 IoT devices can be connected to one base station, which is in large excess over the range of existing mobile communication standards. The technical data from the devices are stored in a secure cloud storage and are accessible to the user through a web interface and a mobile application.

The Smart H&U solution allows the supplying and managing organizations to ensure transparency of mutual payments, proper control over the supply and distribution of resources, remote control of meters, as well as reduction of energy loss and consequent financial loss [10]. "Smart H&U" can be integrated with commonly used systems 1C and SIS H&U. It ensures automatic data collection, without obliging the employees to walk door to door, and the compilation of a clear picture of resource consumption based on such data. At the same time, residents no longer need to manually submit H&U data and pay for the actual consumption of the resources.

“Smart H&U” will undoubtedly be actively implemented nationwide, contributing to the convenience of process management and reducing the loss, which are currently a serious problem for this sector.

We expect a forthcoming transition to the new standard in many urban and consumer services, such as power metering systems, city lighting, road infrastructure and intelligent H&U systems. We believe that NB-IoT will take up a serious position in the market in the nearest future. Many organizations in the industry, such as water utilities, power suppliers etc., striving to improve their service and reduce their labor costs, have long been introducing various automated control systems and technological and commercial accounting solutions.

The activities of managing organizations are gradually digitalized, promoting introduction of advanced technologies into their work: mobile applications for communication with house residents and services are being developed in order to simplify the work of H&U organizations. Residents of apartment buildings are involved in the process of managing their houses using mobile applications.

IOP Conf. Series: Materials Science and Engineering **828** (2020) 012028 doi:10.1088/1757-899X/828/1/012028

They can communicate with the managing organization, evaluate the quality of the works it performs, and solve problems in apartment buildings.

For example, management company “H&U League” (Yekaterinburg) successfully applies new technologies in its work: it communicates with residents of apartment buildings online through its own mobile application and repairs entrances and stairwells using a special block material.

Special calendars tell the management company when it is time to make repairs in the building, help to meet the deadlines for responding to requests. A mobile application also works for employees: it solves the issue of internal communication and work organization within the team. Through the mobile application, employees see all tasks received during the day, for each of them they can correspond with colleagues or with residents. The application works offline. This means that even without active connection, an employee can leave comments and change the status of the task. When the connection appears, it will automatically synchronize with the server and everyone will see the updated description of the task.

The IT system of the “H&U League” has a recognizable interface, it is based on the chessboard principle. This is a kind of house map, showing the number of floors and the number of apartments on each floor. The icons indicate which apartments have unresolved issues.

The internal quality control system is an important part of the work of a managing organization. “H&U League” asks its customers three questions to find out if they are satisfied with the company:

1. Do you like our work?
2. Do we like our work? Interestingly, the residents of the building are encouraged to check the work of the managing organization, so the company receives unbiased assessment.
3. Do we have enough money for this?

These three simple questions will help management organizations to promptly resolve any conflict situations.

The "League" also has a special website "Entrance Constructions". On this site you can choose the materials and design for the entrance and stairwell interior decoration and immediately calculate how much the repairs will cost.

A resident of the house can use a special mobile application and a personal cabinet on the organization’s website for paying bills, leaving submitting requests for a specialist’s visit and evaluating the work of the management company.

The “State Information System of Housing and Utility Services” (SIS H&U) is an important electronic tool for public administration in the H&U area.

The State Information System for Housing and Utility Services (hereinafter - SIS H&U) is a comprehensive centralized federal information system ensuring collection, processing, storage, provision, placement and use of information on housing and utility services of the constituent entities of the Russian Federation.

The system contains the following information:

- management and resources supply contracts;
- regulations and municipal programs for H&U;
- readings of whole-building metering devices;
- planning and performance of maintenance and repair works for common property in apartment buildings;
- tariffs and payment documents for housing and utility services;
- citizens' requests on H&U issues and results of addressing such;
- financial documentation.

SIS H&U is posted online at: https://dom.gosuslugi.ru.

The activity of SIS H&U is regulated by the Federal Law No. 209-FZ dated July 21, 2014 “On the State Information System for Housing and Utility Services”, which establishes the requirements for creating the system, types of information it is to contain, and liability for the failure to enter information in the SIS H&U.

Effective July 1, 2019, all managing organizations in Russia are to enter data into the SIS H&U, including companies from Moscow, St. Petersburg and Sevastopol. Earlier, according to the Federal Law of July 21, 2014 No. 209-FZ, managing organizations operating in cities of federal significance could not do this [11].

SIS H&U is integrated with other federal information systems: Unified State Register of Property Title to Real Estate, Federal Information Address System and the "H&U Reform" portal, which accelerates and updates information flows.

SIS H&U is to facilitate the interaction between owners of residential premises and managing organizations, service providers and authorities: this system will help to receive payment documents for settling of housing and utility services, pay utility bills, hold general meetings of homeowners, execute contracts for utility provision, etc.

The implementation of this system will allow H&U to become more open and transparent, to eliminate data duplication and expedite its search, as well as increase its completeness, reliability, and relevance, as the citizens - consumers of H&U services will be able to use this system to access current information on the activities of managing organizations and authorities in this area, to control their expenses and the accuracy of charged fees for housing and utility services. This allows citizens to see the results of the management organization's work, which promotes honesty and fair competition, contributing to the dynamic development of the economy and society.

SIS H&U will also make it possible to optimize the public control mechanisms by providing an opportunity to compare the cost of the works and services of various H&U organizations and the opportunity to review the results of inspections and audits performed by the authorized bodies.

SIS H&U contains H&U regulations, a list of licenses of managing organizations and housing facilities and the results of inspections performed by regulatory authorities.

This is but an example of the modern electronic technologies contributing to the improvement of the efficiency of public administration of housing and utility services. Competent application of modern electronic technologies in the government regulation system for H&U will make it possible to bring Russian housing and utility services to a new level improving the quality of public services ensuring significant energy saving and facilitating interaction between the homeowners, managing organizations and state authorities.

The introduction of electronic technology improves the public administration system for H&U along with the system of interaction between the homeowners, managing organizations and state authorities.

4. Conclusion

Based on the results of the study, the following conclusions have been made.

The state is implementing the "Smart City" project, which includes the "Smart H&U" system. The main objective of Smart H&U is full control over the automatic operation of the housing and utility services. The idea is to provide for online tracking of all information on the technical condition of the house, engineering networks, as well as the history of all previous problems and their solutions.

Nowadays, the digitalization of the management organizations activities is universal, the introduction of advanced technologies is highly promoted: mobile applications for communication with the residents and services which are to simplify the work of H&U organizations are being developed.

SIS H&U is an Internet portal for the collection, processing, storage, provision, placement and use of H&U information in the constituent entities of the Russian Federation.

The implementation of this system will allow H&U to become more open and transparent, eliminate duplication of information and expedite its search, as well as contribute to its completeness, accuracy and relevance.

Modern electronic technologies work towards improving the efficiency of public administration of housing and utility services. Competent application of modern electronic technologies in the government regulation system for H&U will make it possible to bring Russian housing and utility services to a new level improving the quality of public services, ensuring significant resources saving and facilitating the interaction between the homeowners, managing organizations and state authorities.

References

[1] 2005 *Housing Code of the Russian Federation No. 188-FZ dated December 29, 2004* (RF Statute-Roll) **1** part 1 art 14
[2] *Basic and Supplementary Requirements for Smart Cities: “Smart City” Standard* URL: http://www.minstroyrf.ru/docs/18039
[3] *The concept of the "Smart City" project for digitalization of urban economy, by the Ministry of Construction and Housing and Utility Services of the Russian Federation* URL: https://www.minstroyrf.ru/trades/gorodskaya-sreda/proekt-tsifrovizatsii-gorodskogo-khozyaystva-umnyy-gorod/
[4] *Over 80% of apartment buildings have already provided basic information for SIS H*&U URL: http://www.minstroyrf.ru/press/bolee-80-mkd-uzhe-predostavili-bazovuyu-informatsiyu-dlya-GIS-zhkkh
[5] Kovalev A S 2019 *Electronic Scientific Journal "GosReg"* **1**
[6] Kolokolova E O 2018 *Bulletin of the Russian Cooperation University* **1(31)**
[7] *MegaFon will help H&U* URL: https://www.comnews.ru/content/120877/2019-07-18/megafon-pomozhet-zhkh
[8] *RosKvartal® - Internet service No. 1 for Managing Organizations* URL: https://roskvartal.ru/tehnologii-v-zhkh/9995/upravlyayuschie-organizacii-v-umnom-gorode-novye-tehnologii-v-zhkh
[9] *Smart H&U: Seven Russian solutions* URL: https://vc.ru/flood/37565-umnoe-zhkh-7-otechestvennyh-resheniy
[10] Shadrina T 2017 *Rossiiskaya Gazeta* **7392** URL: http://rg.profkiosk.ru/article.aspx?aid=594645
[11] 2014 *Federal Law No. 209-FZ Dated July 21, 2014 On the State Information System for Housing and Utility Services* (RF Statute-Roll) **30** part I art 4210

IOP Conf. Series: Materials Science and Engineering **828** (2020) 012029 doi:10.1088/1757-899X/828/1/012029

Formation of professional competencies of future builders in technical school

O I Vaganova [1], Zh V Smirnova [1], O P Kozlova [2], N A Vishnevetskaya [3] and A V Prostov [4]

[1] Minin Nizhny Novgorod State Pedagogical University, 603004, Chelyuskin str., 9, Nizhny Novgorod, Russia, z.v.smirnova@mininuniver.ru

[2] Volgograd State Technical University Institute of Architecture and Building Construction, 400005, Chair of Linguistics and Intercultural communication, Lenina prosp, 28, Volgograd, Russia, kornienko_o@list.ru

[3] Volgograd State Technical University Institute of Architecture and Building Construction, 400005, Chair of Linguistics and Intercultural communication, Lenina prosp, 28, Volgograd, Russia, nattimsweet@mail.ru

[4] Volgograd State University, 400062, Universitetsky prosp, 100, Volgograd, Russia, sacha_prost@volsu.ru

E-mail: z.v.smirnova@mininuniver.ru, kornienko_o@list.ru, nattimsweet@mail.ru, sacha_prost@volsu.ru

Abstract. The conditions of the modern labor market in the construction sector, the new technology in the construction industry, changes due to scientific and technological progress and the implementation of modern information technologies require special training of mid-career professionals. The society needs graduates who are ready to independent inclusion in production processes, who are able to solve life and professional tasks that they come across. It is necessary to update the content-technological support for the formation of professional competencies of future builders to a higher level because of the transition to competence-oriented training in the system of secondary vocational school. The authors have developed a model of professional competencies development to improve the training of builders within the framework of secondary vocational schools. The model can be realised in any construction college taking into account current requirements of Russian employers and the All-Russian Intersectoral Association of Employers "Russian Union of Builders" to the level of professional knowledge of a modern graduate of secondary vocational schools and training features of modern builders.

1. Introduction

Nowadays there are a lot of global changes in modern education which are connected with the adoption of new Federal state educational standards [1].

These transformations are the answer to the technical updating, computerization and expansion of communication channels and flows, changing needs of the country and society [2].

It is obvious that some specialties are not much in demand among young people that's why it is necessary to update them, to create new specialties which will allow developing professional competence on higher level [3]. The training of a competent general professional is becoming an increasingly complex multi-stage task [4,5,6]. O.V. Akulova, V.A. Bodrov, N.F. Rodionova, N.N. Surtaeva, A.P. Tryapitsina, N.V. Chekalev have contributed to the development of the issue of training students under the competence approach [7].

The competence-based approach focuses on such educational goals such as:

- the quality of education;
- learnability [8,9];
- self-determination [10];
- self-actualization;
- socialization [11];
- the development of the individuality of the student [12].

Today, due to the changes in society we have indicated the needs of society and country. Thus, it has become necessary to improve the quality of professional training for mid-career professionals [13,14]. A modern graduate of technical school should solve any professional tasks, apply knowledge and skills in practice, be ready for continuous self-improvement and professional growth [15].

Mid-career professionals have to have special training to work with modern information technologies in the construction sector. Besides, if the developer wants to be competitive it is necessary to update the ways of building.

2. Methodology

Training of future builders in institutions of secondary vocational education is carried out within the framework of the competence approach. In the new conditions, the process of forming competencies should take into account the requirements of modern employers. Therefore, the authors have developed a model consisting of several blocks, the purpose of which is the formation of professional competencies in technical school of future builders. This model can be applied in any of the construction technical schools since it meets the modern requirements of employers and takes into account the specifics of the training of builders. The model contains: targeted, informative, technological and efficient blocks.

3. Results and discussion

Formation of professional competencies in students of secondary vocational schools is a process that involves the creation of forms and conditions for the student to achieve a certain result, that is, the generated competencies necessary for the implementation of future professional activities in accordance with the requests of employers and opportunities for further education. The requirements of the Russian Union of Builders are reflected in Table 1.

Table 1. Requirements of Russian employers and the All-Russian Intersectoral Association of Employers "Russian Union of Builders" to the level of professional knowledge of a modern graduate.

№	Content requirements
1.	Know the theoretical foundations of special disciplines.
2.	To be able to transform acquired knowledge into innovative technologies.
3.	Be able to independently find, perceive and analyze new information, taking into account existing legislative and social legal regulations.
4.	To be able to work with global information sources based on computer and information and communication tools.

5. To be able to make decisions in uncertain, suddenly complicated professional conditions.
6. To own computer-aided design systems.
7. Master the methods of system analysis.
8. Master the methods of conducting scientific research.
9. Own software systems design technical solutions.
10. To possess adequate methods of professional communication and behavior, ability to cooperate and work in a group for a common result.
11. Have the ability to be creative and innovate when dealing with professional tasks.
12. Understand the trends and main directions of development of science and technology.
13. Identify information needs, use and give them a professional assessment for solving specific problems.

Requirements analysis showed that employers consider the ability of a graduate to solve non-standard professional tasks, apply computer tools, information and communication technologies possess such personal qualities as initiative, communication, reflexivity, creativity [16,17]. Preparation of students of a technical college should meet modern requirements associated with changes in modern construction [18]. If in the 1950s - 1980s, Soviet typical panel or brick residential buildings were built mainly five-story with small apartments, today the construction of high-rise buildings, townhouses and duplexes, cottages and private households is developing [19]. Therefore, future professionals require the acquisition of relevant competencies. Based on the technical specification, architects and designers create drawings in the corresponding computer programs [20,21,22]. Their list is quite large and Archi-CAD remains the most famous [23]. Virtual drawings have a major advantage. Any change is made on a computer and does not require processing dozens of paper schemes [24]. That is, the software allows you to freely change the parameters and create copies of buildings with different layouts. After checking the drawings for compliance with state standards and the wishes of the customer, a three-dimensional model is created [25]. For this, computer programs are also used [26].

Generally recognized among professionals is 3dsMax. It allows you to create a three-dimensional layout based on computer drawings and provides the client with the opportunity to see what the future house will look like. After the approval of the layout, the designers create a paper version of the project in order to transfer it for inspection to the city architecture, contractors, foremen and builders. This system is actively developed and proves its effectiveness. Modern design is impossible without the full use of computer technology and the latest programs. Today, students of a construction technical school should master the necessary computer programs corresponding to the modern level of construction technology.

The specificity of the competence-based approach of modern vocational education consists in training specialists at a functional level, which implies not only arming students with a certain set of knowledge and skills in the chosen field, but also the formation of the personality of a future professional who is capable of individual creative solutions and learning. The formation of professional competencies of future builders in secondary vocational education [27,28,29] is aimed at this. The composition of professional competencies is structured in such a way that the qualifications of a graduate of technical school meet the requirements of employers and the regional labor market and allow, after a short adaptation period, to fully engage in the work of an enterprise at an appropriate level.

For the successful formation of professional competencies of future builders, we developed a model that can be applied in any construction technical school, since it takes into account all the features of the training of builders, as well as meets the modern requirements of employers specified by us earlier.

The target component of the model includes goal, methodological approaches (cognitive; activity; competence; personality-oriented), principles (science, continuity, subjectivity, variability, systematic, interactivity). The content block reflects the content of educational disciplines that implement the development of several activities: survey, design and engineering, production and technology, experimental research, installation and commissioning, and service and operational. In the technical unit, the following methods are used: methods (knowledge transfer, design, implementation of social partnership technologies, master classes, workshops); forms (lectures, seminars, individual lessons, internships at enterprises, project implementation); means (electronic textbooks, interactive boards, theodolites, levels, range finders, electronic total stations). The effective block reflects pedagogical conditions (gradual and sequential complication of the process of formation of competencies, strengthening practice-oriented orientation, the use of modern educational technologies). Also, the effective block contains the criteria for competencies development (motivational, epistemological, motivational, organizational, corrective-reflexive).

4. Conclusion

The competence-based approach in education allows you to prepare creative-minded professionals who are able to approach outside the standard to solve new and well-known problems that can take responsibility for the decision. The developed model can improve the conditions for the formation of professional competencies and increase their level of improvement among future builders studying in secondary vocational schools taking into account the basic requirements of modern employers.

References

[1] Vaganova O I, Smirnova ZH V, and Abramova N S 2019 *Baltic Humanitarian Journal* **8 (1 (26))** 277-280.
[2] Ajeenkya D, Patil Y, Dr. Gagandeep Nagra and Dr. Gopal R A 2014 *International Journal of Management* **5(5)** 1-6.
[3] Arkhipova M V, Belova E E, Gavrikova Y A, Lyulyaeva N A, and Shapiro E D 2018 *Advances in Intelligent Systems and Computing* **622** 380-386. DOI: 10.1007/978-3-319-75383-6_49.
[4] Ibatova Aygul Zufarovna, and Ippolitova Natalia Viktorovna 2018 *International Journal of Civil Engineering and Technology* **9(3)** 394–399.
[5] Barber M, Donnelly K, Rizvi S, & Summers L 2013 *An avalanche is coming. Higher Education and the revolution ahead* **73.** URL: http://www.studynet2.herts.ac.uk/intranet/lti.nsf/0/684431DD8106AF1680257B560052BCCC/$FILE/avalanche-is-coming_Mar2013_10432.pdf
[6] Abramova N S, Vaganova O I, and Smirnova Zh V 2019 *The azimuth of scientific research: pedagogy and psychology* **8 (1 (26))** 13-15
[7] Braine G 2013 *Non-native educators in English language teaching. Routledge.* URL: https://www.taylorfrancis.com/books/9781135461867
[8] Ergunova O T , Lizunkov V G, Malushko E Yu, Marchuk V I, Ignatenko A Yu 2017 *OP Conference Series: Materials Science and Engineering* **177(1)** 012046. DOI: 10.1088/1757-899X/177/1/012046
[9] Smirnova Z V., Vaganova O I, Loshkareva D A, Konyaeva E A, and Gladkova M N 2019 *IOP Conference Series: Materials Science and Engineering* **483 (1)** 012003
[10] Vaganova O I, and Ilyashenko L K 2018 *Vestnik of Minin University* **6 (3)** 2. DOI: 10.26795/2307-1281-2018-6-3-2.
[11] Gruzdeva M L, Smirnova Z V, Chaikina Z V, Golubeva O V, and Cherney O T 2019 *Lecture Notes in Networks and Systems* **57** 1193-1199.
[12] Prokhorov M P, and Vaganova O I 2019 *Bulletin of the Minin University* **7 (1 (26))** 4.
[13] Ilyashenko L K, Smirnova Z V, Vaganova O I, Prokhorova M P, and Abramova N S 2018 *International Journal of Mechanical Engineering and Technology* (IJMET) **9 (4)** 1097-1105.
[14] Malushko E, Bolsunovskaya L, and Martyushev N 2019 *Asian EFL Journal* **23 (3.2)** 315-328.

[15] Ilyashenko L K, Vaganova O I, Smirnova Z V, Sedykh E P, and Shagalova O G 2018 *International Journal of Mechanical Engineering and Technology* **9 (4)** 1029-1035.
[16] Ilyashenko L K, Smirnova Z V, Vaganova O I, Chelnokova E A, and Kaznacheeva S N 2019 *International Journal of Mechanical Engineering and Technology* **10 (2)** 908-917.
[17] Ilyashenko Lubov Kiryalovna 2018 *International Journal of Civil Engineering and Technology* **9(3)** 607-616.
[18] Manikandan, and M. Muthumeenakshi 2018 *International Journal of Mechanical Engineering and Technology* **9(3)** 706-710.
[19] Markova S M, Sedykh E P, Tsyplakova S A, and Polunin V Y 2018 *Advances in Intelligent Systems and Computing* **622** 129-135. DOI: https://doi.org/10.1007/978-3-319-75383-6_17.
[20] Kamenez N V, Vaganova O I, Smirnova Z V, Bulayeva M N, Kuznetsova E A, & Maseleno A 2018 *International Journal of Engineering and Technology (UAE)* **7(4)** 4085-4089.
[21] Frolova O A, Chanchina A V, Frolova N V, Shevchenko S M, Chelnokova E A, & Bystrova N V 2017 *European Research Studies Journal* **20** 549-556.
[22] Fedorov A, Zakablukovskiy E, & Galushkina A 2018 *E3S Web of Conferences* **48** Doi: 10.1051/e3sconf/20184802007
[23] Smirnova Zh V, and Krasikova O G 2018 *Vestnik of Minin University* **6 (3)** 9. DOI: 10.26795/2307-1281-2018-6-3-9.
[24] Fedorov A A, Ilaltdinova E Y, & Frolova S V 2018 *Vysshee Obrazovanie v Rossii* **27(7)** 28-38. Doi: 10.31992/0869-3617-2018-27-7-28-38
[25] Smirnova Zhanna V, Mukhina M V, Kutepova L I, Kutepov M M, and Vaganova O I 2018 *Advances in Intelligent Systems and Computing* **622** 187-193.
[26] Fedorov A A, Ilaltdinova E Y, & Frolova S V 2019 *Perspektivy Nauki i Obrazovania* **37(1)** 262-274. DOI: 10.32744/pse.2019.1.19
[27] Kamenez Natalie V, Smirnova Zhanna V, Vaganova Olga I, Bystrova Natalia V, and Tsarapkina Julia M 2019 *International Journal of Mechanical Engineering and Technology* **10(02)** 899–907
[28] Ilyashenko Lubov K, Smirnova Zhanna V, Vaganova Olga I, Chelnokova Elena A, and Kaznacheeva Svetlana N 2019 *International Journal of Mechanical Engineering and Technology* **10(02)** 908–917
[29] Vaganova O I, Smirnova Zh V, Markova S M, Chaikina Zh V, and Bulaeva M N 2019 *Perspectives of Science and Education* **39 (3)** 500-514. DOI: 10.32744/pse.2019.3.38

IOP Conf. Series: Materials Science and Engineering **828** (2020) 012030 doi:10.1088/1757-899X/828/1/012030

Intelligent decision support for information protection system's design

E A Vitenburg[1], A V Nikishova[1], M Yu Umnitsyn[1] and T A Omelchenko[1]

[1]Volgograd State University, 100 prospect Universitetskii, Volgograd, 400062, Russia

E-mail: nikishova.arina@volsu.ru

Abstract. The article defines the need to protect enterprise information systems. To ensure the security of enterprise information system, depending on the stage of a life cycle, it is necessary to design, control and upgrade the information protection system. We propose the approach to designing enterprise information system's protection based on the method of intelligent decision support, which uses a neural network. To do so we form the mathematical model of intelligent decision support for designing of enterprise information system's protection. It takes into account the data of monitoring and analysis of security events.

1. Introduction

The development of information technologies in the modern world allows not only to minimize human labor expenditures, but also to optimize the production process and to establish full control over it. Modern industry started using information technology relatively recently. Integration of industrial systems and corporate systems provided users with extensive opportunities to maintain technological and support processes in the enterprise. However, from the information security point of view enterprise information systems have become less isolated, which has led to an increase in various kinds of malicious attacks. This factor determines necessitates of the timely design of an effective protection system.

The variety of protection subsystems that can be used to protect against several types of threats and the territorial distribution of enterprise information system's components complicates the task of selecting a set of protection subsystems that form a protection system.

The choice of protection subsystems can be carried out several times during the life cycle of information system. Thus the important characteristics of the selection process are the speed of result's reception, and the reduction of the residual risk.

Therefore, it is important to use artificial intelligence systems in selection for the best set of protection subsystems. It is important to take into consideration that information protection system's forming can be carried out at different stages of an enterprise information system's life cycle. Also the formation of an effective existing information protection system should take into account the statistics of information security incidents that occurred at the enterprise in recent years, as well as forecasted indicators.

The international standard ISO 27001:2013 [1] points out the need to apply a procedure for information security incidents' managing. Since without timely response to information security incidents, eliminating their consequences and possible causes, it is impossible to effectively manage information security. Therefore, the actual direction of countering modern security incidents in information systems is the creation and application of security monitoring and management systems.

IOP Conf. Series: Materials Science and Engineering **828** (2020) 012030 doi:10.1088/1757-899X/828/1/012030

Their functions are monitoring and correlation of security events, security control, timely detection of unauthorized changes and their consequences, as well as decision support for design or modification of information protection system.

The main theoretical and practical aspects of the problem of information security monitoring and management systems' creation are explored in the works of modern Russian scientists and specialists: Tsaregorodtsev A.V. [2], Makarevich O.B. [3], Zefirov S.L. [4], Astakhova L.V. [5], Livshits I.I. [6], Vasilyev V.I. [7], Bedran A. [8], Kotenko I.V. [9] and others. The analysis of these scientists' works shows that research in this area can be divided into several groups:

- creation and improvement of security events' data collection and correlation;
- security events' analysis for attacks' detection;
- methods' development for risk assessment and security events' ranking in order to identify and predict information security incidents and timely form an effective information protection system;
- methods' development for countermeasures' selection based on the analysis of information security events and incidents;
- development of new and improvement of existing algorithms for decision support in information security management and/or security system's design.

At present leading vendors in the field of information security offer a wide range of software products for security events' monitoring and analyzing [10]. Events monitoring and analysis tools are provided by the following software: MaxPatrol SIEM, HP ArcSight, Security Capsule SIEM, IBM Qradar. Each of these programs has a number of advantages and disadvantages. MaxPatrol SIEM, developed by Russian company Positive Technologies, as well as Security Capsule SIEM, developed in Russia by "ITB", has a certificate of conformity of FSTEC of Russia on the level of control of undocumented features absence. In addition, MaxPatrol SIEM and Security Capsule SIEM provide the use of a large number of connectors, which allows to receive and analyze events from various sources. Unlike their foreign counterparts, HP ArcSight and IBM Qradar, MaxPatrol SIEM and Security Capsule SIEM provide file and software integrity control.

As a result, each software presents a detailed report on the events' analysis to the user [10]. Based on this report, the user can make a conclusion about the security degree and absence of incidents in the information system. If the user observes signs of attack's implementation or occurrence of information security incident in the report, he can make the decision on correction of protective measures. So he can strengthen the information system's protection and counter the malicious effect.

The decision process, in this case, depends entirely on the decision-maker. Especially it depends on his level of professionalism as the information security specialist, his subjective views on the process of information security ensuring, and psycho-emotional component. In addition, from the beginning of attack's identification to the protection measures' application, a long period of time can pass. And so the decisions aimed at providing protection against a particular malicious impact will be taken untimely.

This problem can be eliminated if the decision process is automated. However, nowadays in the field of information security there are no ready-made software solutions that allow not only to monitor and analyze security events, but also to generate management solutions aimed at enterprise information system's protection improvement based on security events' analysis. These management solutions can include changing in information protection system's structure or correction of protection tools' settings.

A large number of works on security events' monitoring and decision models based on monitoring data for information protection system's design shows relevance and necessity of this research. The purpose of the research is to automate the process of information protection system's design and/or modifying taking into account data of security events' monitoring and analysis.

IOP Conf. Series: Materials Science and Engineering **828** (2020) 012030 doi:10.1088/1757-899X/828/1/012030

2. Proposed approach

We propose a design methodology that takes into account security events generated by the existing enterprise information system. This allows the information security specialist to use exactly those protection means that belong to the most important protection subsystems. These means allow protecting against the most relevant classes of threats. To calculate the importance of information protection subsystems we use formula (1):

$$I=N(A), \tag{1}$$

where $I=(I_1,\ldots,I_9)$ – vector of information protection subsystems' importance; $A=(A_1,\ldots,A_6)$ – vector of threat classes relevance; N – functional dependence represented by a neural network.

Moreover, in order to determine the vector of threat classes relevance based on security events' monitoring data, we define two compliance matrices:

1) matrix MTh of compliance between a set of threats Thr and a set of threat classes CTh. $MTh=Thr\times CTh=(mth_{ij})$, where

$$mth_{ij}=\begin{cases}\frac{1}{n}, if\ threat\ belongs\ tothreat\ class\\ 0, otherwise\end{cases} \tag{2}$$

where n – number of threat classes. Meanwhile $\forall i, \sum_j mth_{ij}=1$.

2) matrix MEv of compliance between a set of security events Ev and a set of threat classes CTh. $MEv= Ev\times CTh=(mev_{ij})$, where

$$mev_{ij}=\begin{cases}1, if\ event\ occur\ during\ implementation\ of\ threat\ of\ class\\ 0, otherwise\end{cases} \tag{3}$$

The used neural network is a multilayer perceptron, which functions according to (4).

$$\begin{cases} In_{0k}=v_k\\ Out_{ij}=f(\sum_l w_{ijl}In_{ijl}-\theta_{ij})\\ In_{ijl}=Out_{i-1l}\end{cases} \tag{4}$$

where In_{0k} – the k–th neuron in the input layer; v_k – the k–th element of the input vector; Out_{ij} – the output value of the j-th neuron of the i-th layer; f – activation function of the neuron; w_{ijl} – the weight of l-th entry of the j-th neuron of the i-th layer; In_{ijl} – value of l-th entry of the j-th neuron of the i-th layer; θij – the activation level of the j-th neuron of the i-th layer; Out_{i-1l} – output value l-th neuron of (i–1)-th layer.

So we present the functional model of information protection subsystems' intellectual selection as following (figure 1).

The inputs to the security event analysis phase are the event sets generated by the enterprise information system. We use the classification of security events proposed by MicroSoft in this research [11]. The set of security events includes the following types (5):

$$Events=\{EnterEv,\ ManagementSubEv,\ AccessObjEv,\ PolicyChangeEv,\ UsePrivilegesEv,\ ISProcessesEv,\ LevelISEv\}. \tag{5}$$

where EnterEv – set of events of "account logon" type; ManagementSubEv – set of events of "account management" type; AccessObjEv – set of events of "object access" type; PolicyChangeEv – set of events of "policy change" type; UsePrivilegesEv – set of events of "privilege use" type; ISProcessesEv – set of events of "process tracking" type; LevelISEv – set of events of "system" type.

We selected dangerous events from a variety of security events. We also mapped dangerous security events to threat classes using compliance matrixes. As a result, we determined the vector of relevance of threat classes. We use a neural network to determine the compliance between the relevance of threat classes and the importance of information security subsystems.

IOP Conf. Series: Materials Science and Engineering **828** (2020) 012030 doi:10.1088/1757-899X/828/1/012030

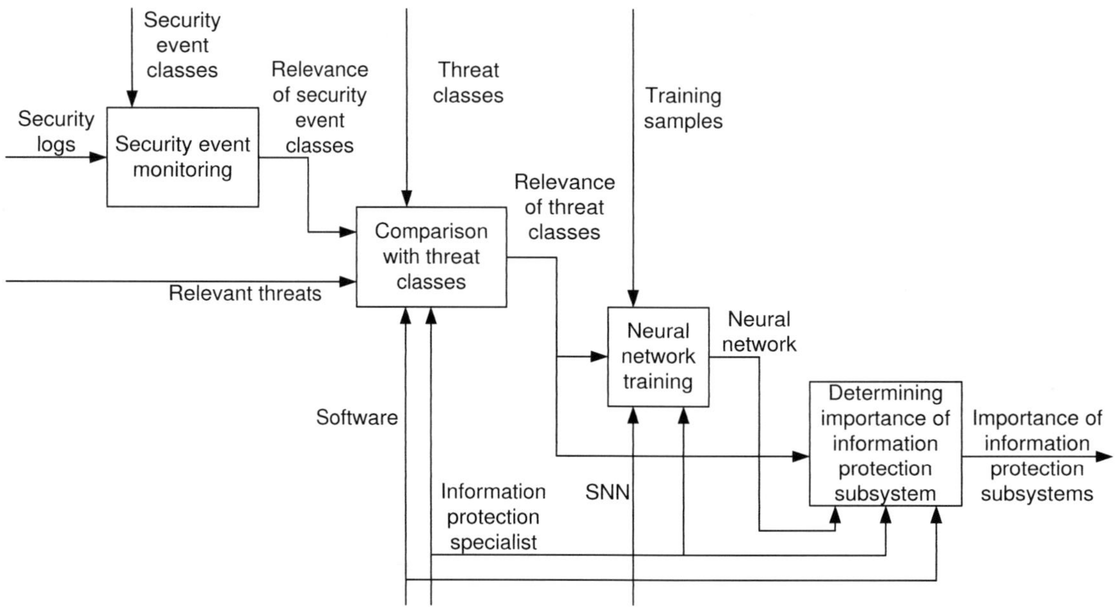

Figure 1. Intellectual choice of information protection subsystems

It is worth considering the fact that the number of event sources increases with the growth of the number of users, servers, workstations and software and hardware elements of enterprise information system and/or reconfiguration of enterprise information system. Taking into account a large number of events generated by enterprise information system during its functioning, it is important to conduct automatic collection and analysis of events' data from the registration subsystems of information system's objects. In this regard, we gave the most detailed consideration of the stage of security events' analysis. Because the correctness of the conclusions obtained during this stage affects the correctness of the generated design decisions.

According to the result of sources' [12-14] analysis, we determined that the current state of the enterprise information system – S_T at a certain time T, $S_T \in$ State, State={Snorm, Sdang, Sanorm} can belong to one of three possible types:

- normal Snorm – regular functioning of information system in accordance with its tasks and according to the documents regulating its work;
- dangerous Sdang – normal functioning is disrupted, violations of information system's functioning associated with attacks, failures of software and/or hardware of information system occurs;
- abnormal Sanorm – characterized by a temporary change in the normal mode of information system's functioning and a surge of abnormal activity of users, software and network traffic.

We described each distinguished type of information system's state by its own set of events and parameters, which according to the states we also divided into normal, dangerous and abnormal events.

Events related to dangerous events and anomalous events require more detailed analysis. The set of such events is the input data of the threat class comparison stage. These events indicate the implementation of security threats.

Since the methods of attacks' implementing can change over time, it is more convenient to operate with classes of threats of information security violations for the purposes of scientific research [15].

The set of threats of information security violation Threat can be conveniently divided into classes CTh, which we defined in the formula (6).

$$\text{Threat} = \{\text{Breaking, Leak, Distortion, Loss, Blocking, Abuse}\}, \quad (6)$$

where Breaking – set of threats that fall into the "breaking" category, Leak – set of threats that fall into the "leakage" category, Distortion – set of threats that fall into the "distortion" category, Loss – set of threats that fall into the "loss" category, Blocking – set of threats that fall into the "blocking" category, Abuse – set of threats that fall into the "abuse" category.

Then the mapping of threats' classes and security events' classes we defined as in table 1.

Table 1. Compliance of threats' classes and security events' classes

Threats' class	**Security events' class**
Threats of breaking category	Events of object access type
Threats of leakage category	Events of object access type; events of process tracking type
Threats of distortion category	Events of object access type; events of process tracking type; events of system type
Threats of loss category	Events of system type; events of privilege use type
Threats of blocking category	Events of account logon type
Threats of abuse category	Events of account management type; events of privilege use type; events of policy change type

Therefore, the input data for the neural network is:

- results of enterprise information system's events' monitoring – MIS;
- classification of events and evaluation of information system's private security indicators.

Data about each event $EventIS_i$ are collected from the log during monitoring of information system within the specified time interval ΔT. These events form a set of information system's events EventIS and are assessed. We use cortege (7) to describe event attributes.

$$EventIS_i = \langle ID, Date, Level, Source, EventType, EventState, SecureParams \rangle, \quad (7)$$

where ID – event's code; Date – time and date of event's occurrence; Level – event's level; Source – event's source; EventType – event's type; EventState – event's state; SecureParams – vector of event's security parameters (8).

$$SecureParams = \langle h, u, risk \rangle, \quad (8)$$

where h – frequency of event's occurrence; u – potential damage of event's consequences; risk – amount of information security risk of the event.

We determine the frequency of event's occurrence h by the ratio of the number of events of certain code to the total number of events within the specified time interval ΔT (9).

$$h = NEventID/NEvent, \quad (9)$$

where NEventID – number of events of certain code; NEvent – total number of events within the specified time interval ΔT.

The potential damage of event's consequences u depends on the value level of each event Level and we describe it by a quantitative – qualitative scale of levels (table 2).

We calculate information security risk of each event, which is a function of frequency of event's occurrence and potential damage of event's consequences, by the formula (10).

IOP Conf. Series: Materials Science and Engineering **828** (2020) 012030 doi:10.1088/1757-899X/828/1/012030

$$risk=h*u. \quad (10)$$

We calculate the total risk RiskSum for all events as the sum of the private risks according to (11).

$$RiskSum = \sum_{i=1}^{EventID} risk_i. \quad (11)$$

Table 2. Scale of compliance between value of potential damage and value level of event

Potential damage u	Value level of event Level
u=0 low	Information Success audit
u=3 medium	Warning
u=6 high	Mistake Failure audit
u=10 critical	Critical error

We describe the set that provides classification of events and estimation of private security indicators by subsets of elements (12).

$$ACE=\{\{EventType\}, \{EventState\}, \{ISState\}\}, \quad (12)$$

where EventType – set of detected events' types; EventState – set of possible events' states EventState=$\{S^n, S^d, S^a\}$=$\{0;1;0,5\}$, where S^n – normal events that are characteristic for the normal mode of enterprise information system's functioning; S^d – dangerous events that are characteristic for enterprise information system where security breach (attack or security incident) is located; S^a – abnormal events characterizing deviations of information system from the normal mode of functioning. These events need additional analysis.

We reduced the process of determining whether an event $EventIS_i$ belongs to one of the three states to solving the classification problem. For this purpose we use the set StatePatern=DamgeEvent∪NormEvent. We divide the set into two basic subsets which are formed by an expert group based on data on typical events of normal enterprise information system's functioning and previously detected information system attacks and incidents. So we formed the template or signature of events typical for this information system and dangerous events:

- set of typical security breach events DamgeEvent includes events that are known signs of an attack or are included in the events describing the scenario of a security incident;
- set of events NormEvent typical for information system's normal functioning.

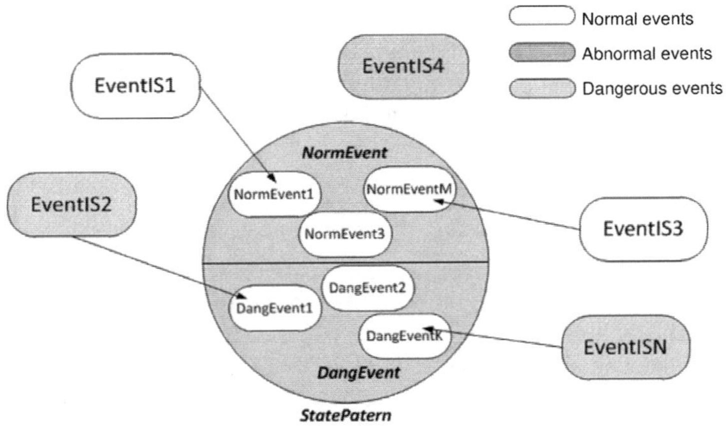

Figure 2. Event classification based on basic templates

We implement classification F(EventIS$_i$(EventState)→{StatePatern}$_j$) as in (13) (figure 2):

$$EventIS_i(EventState) = \begin{cases} S^n, EventIS_i \in NormEvent \\ S^d, EventIS_i \in DamageEvent \\ S^a, EventIS_i \notin NormEvent \cup DamageEvent \end{cases}, \quad (13)$$

where EventIS$_i$(EventState) – information system's events with state attribute, each of which corresponds to a set of relationships StatePatern$_j$ with typical events from sets of events of normal functioning and security breach.

Thus, if such an event is not in the template of known regular events or security breach events, it is considered abnormal and may indicate abnormal activity, caused either by the actions of an attacker, or a temporary authorized deviation from the normal mode of functioning.

At the second stage according to table 1 we carry out the comparison of detected dangerous events DamageEvent to the set of threats defined in the formula (6).

And then the structure of information protection subsystems corresponding to the identified relevant classes of threats is automatically proposed.

3. Results and Discussion

On the basis of the proposed model, we developed software that allows both designers of information protection systems and information security specialists at the enterprise to form the most optimal set of information security tools – projects.

A screen copy of the results of software for intelligent decision support for information protection system's design functioning is shown in figures 3-4.

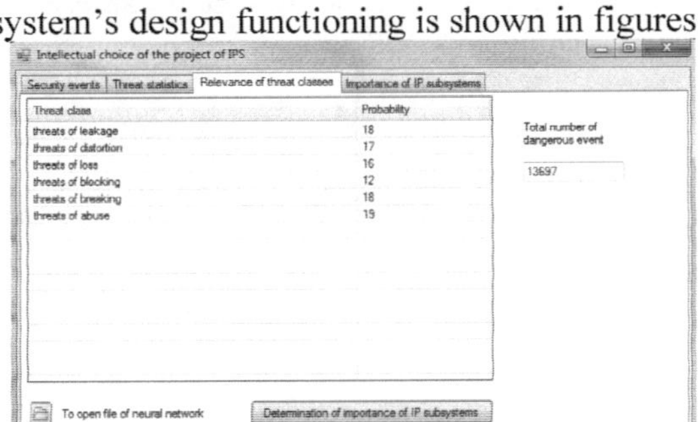

Figure 3. Determining the relevance of threat classes

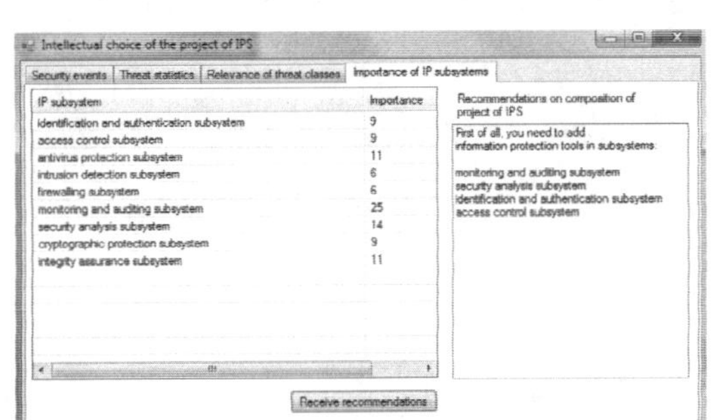

Figure 4. Determining the importance of protection subsystems

Implementation of proposed projects increases the efficiency of information protection system in comparison with the process of information protection system's design without decision support tools. This conclusion follows from the fact that the design will take into account the whole set of data about the protected system, which is difficult in "manual" design [16].

4. Conclusion

It's important to note that developed software minimizes the subjective factors included in the design process by the protection system's developer. Such factors include mistakes of the developer in the analysis of initial data, short terms of modernization and/or development of protection system, taking into account a large amount of information about the protected system, the inability to carry out a deep and comprehensive analysis of security event logs or statistics in a short time.

The application of the developed software allows designing optimal protection for a particular information system, taking into account the peculiarities of its functioning and its relevant threats.

Acknowledgment

The reported study was funded by the Council for grants of Russian Federation President, according to the research project No. MK-6404.2018.9.

References

[1] Sokolov S S, Alimov O M, Golubeva M G, Burlov V G and Vikhrov N M 2018 *IEEE Conference of Russian Young Researchers in Electrical and Electronic Engineering (EIConRus)* **5** 124–7
[2] TSaregorodtsev A V 2009 *Methods, Models and Algorithms of Synthesis of Protected Information Systems* (Moscow: Tax Academy of the Russian Federation Press) p 208
[3] Makarevich O B and Sheludko I A 2003 *Izvestiya SFedU. Engineering Sciences* **4** 211–6
[4] Zefirov S L and Shcherbakova A IU 2014 *TUSUR Reports* **2(32)** 77–81
[5] Astakhova L V and Tsimbol V I 2016 *Bulletin of the South Ural State University. Ser. Computer Technologies, Automatic Control, Radio Electronics* **16(1)** 165–9
[6] Livshits I I and Lontsikh P A 2015 *IrSTU Bulletin* **98(3)** 268–73
[7] Vasilev V I, Pestrikov V A and Krasko A S *Izvestiya SFedU. Engineering Sciences* **8** 7–14
[8] Bedran A 2009 *T-Comm — Telecommunications and Transport* **2** 10–3
[9] Kotenko I V, Saenko I B and Iusupov R M 2014 *St.Petersburg State Polytechnical University Journal: Computer Science. Telecommunications and Control Systems* **198(3**) 7–18
[10] Tarasenko N 2018 *SecurityLab.ru by Positive Technologies*
[11] Dwyer J and Truta T M 2013 *9th IEEE International Conference on Collaborative Computing: Networking, Applications and Worksharing* **9** 563–70
[12] Mashkina I V, Sentsova A IU, Guzairov M N and Kladov V E 2011 *Izvestiya SFedU. Engineering Sciences* **125(12)** 25–35
[13] Astrakhov A V. Klimov S M and Sychev M P 2013 *Countering Cyber Attack. Technological Base* (Moscow: BMSTU Press) p 70
[14] Kazimir V V and Seraia A A 2010 *Journal Mathematical Machines and Systems* **4** 52–61
[15] Vitenburg E A, Pushkarskaia A I and Oladko V S 2017 *Information Systems and Technologies* **101(3)** 21–30
[16] Vitenburg E A 2018 *Proc. of VolSU Scientific Session* (Volgograd: VolSU Publishing) pp 295–9

Optimizing cargo delivery employing vehicles of different carrying capacity

V M Volchkov, V A Korolyash, E S Rayushkin, O U Soloveva and E G Shvedov

Volgograd State Technical University, 28 V. I. Lenin prospect, Volgograd, 400005, Russia

E-mail: soloveva8@yandex.ru

Abstract. Consider the problem of mathematical modeling. Making optimal use of vehicles of different carrying capacity can help minimize delivery costs. This article presents the models for finding the optimal delivery solution for ring roads and pendulum routes. The models were created for several car parks (depot) with vehicles of various carrying capacity. We also have taken into consideration that some routes are unavailable for particular types of transport.

1. Introduction

In many situations, low-tonnage vehicles are more appropriate to deliver goods [1]. This is due to some obstacles with delivering goods within the city. These obstacles occur due to the heavy traffic in the main highways, courtyards infrastructure, the need to leave a car in a small free area near the ramp or on the "patch" near the house. The situation is the same when it comes to small consignments within agricultural areas or factory territories where the manageability of heavy vehicles or trailer-trucks is extremely limited.

This paper reviews some routing schemes in terms of optimizing cargo delivery:

– when using freight transport of various carrying capacity;

– when it is impossible to use certain types of freight transport on certain routes.

2. The issue of cargo delivery (overview)

The issue of optimizing, namely, minimizing, delivery costs is a multidimensional task. Such optimization has various goals and applications:

– from a common transportation problem in which the routes between suppliers and consumers are clearly defined; while the task is reduced to determining the amount of goods transported along these routes to optimize the total costs,

– to complex logistic tasks which take into account the capabilities of suppliers (including the production structures) and customers (which may be associated with further transportation systems, e.g., a seaport, or, for a factory, the production capacity).

Nowadays, a great number of different methods have been developed to find the optimal solution with regard to the nature and goals of cargo delivery in various conditions. An overview of various goal statements and methods to achieve them can be found in [2, 3, 4, 5].

2.1. Routing tasks

The routing is a part of these big logistical tasks. Routing is regarded as the search for an optimal route scheme that minimizes the delivery cost from suppliers (manufacturers, from warehouses, etc.) to

IOP Conf. Series: Materials Science and Engineering **828** (2020) 012031 doi:10.1088/1757-899X/828/1/012031

customers (consumers). A detailed classification of routing problems and a review of methods for solving them can be found in [6]. In routing, there are many wording options that take into account the location peculiarities of routes and possible requirements for goods delivery. There are also methods to address the aforementioned issues that are determined by:

1. Customer requirements are characterized by:
a) the location of consumers (a graph of paths and endpoints of routes);
b) consumer requests for the volume of goods (e.g. goods of various types, or the so-called multi-product tasks),
c) the time periods by which the consumer should receive goods (tasks with "time windows"),
d) time limits for handling charges.

2. The distribution network (routes) that is characterized by pendulum, cyclic, and tree-like schemes of route networks.

3. The fleet of vehicles that is characterized by:
a) the places of departure and destination (transport companies) and the location of consignors and consignees,
b) the carrying capacity of vehicles (expressed as the maximum volume of transported goods or the accommodation capacity (e.g. trays)),
c) devices for handling charges,
d) the roads trafficability for the particular vehicle,
e) vehicle maintenance costs.

We can also distinguish a specific class of routing tasks. In this class, the delivery tasks and exporting other goods from other consumers are simultaneously solved (*pickup and delivery problems*).

Usually, the goal of optimization is to minimize the total cost of transportation, also by choosing the optimal route. However, sometimes there can be a multicriteria problem (with different, often conflicting, goals):

– minimization of total costs that usually depends on the total distance traveled by cars or the total travel time, and on the costs of using particular vehicles;
– reducing the number of vehicles (or drivers);
– correlation of routes lengths with travel time and vehicle loading;
– minimization of fines related to customer service quality.

Also, sometimes it is necessary to take into account that some factors are time-dependent and are possible to occur.

2.2. The mathematical formulation of routing tasks

The mathematical formulation of tasks depends on the aforementioned routing types. Therefore, we will illustrate only one of the possible tasks. It is the so-called routing problem with partial unloading of goods for each consumer and the possible loading of different product from this consumer (*VRP with Pickup and Delivery – (VRPPD)*).

This task can be regarded as a search for some simple ring routes that require minimal costs with the following restrictions:

– each ring route begins and ends at the starting point ("depot");
– each consumer is attended to in only one ring route;
– the value of the cargo along each ring route should be non-negative and should not exceed the carrying capacity C;
– for each consumer i (different from the "depot"), consumer D_i should be attended to on the same ring route as consumer i [5].

There are many variations of restrictions. Restrictions appear in tasks with partial handling charge at consumers' or with the imposition of time restrictions on the delivery of goods. Options with several "depots" or several types of vehicles are possible. Also, the task can be temporary in nature when delivery can take several days. In addition, some task components (number and customer requests, path length, etc.) may be random.

Various approaches to modeling and mathematical formulation of these problems can be found in [5, 7, 8].

2.3. Methods for solving routing tasks

The classical transportation task belongs to the class of linear programming problems. Therefore, various modifications of the so-called simplex method are usually used to solve it [9, 10].

Methods for solving routing tasks, in contrast to the simplest transport task, as a rule, include integer constraints. Therefore, integer programming methods are the basis for their solution [11] (e.g. the "branches and bounds with clipping" method).

Methods for solving routing problems can be divided into “exact” and “heuristic”.

“Exact” methods are usually various versions of the “branches and borders with clipping” method [5, 12]. Since the computational costs for implementing such methods can increase significantly with increasing degree of the task (a large number of consumers, routes, or restrictions), various heuristic approaches are often used [13]. The use of heuristics does not allow obtaining “exact” solutions but only a so-called good solution. However, it can significantly reduce computational costs.

In more complex logistics tasks, a strict mathematical formulation can be difficult to use due to a large number of additional conditions that describe the behavior of suppliers and consumers. In such cases, simulation methods can be used [14, 15]. Sometimes, genetic optimization algorithms are used to solve transportation tasks [8].

3. Setting the task

According to the aim of this study, we will determine the optimal composition of the vehicle fleet for the delivery of goods to consumers.

We will consider three cases:

a) a fleet with one type of heavy goods vehicle,
b) a fleet with several types of vehicles of different carrying capacity,
c) several fleets with vehicles of various carrying capacity.

The option with a radial route diagram does not require any routing. In the radial diagram, routes are defined by the distance between the vehicle fleet (“depot”) and consumers. In addition, having one type of vehicle does not require optimization since the number of vehicles on each route is determined by the volume of goods transported by the consignor (consignee). The only parameter to be determined is the number of vehicles of a certain type. This will depend on customers’ needs, vehicle carrying capacity, route availability, and fuel consumption.

Nevertheless, this structure of the model will be preserved in more complex cases. Basically, this is a common transportation task (with one supplier) but with different types of vehicles.

The formal statement of the problem (its mathematical formulation) can be written as follows.

X_{ij} is the quantity of the product/goods, $Cost_{ij}$ is costs associated with delivery along route i (to consumer i) by vehicle j.

Consumer (route) i is characterized by distance $Distance_i$ and demand $Demand_i$.

Vehicle $Auto_j$ is characterized by $Auto\_Capacity_j$ and average fuel consumption per 100 km $Fuel\_Consumption_j$.

Minimized:

$$\sum_{i\,j} X_{ij} \cdot Cost_{ij}$$

under restrictions:

for each route (consumer), i number of deliveries to the i-consumer by all types of j vehicles should attend to their needs:

IOP Conf. Series: Materials Science and Engineering **828** (2020) 012031 doi:10.1088/1757-899X/828/1/012031

$$\sum_{j} X_{ij} \geq Demand_i$$

the amount of goods exported by all types of *j* transport to the *i*-consumer should not exceed stocks in the warehouse; for simplicity, we will take stocks equal to the sum of all consumer requests:

$$\sum_{j} X_{ij} \geq Depot\_capasiti$$

In addition to the standard restriction system of the transportation task, we will introduce $Avaiable_{ij}$ parameter for the availability of *i*-route for the *j*-type of vehicle.

After optimization, the number of cars for a given direction is determined by dividing the amount of cargo transported in this direction by the vehicle capacity (the total value is rounded to the nearest upper integer).

4. The case of several suppliers with vehicles of different carrying capacity and restrictions on the availability of the delivery route

This case will represent a general transport task with any number of suppliers, any number of vehicle types, and restrictions on the availability of the delivery route.

In order not to complicate the model with a large amount of source data, we will use the case of two suppliers and four types of vehicles. Generalisation for a larger number of suppliers and vehicles will only require adding the source data which will not affect the main part of the model.

Figure 1 shows the distribution network that we will use as an example.

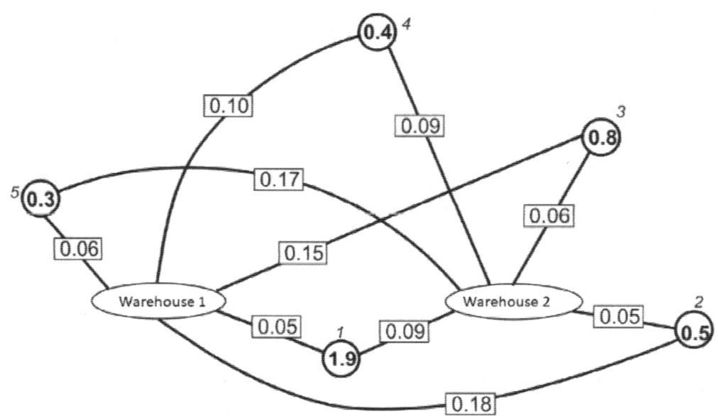

Figure 1. Two-supplier transportation scheme with five consumers.

Such a scheme provides a large number of options by changing the number of stocks in the 1st and 2nd warehouses, by closing various routes for any vehicles, by changing the consumers' needs, the distance between warehouses and consumers. Given the additional restrictions on the availability of certain routes, this scheme can be considered relevant to the goal of this study.

In this model, the objective function and constraint system are the same as in the aforementioned one. Now, it is necessary to enter another index to iterate over all the options with suppliers. In particular, the main operator with the objective function should now be written as:

MIN = @SUM(PATH(I,J,K)| Available(I,J,K)#NE#0:COST(I,J,K)).

Similar changes to the third index should be made for other operators. Naturally, all source data should contain full source information with regard to several warehouses/suppliers.

Although all the routes are available for all types of vehicles, unlike the model with one warehouse (fleet), the optimal distribution of goods along routes is not known in advance.

The software was developed to determine the optimal use of vehicles.

With the developed model, it is possible to vary the different availability of routes for different types of vehicles:

- to close some routes for all types of vehicles (because there is no such road or it is under repair),
- to close some routes for multi-ton trucks (e.g. because there are bridges in the countryside that do not support the weight of the truck or the dimensions of the bridge do not match the track of the truck),
- to close some routes for small-capacity freight transport with a trailer due to their low manoeuvrability in certain parts of the route.

As an example, we consider the following scheme (Figure 2).

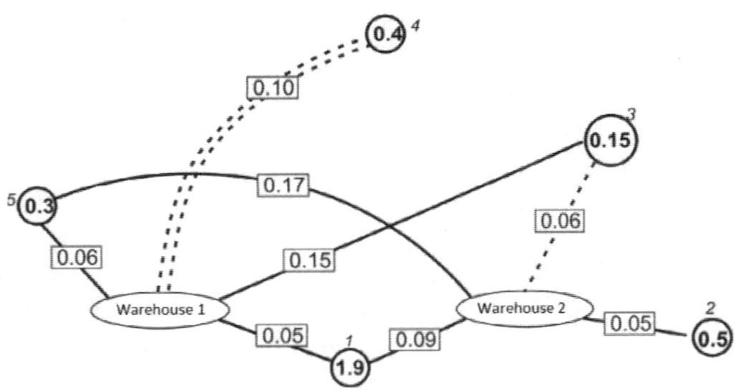

Figure 2. Scheme of routes with two warehouses (some roads are missing, two routes are partially accessible).

Compared to the previous scheme (Figure 2), in which all suppliers were available for all modes of transport from both depots, now

- there is no route from the *first* warehouse to the second consumer (the *second* row in the *Available* matrix below),
- there is no route from the *second* warehouse to the *fourth* consumer (the last row in the *Available* matrix below),
- the route from the *second* warehouse to the *third* consumer is not available for multi-ton trucks (ninth line),
- the route from the *first* warehouse to the *fourth* consumer is not available for freight transport with a trailer (fourth line).

In the model, it is reflected in the data matrix *Available*:

Available = (1, 1, 1, 1
0, 0, 0, 0,
1, 1, 1, 1,
1, 1, 1, 0,
1, 1, 1, 1,

1, 1, 1, 1,
1, 1, 1, 1,
0, 1, 1, 1,
1, 1, 1, 1,
0, 0, 0, 0);

It is easy to make other changes to the initial data of the model and promptly determine the optimal distribution of vehicles for various situations.

5. Conclusion

IOP Conf. Series: Materials Science and Engineering **828** (2020) 012031 doi:10.1088/1757-899X/828/1/012031

1. We have tried to justify the prospects of using vehicles of various carrying capacities for the carriage of goods under various conditions:

– inside the city in terms of heavy traffic in the main highways, difficulties when driving through domestic territories, the need to maneuver in a small area;

– in rural areas;

– in-plant territories.

2. We have studied some variants of transportation tasks with the use of different types of vehicles of various carrying capacity.

3. The developed software helps quickly assess the fleet of transport vehicles depending on the customers' needs and road traffic for various types of vehicles.

References

[1] *The use of small vehicles for the organization of cargo transportation within the city*. URL: http://gengruz.com/transportno-logisticheskie-uslugi/2385-ispolzovanie-malotonnazhnogo-transporta-dlya-organizacii-gruzoperevozok-v-cherte-goroda.html

[2] Volchkov V M, Godenko A E, Kalinin Y V and Tarasova I A 2018 *Optimisation in engineering and management tasks* (Volgograd: VSTU) p 136

[3] Mirotin L B, Balalaev A S, Gudkov V A, Zyryanov V V, Kapustin S N and Lyapina T V 2014 *Transport logistics: students book. Additional educational methodical association of universities for education in the field of transport vehicles and transport and technological complexes,* ed K L.B.Mirotin (Moscow: Hotline– Telekom) p 301

[4] Mirotin L B, Gudkov V A, Zyryanov V V, Nekrasov A G, Balalaev A S, Larin O N, Pokrovsky A K, Belyaev V M, Velmojin A V, Mashevets N U, Gronin D P, Shiryaev S A, Keridy P G, Melanich E V, Voronejtsev S U and Kuznetsov S A 2010 *Cargo management in transport and logistics systems: a monograph,* ed K Mirotin L B (Moscow: Hotline– Telekom) p 702

[5] Toth P and Vigo D 2002 *The Vehicle Routing Problem* (Philadelphia: SIAM Society for Industrial and Applied Mathematics) p 367

[6] *Schrage Linus. Optimization Modeling with LINGO* 2015 (Chicago: LINDO System Inc) 620 p

[7] Díaz-Parra O, Ruiz-Vanoye J Λ, Bernábe L B, Fuentes-Penna Λ and Barrera-Cámara R Λ 2014 *Journal of Applied Mathematics* 848129

[8] Ghani N E A, Shariff S S and Zahari S M 2016 *Advances in Pure Mathematics* **06(05)** 657 71

[9] Bandy B 1989 *Linear programming basics:* (Moscow: Radio and communication)

[10] Taha H 2005 *Introduction to Operations Research* (Moscow: Publishing House «Williams»)

[11] Vasilyev I L and Klimentova K B 2010 *J. Appl. Industr. Math.* **4:3** 441–4

[12] Gadjinsky A M 2007 *Logistic*: *Studensbook* (Moscow: Publishing and Trading Corporation «Dashkov and Co») 472 p

[13] El-Sherbeny N A 2010 *Journal of King Saud University* **3** 123-31

[14] Low A M and Kelton V D 2004 *Simulation Classic CS* (St Petersburg: Peter; Kiev: Publishing Group BHV)

[15] Toluev U I 2008 *Logistics and Supply Chain Management* **2 (25)**

IOP Conf. Series: Materials Science and Engineering **828** (2020) 012032 doi:10.1088/1757-899X/828/1/012032

Effect of mineral and organic fertilizers on potassium leaching in sandy loam soils

L V Yakovleva [1], D A Danilov [1,2], E A Nikolaeva [1]

[1] Leningrad research institute for agricultural science «Belogorka», 1, Institutskiy st., Belogorka, 188338, Russia

[2] Saint-Petersburg State Forest Technical University named after S.M. Kirov, 5 U, Institutskiy per., Saint-Petersburg, 194021, Russia

E-mail: livlaya@mail.ru

Abstract. In a model experiment, a study was carried out on potassium leaching from the soil. Soil samples were taken from a protective strip of long-term experiment (established in 1981). Various fertilizers and their mixtures were introduced. The effect of organic matter on the capacity formation of the cationic soil exchange was considered. Significant differences were noted in the potassium migration during various fertilizers application. A significant role is played by liming of acidic soils. Potassium ions displace calcium ions more easily calcium ions from the soil absorption complex than aluminum. Therefore, when liming, leaching of potassium may be reduced. With the introduction of peat and peat-based fertilizers, the cation exchange capacity of the soil increased significantly. The leaching of potassium did not increase. Introduction of cattle manure without heat treatment significantly increased the migration of potassium in the soil and its leaching to the lower lying horizons. In general, the migration and leaching of substances in arable soil rely upon a number of factors and their combinations: natural and climatic factors, physical and chemical properties of the soil, the intensity of agricultural use of arable land, the type and composition of fertilizers, and the mineralogical composition of the soil. When growing plants on calcareous, fertilized with potash fertilizers soils, the loss of potassium does not exceed its leaching from non-fertilized soils.

1. Introduction

The role of potassium and minerals including K for the soil complex is enormous. Of the many nutrients for plant species, soil potassium and minerals including K are important. In his experiments, J. R. Glauker (17th century) received the greatest increase in plant growth from the addition of nitrate (KNO_3) to the soil, and potassium was found to be useful for plant growth. Since the works of von Liebig published in 1840, the importance of K for plant growth has been confirmed [1, 2].

The total K content in soils ranges from 3,000 to 100,000 kg ha^{-1} in the upper 0.2 m soil profile. Of this total content, 98 % of potassium is bound in mineral form, and only 2 % is in soil solution and metabolic phases [3].

Plants need in potassium is second only to the need for nitrogen. An increase in nitrogen in the soil can be achieved by including legumes into organic crop rotations, while potassium needs to be introduced into the fields. A lack of potassium in legumes can reduce nitrogen accumulation. Due to the low cation exchange capacity on coarse-grained sandy soils, there is a potentially high risk of losses in potassium leaching, which, according to various estimates, ranges from 20-50 kg ha^{-1} per year [4].

Potassium, along with nitrogen and phosphorus, is an important macronutrient for plants and plays an important role in the development of plants. This element has a wide range of functions in plant nutrition, through the cell membranes it provides gradients of electric potential, maintaining turgor and activating enzymes. Potassium is necessary for plants for photosynthesis, protein synthesis and regulation of stomata and is the main cation in maintaining anion-cation balances [1]. Methods of cultivation, types of crops, type of soil-hydrological and climatic conditions affect the provision of potassium to plants. Compared to nitrogen, in many developing countries, the use of potash fertilizers was ignored, which led to the constant deficiency of the soil this element [5, 6].

Rate and timing of fertilizer and organic fertilizers are often based on the optimum nitrogen, not potassium requirements of plants. This often leads to potassium deficiency or excess depending on the size of the harvest and crop rotation. In the agricultural crop rotations with a predominance of perennial herbs, including many organic farming systems, the potassium was an important element, especially in areas dominated by sod-podzolic or peaty soils. One needs to consider managing potassium regime in relation to the long-term sustainability of soil as resource, and the production of crops of high yield and quality. Some regions, where clay-rich soils dominate, have the potential to provide a sufficient amount of K, while other areas with poor sandy or peaty soils have a very low potential for weathering and, thus, will require the introduction of potassium [7].

In Lithuania, the study on the comparison of several systems of crop rotations to reduce leaching losses of potassium from soils was performed [8].

In several studies it was noted that the total losses in a clay soil were more than twice the losses in sandy soil (13 and 6 kg ha^{-1}, respectively) due to the development of preferential flow in clay soil [9].

A serious problem in areas of intensive agricultural production is the use of nitrogen with consequent large nitrogen losses and environmental pollution. A serious problem in areas of intensive agricultural production is the use of nitrogen with its subsequent large losses and pollution. Nitrogen-potassium interaction is currently of interest in many studies, however, focuses on plant nutrition with potassium at different doses of nitrogen. Fixation and release of potassium from the soil, as well as absorption, transport, and reuse in crops, are affected by the form and rate and time of nitrogen use. High-yielding crops can be obtained in optimal nutrient ratios N: K. High rates of N and K application do not necessarily lead to increasing in increments of productivity and can even reduce the yield. A better understanding of the mechanism of nitrogen - potassium interaction may be a useful guide to better management of nutrients in agricultural practices to achieve high yields with high efficiency of nutrient use [10]. Years of experience in agriculture shows that it is impossible to achieve high yields in sod-podzolic soils of light granulometric composition without regular use of mineral fertilizers, including potash, [11,12,13]. However, this significantly increases the migration of bases in soil. Leaching of potassium in arable soils of light granulometric composition can reach 20 kg ha^{-1} according to our linemetrics studies [14,15]. On acidic soils in the conditions of washing water regime it is very important to preserve the bases in the soil. Liming of acidic soils plays a special role in potassium migration. In nature, everything is interconnected, so leaching is the result of the interaction of a number of factors. We have tried to simulate the impact on the soil of various fertilizers and their combinations in the laboratory, taking into account the average annual rainfall for the research region.

2. Methods & objects

The object of study is acid sod-podzolic sandy soil. Agrochemical research of soils was carried out according to standard procedures [16, 17, 18]. The soil was taken from the protective strip of a long field experiment. The experiment examines the effect of seven doses of lime at four levels of application of mineral fertilizers. The model experiment was carried out in lysimetric columns. The scheme of the experiment included the use of fertilizers on a limed and non-limed background [19, 20]. We tested the effect of mineral and organic fertilizers. The experiment was carried out without plants. The soil was washed with distilled water. Five washings have been done. In total, the average annual rainfall for the conditions of the Leningrad region was passed through the soil. Potassium in the lavage was determined by flame photometric method [21]; organic matter by phenylantranilic acid titration after wet combustion in a mixture of sulfuric acid (H_2SO_4) and potassium dichromate ($K_2Cr_2O_7$) according to the method of Tyurin. The cation exchange capacity was determined by the Bobko-Askinazi-Aleshin in the modification of the TSINAO. The soil was saturated with a buffer solution of barium chloride ($BaCl_2$), washed with sulfuric acid (H_2SO_4) and titrated with sodium hydroxide (NaOH).

3. Results & Discussion

Migration and leaching of substances in arable soils depend on a number of factors: climatic, physico-chemical properties of soil, intensity of agricultural use of arable land, the mineralogical composition of the soil. One of the most important properties of soil is the cation exchange capacity. It is also the result of the interaction of a number of factors. An important role in shaping the capacity of cationic exchange plays the contents in soil organic matter.

Table 1 shows a rather strong dependence of capacity of cation exchange from the organic matter in the soil when applying various fertilizers and their combinations. Rank analysis and Spearman correlation coefficient at the 95% significance level was used to determine the relationships. The results showed a wide range of fluctuations in the potassium concentration in the wash water.

Table 1. The value of the capacity of cation exchange (ECO) of the soil at different content of organic matter

Options	Organic matter content, % (X)	ECO, mg-eq. per 100 g of soil (Y)
Control	2.46	4.88
NPK	2.19	12.59
Lime	1.92	15.23
Peat	5.17	19.65
Manure	2.86	15.93
NPK+ lime	2.23	18.38
NPK + peat	5.23	18.98
NPK+ peat + lime	4.99	22.10
NPK + manure	2.28	16.59
NPK + manure + lime	2.76	18.40
The accuracy of the experiment	6.8 %	1.23%
NSR05	0.65	0.62
Correlation coefficient (Rs)	0.745	

There were no clear relationships between the doses of lime and mineral fertilizers with the concentration of potassium in the washing waters. The effect of liming on potassium leaching was twofold: in some cases, it did not have a significant impact on potassium migration, in others it reduced leaching. This may be due to the following reasons:

1. liming leads to the replacement of exchange aluminium in the soil absorbing complex with calcium. Potassium in dilute solution displaces calcium from the exchange complex of the soil absorption easier than exchange aluminium. This keeps the potassium from leaching out.
2. liming increases the cation exchange absorption capacity and helps to keep potassium in the exchange form from leaching. Previous lysimetric experiments have shown that an increase in the dose of potash fertilizers by 10 kg per 1 ha leads on average to an increase in potassium leaching on light soils by 0.52 kg ha per year. The use of peat did not increase the leaching of potassium in comparison with mineral fertilizers. When introducing cattle manure (KRS) without heat treatment, potassium leaching increased.

IOP Conf. Series: Materials Science and Engineering **828** (2020) 012032 doi:10.1088/1757-899X/828/1/012032

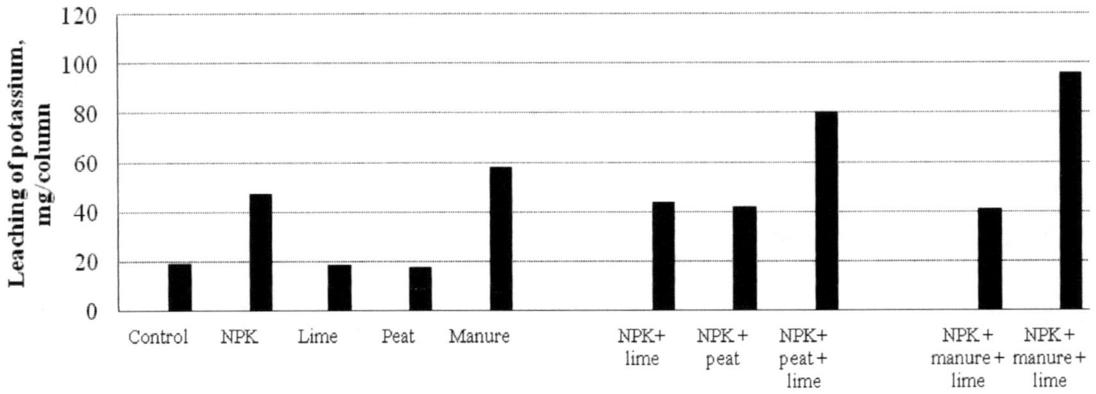

Figure 1. Effect of fertilizers on potassium leaching

It is important what the value of the absorption capacity is related to. In figure 2 it is possible to allocate two zones: The 1st, upper zone, is associated with the presence of organic components of manure in the soil. The lower zone characterizes the leaching of potassium when applied to the soil minerals and peat. The introduction of manure stimulates increasing migration of bases, including potassium. This is due, in our opinion, to the presence of organic substances in the dissolved form and the alkaline reaction of the medium due to the content of ammonia salts and ammonium hydroxide in the manure.

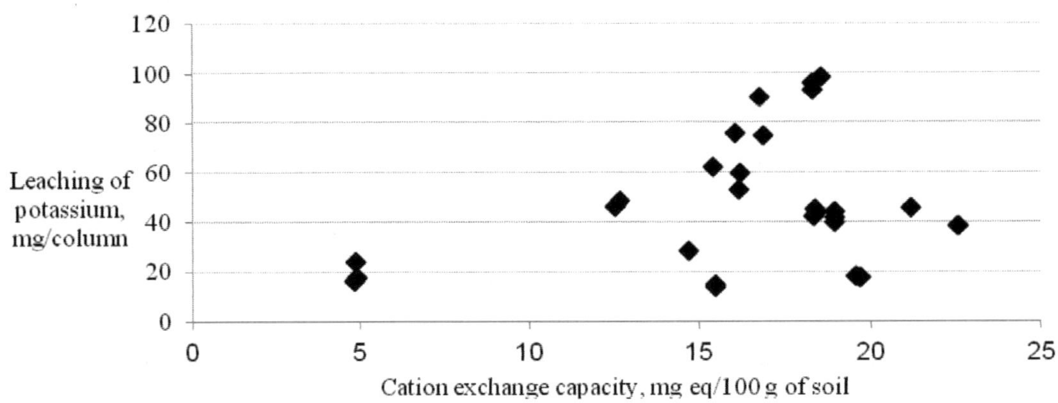

Figure 2. The leaching of potassium, depending on the ECO

The results of the study of the migration dynamics of substances in the soil showed that when cattle manure is applied to the produced, fertilized soil with mineral fertilizers, the migration ability of the bases, including potassium, increases approximately twice as much as the option of applying mineral fertilizers and lime with peat (Figure 3). The cation exchange capacity of the soil and the organic matter content were higher on the options of peat use than on the options of manure introduction.

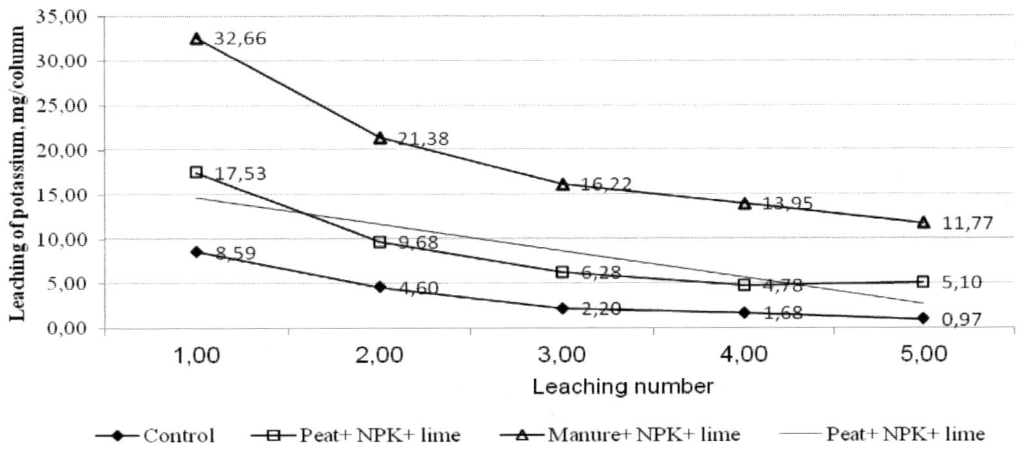

Figure 3. The dynamics of potassium migration during the application of various fertilizers

4. Conclusion

When characterizing the potassium regime of sod-podzolic soils, it is necessary to take into account the granulometric and mineralogical composition of soils, the level of liming and the availability of soil with potassium, as well as the composition of cations in the soil. An increase or decrease in fixation or migration of potassium in the soil is not a general pattern, but depends on a combination of many factors. The leaching of potassium largely depends on the granulometric composition of the soil, the composition of the cations of the soil absorbing complex, and the presence of mobile forms of organic substances. The introduction of potassium-containing fertilizers increases the migratory capacity of potassium. When growing plants on calcareous, fertilized with potash fertilizers soils, the loss of potassium does not exceed its leaching from non-fertilized soils.

Acknowledgment

The work was performed as part of the State Assignment of the Ministry of Higher Education and Science of the Russian Federation on the topic: "To study the patterns of changes in the basic properties of sod-podzolic soils of the North-West of the Russian Federation in agrolandscapes".

References

[1] Cakmak I and Engels C 1999 *Mineral Nutrition of Crops—Fundamental Mechanisms and Implications: Role of Minerals in Photosynthesis and Yield Formation* ed Rengel Z (New York: The Haworth Press) pp 141–68
[2] Römheld V and Kirkby E A 2010 *Plant Soil* **335** 55-80
[3] Rahmatullah C and Mengel K 2000 *Geoderma* **96** 291-305
[4] Askegaard M and Eriksen J 2008 *Agriculture, Ecosystems and Environment* **123(1–3)** 99–108
[5] Regmi A P, Ladha J K, Pathak H, Pasuquin E, Bueno C, Dawe D, Hobbs P R, Joshy D, Maskey S L and Pandey S P 2002 *Soil Science Society of America Journal* **66** 857-67
[6] Bil L, Xia J, Liu K, Li D and Yu X 2014 *Plant Soil Environ* **12(60)** 537–43
[7] Öborn I, Andrist-Rangel Y, Askegaard M, Grant C A, Watson C A and Edwards A C 2008 *Soil Use and Management* **21(1)** 102–12
[8] Bučienė A, Antanaitis Š, Šlepetienė A, Šimanskaitė D, Mašauskas V and Bučas M 2014 *Zemdirbyste-Agriculture* **1(101)** 3–10 DOI: 10.13080/z-a.2014.101.001
[9] Alfaro M A, Jarvis S C and Gregory P J 2006 *Soil Use and Management* **20(2)** 182–9
[10] Kozak M, Stêpieñ M and Joarder A H 2005 *Polish Journal of Soil Science* XXXVIIII **2** 180-6
[11] Nikitina L V and Volodarskaya I V 2007 *Agrochemistry* **2** 14-8

[12] Zhang F, Niu J, Zhang W, Chen X, Li Ch, Yuan L and Xie J 2010 *Plant and Soil* **335(1)** 21-34
[13] Cakmak I 2005 *Plant .Nutr. Soil Sci.* **168** 521–30
[14] Yakovleva L V 2018 Polyakov V A and Zhdanov S S 2018 *Vladimir Agricolist* **4** 14-20
[15] Yakovleva L V and Nikolaeva E A 2018 *Fertility* **5(104)** 31-5
[16] Bankin M P, Banchina T A and Korobeynikova L P 2005 *Physical-Chemical Methods in Agricultural Chemistry, and Biology of Soils* (St. Petersburg: Publishing house of St. Petersburg State University) p 177
[17] Karpachevsky L O 2010 *Ecological Soil Science* (M.: Kolos) p 687
[18] *Chemical Composition of Soils* URL: http://www.bygeo.ru/materialy/pervyi_kurs/pochvovedi-zem-res-u-chtenie/1767-104 himicheskiy-sostav-pochvmakroelementy-v-pochvah.html (date of access 10.11.2015)
[19] Kilmer V J, Younts S E and Brady N C 1972 *The Role of Potassium in Agriculture* p 503
[20] Mengel K 2016 *Potassium In: Handbook of Plant Nutrition* (CRC Press) pp 107-36
[21] Wallace A 2008 *Communications in Soil Science and Plant Analysis* **25(1-2)** 87-92

IOP Conf. Series: Materials Science and Engineering **828** (2020) 012033 doi:10.1088/1757-899X/828/1/012033

Research and Engineering at Academic Institutions as LPP Factor: Russia Case Study

N V Yudina[1] and O A Seliverstova[2]

[1] Financial University under the Government of the Russian Federation, Moscow 125993, Russia
[2]Vladimir State University, Vladimir 600000, Russia

E-mail: dr.yudina@mail.ru

Abstract. Language policy and planning (LPP) is an increasingly broad field today dealing with a number of complicated issues of utmost importance for the sustainable development of society in general and local communities in particular. In this view, it is necessary to create an efficient network and structure of actors all working towards the same goals. The paper seeks to explore the role of academic institutions as LPP actors with a focus on international cooperation in research and engineering as a major factor that impacts academic LPP at micro-level and shapes foreign language acquisition agenda at the macro-level. The paper presents a case study focused on Russian LPP and considers its current concerns and solutions.

1. Introduction
Language policy and language planning are the two concepts that have existed in sociolinguistics for nearly a century so far. The scope of issues LPP has to deal with has dramatically broadened, making the structure of its branches ever more complicated. With increased diversity of LPP topics, the number of actors has risen too. That is why questions of agency and interaction between the actors need to be considered. Among multiple factors that have an impact on LPP, research and engineering internationalization stands out as the one setting LPP agenda in terms of foreign language acquisition at all levels, from the local to the international one. Russia has always attached great importance to LPP and is currently considering the feasibility of developing a law to set the guidelines of the national language policy and planning. That is why an overview of the current LPP agenda and investigating the Russian LPP model may contribute to exploring the topic of LPP agency.

2. Literature Review
The original idea of language planning as "activity of preparing a normative orthography, grammar and dictionary for the guidance of writers and speakers in non-homogeneous speech communities" is attributed to Einar Haugen, introduced in 1959 [1]. However the idea of language policy can be found in even earlier works of E. Polivanov, a well-known Russian linguist, who claimed, back in 1931, that a true linguist was supposed to be a specialist in 1) a constructing contemporary language cultures; 2) language policy (capable of forecasting the future development of the language); 3) general linguistics, being aware of the previous language development and 4) particular languages with their history including the national and ethnic culture [2]. The two concepts – language policy and language planning - in various interpretations have been extensively used by sociolinguists. The two being

IOP Conf. Series: Materials Science and Engineering **828** (2020) 012033 doi:10.1088/1757-899X/828/1/012033

inseparably connected have come to be blended in a unified notion of language policy and planning [3].

In order to specify the extremely broad array of measures covered, a number of constituent parts were introduced. The traditional framework of LPP includes corpus planning, status planning (Kloss) and acquisition planning (Cooper). These issues are generally treated as internal policy issues. Today however language spread has come to be treated as a separate part or type of LPP which through setting objectives relevant to the government's goals can directly or indirectly influence status and acquisition planning.

In 1997 Ammon defined language spread policy as "a policy which has as its objective the spread of a specific language, or sometimes a set of languages, either to new speakers or to new domains" [4]. Thus, depending on the target domain, he distinguished between internal and external language spread policy types, the former being focused on language spread within a country, while the latter aims to expand the language into new domains, outside a country. The external language spread is a very special type of LPP, since it has to deal with issues in "societies controlled by other governments" (ibid.) which means a very limited array of tools [5]. Hence the governments have to develop and implement their policies via intermediaries in the target polities. Moreover, the success of this LPP depends to a considerable extent on the LPP reception rather than production [6]. This receptive function represents what Haarman calls prestige planning which in turn is affected by the nation's soft power.

Given the existing diversity of LPP types, comprehensive efforts towards the goals set by governments may be fruitful only if they develop an effective network of actors. The issue of agency in LPP has been a major concern for specialists in modern LPP theory seeking to explore "interaction between macro-level LPP decisions and the range of agentive responses and actions at the micro-level" [7]. "Agency in LPP has traditionally been depicted in terms of localized implementers of macro-level policies" (ibid) without any regard to local actors' role in setting LPP agenda and introducing bottom-up initiatives arising out of local language-related needs and concerns. Thus universities and other academic institutions facing similar challenges given the current demands of research and engineering have become powerful actors of LPP. In order to understand the nature of this power in various contexts, multiple research has been carried out across the world with results being presented in "Language Policy and Planning in Universities: Teaching, research and administration" edited by Anthony J. Liddicoat.

Besides the view limiting the role of LPP actors simple implementation of macro-level policies lacks attention to certain issues relevant for understanding LPP. Liddicoat argues that a direct causational link relationship "between decisions made by those who are with power to execute them and the actual results of language planning" appears to be weak since it does not admit of the impact of contextual factors that may bring about unplanned results [8]. Besides this approach "oversimplifies the nature of power as it applies in speech communities", as the coercive and normative power of institutions may appear too straightforward, while changes in the behavior towards language use may be obtained through "subtle operations on the choices of others" (ibid). This statement again brings us close to Nye's idea of soft power, defined as "the ability to attract and co-opt, rather than coerce" or "the ability to shape the preferences of others through appeal and attraction" (Nye). Although soft power is traditionally seen as a tool of international policy, this approach might as well be applicable to lower-level actors. Thus it seems feasible to focus on local contexts and micro-level LPP agents.

Let us look at the bottom-up structure of LPP actors. At the very basic level are individual linguists or groups of linguists and related specialists who undertake to reserve or revive and promote languages. They are either "enthusiasts motivated by a range of different concerns" [9] working on their own or supported by local community groups engaged in promoting the language through informal education. Another type of LPP actors are institutions, involved in language, literature, culture or other related fields. Bigger scale agency is performed by goverrnment or non-government bodies which have a direct or indirect impact on decision making.

The interaction between actors of different levels may be bidirectional. The top-down policy accepted positively at the local level may be successfully implemented. However, in case these decisions fail to address urgent problems or infringe on local communities' or minorities' language rights, the implementation of these policies is likely to be hampered at the micro-level. This may produce reverse bottom-up initiatives with regard to local contexts.

In this view, the actors implementing LPP at the local level have a special role of being mediators and undertake to maintain the balance between decision-makers and communities. Academic institutions appear to face these special tasks being an important link in LPP agency.

One of the recent challenges academic institutions have to face is the need for comprehensive cooperation in the field of research and engineering. Knowledge sharing and exchange of ideas at the international level have become an essential part of academic life which demands high proficiency in English being lingua franca of science and engineering. This brings forth issues of foreign language acquisition to become part of LPP at all levels.

3. Methods

In order to explore the LPP agency as exemplified by academic institutions as well as research and engineering cooperation as a major factor of LPP, we need to describe the framework of guidelines for our study.

First, we will dwell on the basics of the Russian system of higher education which is essential to understand the nature of the relationship between LPP actors of different levels. Second, we will analyze the contribution of various academic institutions to LPP with regard to all the above-mentioned types, namely:

- corpus planning;
- acquisition planning;
- status planning;
- prestige planning;
- language spread policy (both internal and external).

When doing so, we will point out the challenges Russian universities face today in the context of increased demand for cooperation and exchange in the field of research and engineering. In conclusion, we will come up with a characteristic of the Russian LPP model with regard to the role of academic institutions and offer micro-level LPP measures to address the concerns of internationalization arising out of demand relevant for research and engineering.

4. Results and discussion

The system of education in Russia is based on government-funded institutions that have to meet a number of requirements to the conditions of learning, their curricula and education outcomes. Likewise, the work of the Academy of Sciences is funded by the state, making the implementation of LPP a top-down process. Bottom-up initiatives may be supported through grants and scholarships. Besides representatives of major academic institutions are involved in the work of the recently formed Council of the Russian Language which brings together linguists, academics in closely related fields and statesmen to develop recommendations and advice on LPP issues.

The major part of work in the field of *corpus planning* is done by Vinogradov Institute of the Russian Language which is a unit of the Russian Academy of Sciences. Researchers from the Institute launched and have been maintaining the National Corpus of the Russian Language that includes besides the core corpus, that of newspapers, poetry, dialects, multimedia, and parallel corpora, as well as the old Russian corpus and that of ancient Russian manuscripts. The National Corpus has become the basis for a number of dictionaries and studies in descriptive linguistics: Dictionary of Idioms, Grammar Dictionary of New words, Frequency Dictionary of the Russian Word Stock, etc. The Institute also undertakes corpus planning by operating in the field of prescriptive lexicography via introducing and setting language norms and compiling dictionaries and grammar reference books. Another part of its activities is linked with the dissemination of linguistic knowledge, making it

IOP Conf. Series: Materials Science and Engineering **828** (2020) 012033 doi:10.1088/1757-899X/828/1/012033

interesting and appealing, especially for young people. This is done on a regular basis via a series of language and linguistic nationwide contests, clubs, linguistic student societies and summer camps for school children thus contributing to *prestige planning*.

Another big institution that contributes to *corpus planning* is the Institute of Linguistics of the Russian Academy of Sciences. They are mostly engaged in research on languages other than Russian within the country including minority and endangered languages. The main research domains listed on the official web site, are the following: 1) language theory, linguistic typology, comparative linguistics; 2) Romance, Germanic, Celtic, African, Turkic, Mongolian, Iranian, Caucasian, Finno-Ugric and Asian languages; 3) grammars of endangered languages; 4) areal linguistics and ethnolinguistics; 5) sociolinguistics; 6) psycholinguistics; 7) computational linguistics; 8) lexicography; 9) logical analysis of language; 10) experimental phonetics and semantics; 11) translation theory (https://iling-ran.ru/web/en/about). A great deal of the Institute's work involves standardization, description and revitalization of languages. By maintaining and developing minority and endangered languages, the Institute has been a major actor of *status planning* as well as *prestige planning* balancing between the policies aimed at the spread of Russian and preserving linguistic diversity and, through it, the cultural heritage of the country.

Issues related to the Russian language *spread policy* have been high on the agenda in the recent decade. While internal spread LPP deals with problems of status and acquisition of Russian in the republics, where their national languages are recognized as official republics' languages alongside Russian as a national language, the problems of the level of Russian proficiency among immigrants from Central Asia and former USSR republics is also an area of great concern. Another sensitive issue of internal language spread policy is the perceived deterioration of the Russian language in use, involving neglecting language norms even in public domains which are supposed to be examples of literary Russian. To address these concerns, at the recent meeting with the President, held on November 5, 2019, the Council on the Russian Language suggested adopting a comprehensive law on language policy and planning in Russia, which is expected to set the guidelines for subsequent joint work of all actors involved. The areas to be covered besides internal policy issues include the problems and challenges of external LPP.

External Russian language spread policy became a focus of LPP last decade. Initiated first as a way to support compatriots living abroad, it has now come to be seen as an essential tool and asset of soft power which is capable to introduce the image of the country different from the one people abroad perceive from mass media. In order to provide a structured plan of activities in this field the government initiated Federal Special Purpose Program "The Russian Language" which presented targets and tasks for a four-year period of 2011-15 subsequently followed by a similar programme for 2016-2020, both undertaking to lay the basis of reference sources, books, software and on-line resources for teaching and learning Russian abroad.

A major actor here is the Pushkin State Russian Language Institute which "specializes in teaching Russian to international students and in training teachers to teach Russian as a foreign language" (http://www.pushkin.institute/en/). Besides providing support and resources for teaching and learning Russian, the Pushkin Institute has introduced a number of initiatives for language spread. Among them is a volunteer program for students and young people "Russian Language Ambassadors" which travel within Russia and neighboring countries to introduce young people to the diversity of Russian multinational culture and the beauty of the Russian language thus contributing to both external and internal language spread policy and promoting acquisition.

Acquisition planning in Russia is directed by the government. For all levels of education from pre-school through secondary and high school up to higher education, the government develops standards that specify the target outcomes of education at each level in the form of skill sets to be obtained in the course of study mandatory for all education establishments including academic institutions. Thus the latter serve as intermediaries between decision-makers and communities.

On the other hand, universities have to face some specific challenges, the major being internationalization, which makes a big impact on all spheres of academic work including teaching, learning, and research. The trend perfectly described by Liddicoat appears very true for Russian universities. Although the shift to English as language of knowledge within the country is less noticeable than in Europe, there is an increasing trend of the national research evaluation processes "to rank publications as a way of measuring research quality and such rankings have tended to privilege research developed in English" [9].

The scope of issues researchers have to investigate alongside the increasing rate of innovations introduced throughout the world has made international cooperation and joint scientific and engineering projects an essential and inseparable part of academic life. In order to develop and produce cutting-edge technology, one has to follow international trends in the field and be aware of the newest discoveries made. It requires a high English proficiency level on the part of universities' staff. Participation in Erasmus+ projects emphasizes the need for a transition to lingua franca in the academic environment. Thus academics have to be immersed in the universe of shared English language knowledge and be able to introduce their ideas or share research results by communicating in the international academic environment. Likewise, students who get enrolled, especially those in Master programs, should be well prepared to work with English language information and resources. These concerns make *foreign language acquisition* an issue of Russian LLP at the macro-level.

According to the EF English Proficiency Index 2011, Russia featured low English proficiency level ranking 32 with the EF EPI score 45.79. This could in part be accounted for by a relatively late start of English acquisition, beginning from the 5th grade of secondary school (10-11 years old). In 2011 foreign language was introduced in the primary school curriculum in Russia. Other LPP steps related to foreign language acquisition are introducing a second foreign language in the secondary school curriculum and a nation-wide compulsory examination in foreign languages which is due in 2022. The general focus on foreign languages in LPP has to a large extent been caused by the demand for internationalization of research, engineering, and business. The results are yet to be seen when 2011 schoolchildren will enter the student body and the labour market. However today we can observe Russia gradually improving foreign language proficiency with the reported EF EPI average score being 52.96 as of 2018, although being still behind a number of countries ranking 42nd in the list (https://www.ef.com/wwen/epi/).

With the current level of English proficiency, it appears feasible for academic institutions to develop micro-level LPPs tailored to their needs. Academic LPP may be defined as a set of guidelines for teaching and learning languages related to the universities' objectives in the context of internationalization in line with the state LPP demands. The major concern academic LPP should address is significantly varying levels of English proficiency on the part of both students and staff. Hence the model of academic LPP should be bidirectional and aimed at academics and students respectively. The steps involved may include English language support provided by the departments of foreign languages available at all universities, introducing English courses for academics with low proficiency level, increasing opportunities for academic mobility and exchanges in research and engineering as well as learning and teaching, strengthening cooperation with partner universities and companies abroad and extending the network of partners [10]. All these steps will inevitably lead to the need of introducing courses in English in the Russian universities' curriculum.

Another priority in the context of research and engineering internationalization is promoting Russian education abroad to welcome foreign students. Until recently the prevailing approach was to offer a preparatory one-year Russian language course followed by leaning with instruction in Russian [11]. This traditional approach, although serves the aim of language spread, hampers the process of internalization, which Today some universities can offer courses and degree programs (mostly Master programs) in English while others are still making their first steps towards internationalization.

5. Conclusion

To sum up the above we can conclude that Russia features a direct overt top-down model of LPP. Being government-funded, academic institutions mostly implement the language policy and planning initiated by the government. The opportunities for bottom-up initiatives are few but existent. One of them is to act through representatives in the Council of the Russian language Russia. To make this option efficient there should be constant two-way communication between the Academy of Sciences and local universities. The latter being well aware of local contexts can influence decision making by bringing forth problems relevant for the communities, thus acting as intermediaries between the government and the society.

One of the currently pressing concerns in Russian LPP arising out of research and engineering internationalization has brought forth the issue of foreign language acquisition in the nation-wide system education. Although important decisions for encouraging learning English have been made, the results are to be seen in the coming decade. Meanwhile, universities have to develop academic LPP tailored to their specific needs and priorities.

Academic institutions as representatives of local as well as research and engineering communities may be said to have a special place in the structure of LPP agency, being both intermediaries and implementers of top-down policies and agenda-setting force putting forward LPP concerns that require major steps at all LPP levels.

Reference

[1] Haugen E 1959 *Anthropological Linguistics* **1(3)** 8-21
[2] Polivanov E D 1931 *On Marxist Linguistics* (Moscow Federatsiia) p 182
[3] Kloss H *Research Possibilities on Group Bilingualism: A Report" (*Quebec: International Center for Research on Bilingualism) p 94
[4] Cooper R L 1989 *Language Planning and Social Change* (Cambridge: Cambridge University) p 228
[5] Ammon U 1997 *Language problems and Language Planning* **21 (1)** 51-57.
[6] Kamwangamalu N M 14 September 2016 *Language Economics and Prestige Planning for Minority Languages in Postcolonial Settings* (Washington DC)
[7] Bouchard J and Glasgow G P 2019 *Agency in Language Pollicy and Planning: Critical Inquiries* (NY Routledge) p 332
[8] Liddicoat A and Baldauf R 2008 *Language Planning and Policy: Language Planning in Local Contexts: Agents, Contexts and Interactions* (Clevdon: Multilingual Matters) p 287
[9] Nye J S 2004 *Soft Power. The Means to Success in World Politics* (New York Public Affairs) p 208
[10] Liddicoat A J 2016 *Current Issues in Language Planning* **17** 231-41
[11] Shamne N L, Milovanova M V, and Malushko E Yu 2019 *IOP Conference Series: Materials Science and Engineering* **483(1)** 012081. DOI: 10.1088/1757-899X/483/1/012081

IOP Conf. Series: Materials Science and Engineering **828** (2020) 012034 doi:10.1088/1757-899X/828/1/012034

Assessment of the environmental responsibility of industrial enterprises in the region

E A Zaliznyak [1], A A Matveeva [1], A V Kholodenko [1] and S N Kirillov [2]

[1] Volgograd State University, prospect Universitetskyi, 100, Volgograd, 400062, Russia
[2] Lomonosov Moscow State University, Leninskie gory, 1, Moscow, 119991, Russia

E-mail: matveeva@volsu.ru

Abstract. The article discusses the problems of accessibility of information about the impact of enterprises on environmental components for stakeholders. Assessing the environmental component of non-financial reports allows to analyse the environmental activities of any company and draw appropriate conclusions about the effectiveness of the policies of enterprises. Russian companies, unlike foreign ones, have lower trends to disclose environmentally significant information, but there is a positive trend in placing such information in the public domain. The study analyzed the trend of non-financial reporting following the international recommendations of the Global Reporting Initiative (GRI). In the period June-November 2019, the number of Russian companies reports presented in the international GRI database increased by 10%, with a global dynamics of 1.7%. According to the National Register of the Russian Union of Industrialists and Entrepreneurs, 80% of non-financial reporting is the reports of sustainable development field, the most open are the enterprises of the oil and gas industry and the electric power industry. The study covered 107 negative impact facilities of the 1st category, high and significant risk category, located in the Volgograd region. To assess the openness of regional enterprises, we analyzed their official websites and databases of the rating of environmental business performance. Official sites identified for 72% of facilities demonstrate the predominance of environmentally significant declarative information, it is 87% of analyzed sites. A study of companies' readiness to disclose data on anthropogenic impact for stakeholders used copyright checks and showed the commitment of less than 10% of respondents. Regional companies did not support the initiative on the openness of environmentally significant information, which less than 2% of enterprises publish in the open access, and are not ready to prepare reports on sustainable development following international criteria. Since 2018, the exclusion of quantitative data on the negative impact on specific enterprises as a mandatory component of regional reports on the state of the environment has exacerbated the identified trend.

1. Introduction

To solve environmental problems, environmentally responsible companies use various tools in their practice, including social reporting with an environmental component. Developing within the framework of social responsibility, environmental responsibility should exceed the legislatively established requirements in the field of environmental protection, comply with standards and feed into the business plans of companies for future development. The publication of data on the impact of the company on the components of the environment, areas of environmental activity, allows us to assess

IOP Conf. Series: Materials Science and Engineering **828** (2020) 012034 doi:10.1088/1757-899X/828/1/012034

the organization's commitment to sustainable development goals, a modern trend in world development.

2. Research and discussion

The current level of corporate responsibility among the world's largest companies, G250, according to the Corporate Performance Review - Responsibility Report 2017, is more than 90%, which represents a new trend of sustainability accounting [1]. Companies and organizations of all types, sizes and sectors from around the world issue sustainability reports. Thousands of companies in all sectors have published reports that link to the GRI Sustainability Reporting Guidelines [2-4]. Government agencies and nonprofit organizations are also major reporters. The GRI Sustainability Disclosure database contains all the well-known GRI reports. As of November 1, 2019, the GRI database contains 33286 reports executed following the GRI Guidelines, 14269 various organizations, the share of Russian companies is 1.1% - 163 companies [5].

Over the 5 months of this year, the increase in the report number of Russian companies amounted to 10%, with an increase in reporting companies by 1%. Considering that the total number of reports in the GRI database for the period under review increased by 1.7%, the dynamics of the Russian companies' indicator, in our opinion, has a positive trend (Figure 1).

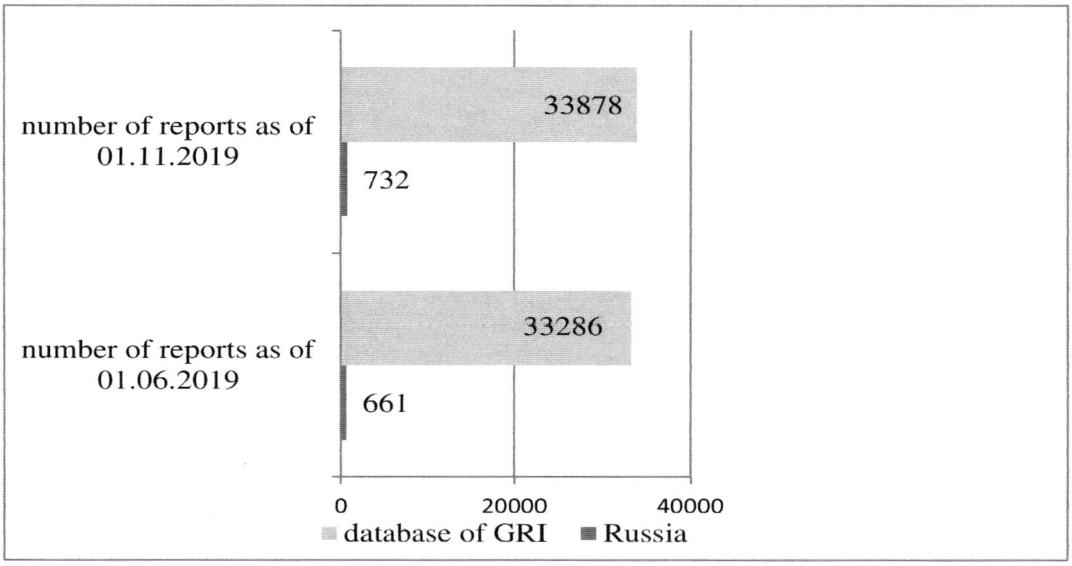

Figure 1. Dynamics of reports placement in GRI system [5].

As a tool for the environmental policy of environmentally responsible companies, non-financial reporting allows deeply analyzing the activities of each organization. Business entities that have the most developed system of environmental measures comply with not only legal acts and regulations but also openly demonstrate the results of both their impact on the natural environment and the measures taken to reduce it.

In Russia, the Russian Union of Industrialists and Entrepreneurs (RUIE) is engaged in the systematization and placement of non-financial reporting in the National Register. Currently, 71% of reports relate to sustainable development (35%) and social orientation (36%) and less than 10% are made by ecology reports [6]. Figure 2 presents the distribution of reports by industry affiliation.

In the framework of non-financial reporting, Russian companies prefer reports in the field of sustainable development, 79.5%, while environmental reports occupy only 20.5%. The distribution by industry is as follows: Oil and gas and energy companies are most interested in the openness of environmental information and provide reports on sustainable development. An interesting fact is a certain imbalance in the provision of non-financial reports, that is, companies are ready to provide either environmental reports or sustainable development reports.

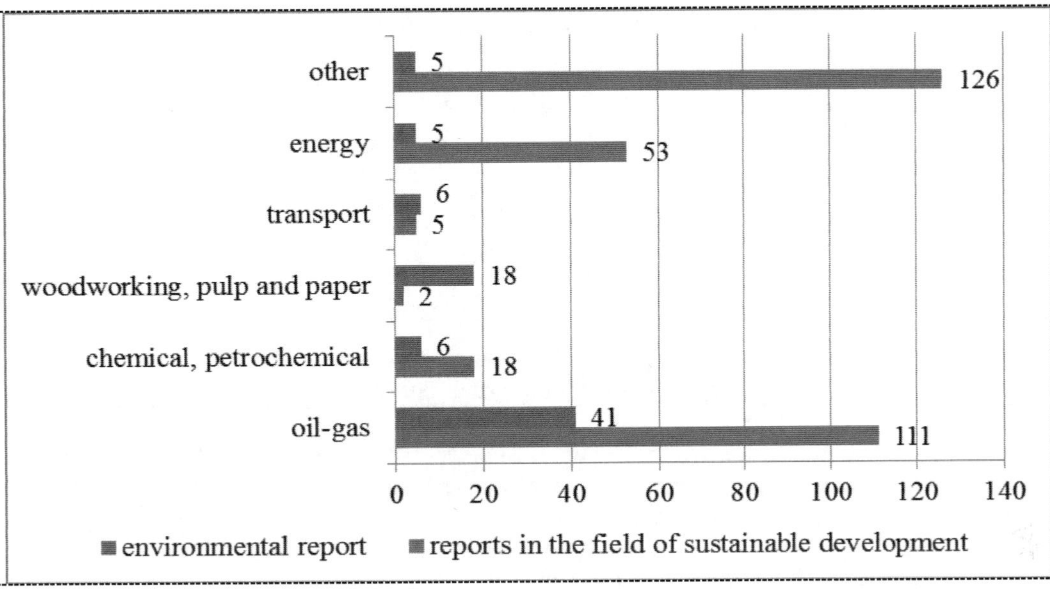

Figure 2. Distribution of reports by industry affiliation of companies [6].

Since 2015, Russia has introduced the categorization of anthropogenic objects that is the division into 4 categories depending on the level of impact [7]. The facilities' categorization is based on types of economic activity, technological production capacity and indicators of negative impact. The level of state supervision depends on the category of the enterprise, its location and other parameters: federal includes all Negative Environmental Impact category 1 facilities [8] and regional. The legislation establishes the classification of impact objects as the following risk categories: low, moderate, medium and significant. Depending on the territorial location, for example, within the boundaries of a specially protected natural territory of federal significance, or in case of legal violations in the field of environmental protection, in case of administrative offence or invalid licence on activities in the field of waste management, facilities are subject to classification to a higher risk category, for example, extremely high [9].

The objects of the study were enterprises of fist category located on the territory of the Volgograd region, of a high and significant risk category, having a significant negative impact on the environment and relating to the areas of application of the best available technologies. As of November 2019, according to the data of the public portal, Software and hardware for an accounting of facilities with a negative impact on the environment, more than 297 thousand objects have registered on the territory of the Russian Federation, of which 7400 belong to the 1st category. On the territory of the Volgograd Region, among the objects subject to federal environmental supervision, it is about 40% of all subjects of regional environmental protection agencies, 6.5% of enterprises belong to 1st category (Figure 3), of which more than 25% are classified as high risk [10].

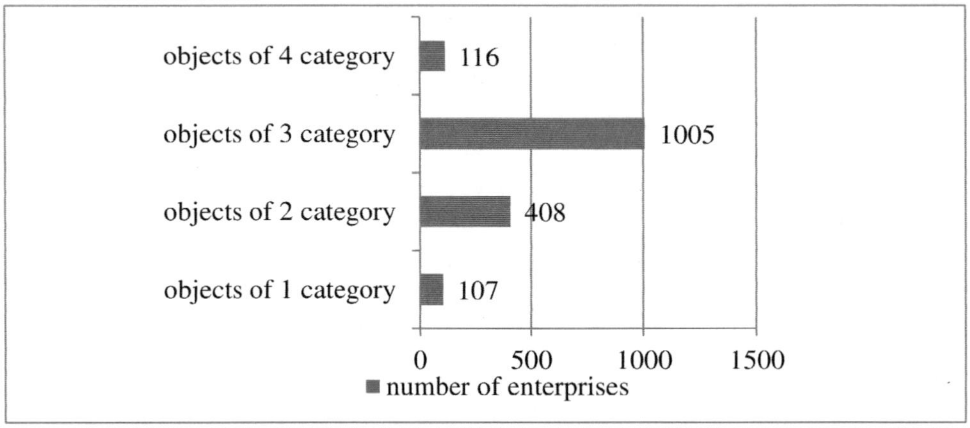

Figure 3. Structure of objects of negative influence of Federal level of supervision in the territory of the Volgograd region [10].

According to the territorial body of the Federal State Statistics Service for the Volgograd Region, the mass of pollutant emissions into the atmosphere in the Volgograd Region from stationary sources amounted to 138.0 thousand tons for 2017. Meanwhile, over 60% of pollutant emissions is in the city of Volgograd, the rest of it fell to the satellite city Volzhsky. In 2017, an average of less than 52% of emissions was captured and neutralized in the region. Pipeline transport enterprises make a significant contribution to the total air pollution since their emissions come without treatment. The territorial production enterprise Volgogradneftegas JSC RITEK, located in several districts of the region, belongs to the first risk category and contribute about 6% to the total air pollution by enterprises of the high-risk category.

Information of 2017 from the annual reports on the state of the environment of the Volgograd Region showed that 34 companies are the main sources of air pollution, 44% are enterprises of the 1st category, 53% is the 2nd category, and the 1st enterprise is the 3rd category (Volgograd branch of Omsktekhuglerod LLC). Figure 4 presents the volume of pollutant emissions for 2017 by categories of negative environmental impact objects.

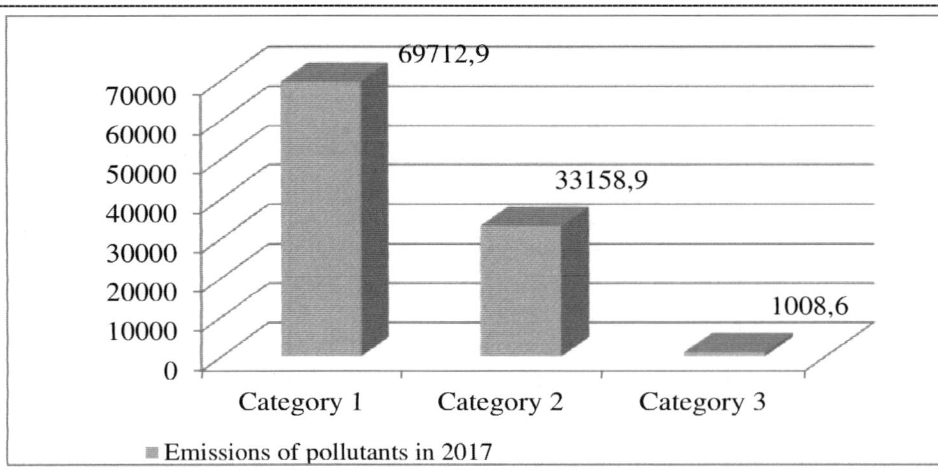

Figure 4. The volume of emissions of pollutants by categories of objects of negative impact in 2017 [11].

The enterprises, the main sources of anthropogenic impact, account for 103.9 thousand tons, that is 75.3% of the total amount of emitted substances from stationary sources. More than 67% of the indicated volume of emissions are the objects of the negative impact of the 1st category (Figure 5).

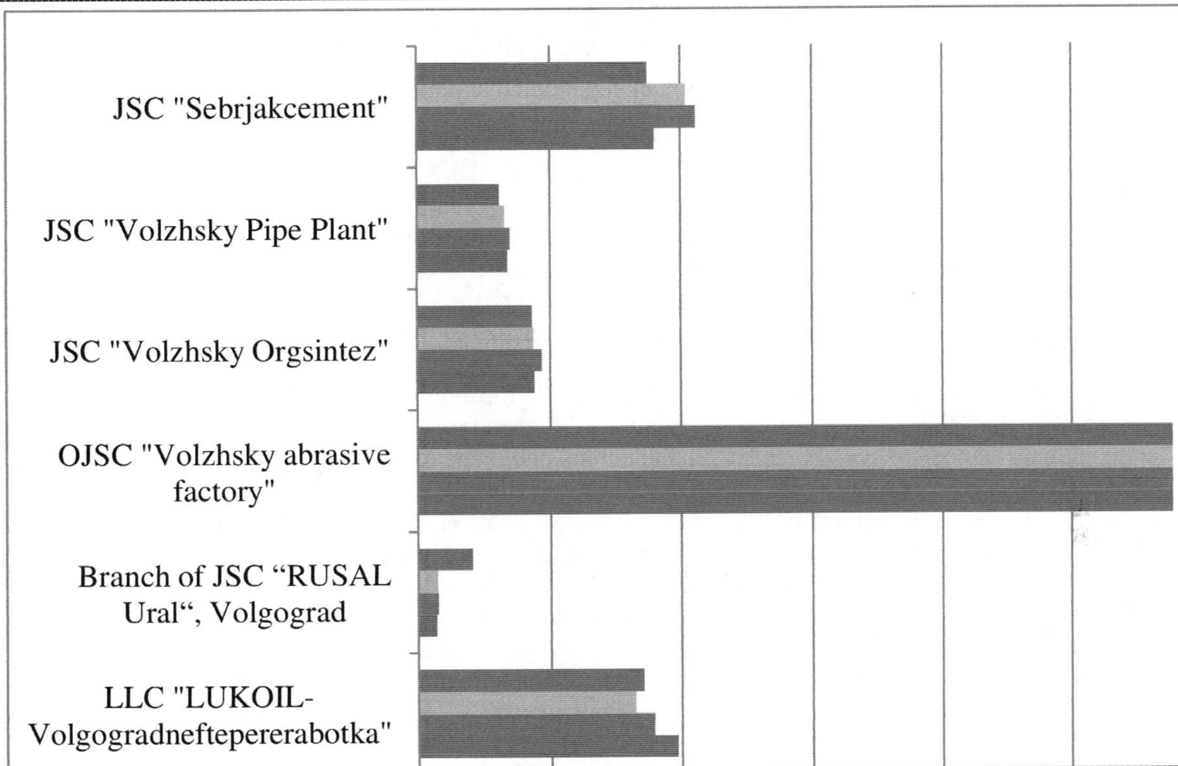

Figure 5. Dynamics of emissions of pollutants by enterprises of 1 category, tons [11-13].

The diagram shows three enterprises with the greatest negative impact on atmospheric air in the region: Volzhsky Abrasive Plant OJSC, LUKOIL-Volgogradneftepererabotka LLC and Sebryakovcement JSC, which account for 45.6% of emissions of category 1st enterprises, 23% of total emissions in the region, 12.2%, 6.1% of the total, and 12.5%, 6.2%of total, respectively. Thus, 3 enterprises of the region form together more than one-third of the total emissions, 35.3%.

In the region, the main sources of wastewater pollution are utilities, chemical industry, building materials industry and the fishing industry. The list of enterprises, the main sources of pollution of water bodies in the region in 2017 included LLC Water Supply Concessions, Volgograd, which belongs to the 1st category of Negative Environmental Impact facilities.

Since 2010, the amount of waste generated in the region has increased by 0.803 million tons or 98.29%. The enterprises, the main sources of environmental pollution in the region engaged in waste management, include Ekomaster LLC and Volga-Business LLC, both organizations are 1st category of Negative Environmental Impact facilities.

In terms of assessing the information openness of environmentally significant information at the regional level, it is necessary to note, starting in 2018, the negative trend of the exclusion of data on the impact on environmental components of individual enterprises, which are the main sources of anthropogenic impact from the annual report on the state of the region's environment.

On the territory of the Volgograd region, there are 4 enterprises included in the List of facilities that harm the environment belonging to 1st category and contribute at least 60% to the total emissions and discharges of pollutants in the country [13].

LLC Concession of water supply as a housing and communal services company engaged in the reconstruction and modernization of the central cold water supply and sanitation system has in its structure sewage treatment plants of the 1st category located on Golodny island [10].

Regional companies of the 1st category also include the oil refinery LLC LUKOIL-Volgogradneftepererabotka which specializes in the production of fuel and oil components, the construction plant of AO Sebryakovcement located in Mikhailovka city of Volgograd Region, and the pipe metallurgical plant OJSC Volzhsky Pipe Plant.

According to the report for 2018, the Volzhsky Abrasive Plant emitted more than 33 thousand tons of pollutants [14] which is more than 65% of the total missions from stationary sources in the city of Volzhsky and 23% of the total contribution to the regional load.

When assessing the transparency of the reporting of the Volgograd region enterprises in the field of environmental impact, using publicly available information, we found that LUKOIL-Volgogradneftepererabotka LLC took 1756 place, Volzhsky Abrasive Plant OJSC for the first time disclosed impact indicators on environmental components in 2018, LLC Water Supply Concessions and JSC Sebryakovcement did not attach importance to participation in the rating of environmental business performance [15].

While a content analysis of the official websites of these enterprises, we found that the official website of LUKOIL-Volgogradneftepererabotka LLC does not contain analytical information about the company's impact on environmental components and the results of environmental measures.

On the official website of JSC Sebryakovcement in the production section, the ecology tab contains information about the equipment of the enterprise with treatment equipment from its foundation until 2013. The development strategy subsection referred to the reduction of negative environmental impacts as a priority in the Technical Upgrade Program. Thus, we can state that the organization's website does not contain any information regarding environmental impact and the presence of environmental policy at the enterprise [16].

The official website of Volzhsky Abrasive Plant OJSC does not contain any information whatsoever about the damage to the environment components and the presence or absence of environmental protection measures.

On its official website, Water Supply Concessions LLC declares support for voluntary disclosure of information by the recommendation of the National Association of Concessionaires and Long-Duration Investors (NACDI) in the field of sustainable development and green investments, but there is no other environmentally significant information on the site. When studying the company's annual report on the corporate information disclosure platform, we noted the declarative nature of the environmental component of the report [17].

According to Interfax-ERA, 87 companies disclose indicators, therefore less than 2% of enterprises in the region registered on the Software and hardware for an accounting of facilities with a negative impact on the environment portal, 87 companies out of 4524, supported the initiative of transparency of environmentally significant information.

In the framework of the study, regional enterprises had the opportunity to show their interest in disclosing environmentally significant information by filling out the author's checklists structured in blocks. We sent formal requests to companies to provide information on emissions, discharges, disposal of production and consumption waste, as well as on energy consumption and energy conservation. The Manage and Responsibility and Openness blocks included issues of environmental policy, environmental protection costs, and the degree of readiness for opening up a business. The Manage block included two blocks: the main one: the presence of an environmental management system and a program of industrial environmental control, and additional: current and capital expenditures for environmental protection, the payments for negative environmental impact. The additional block has an emphasis on the results of state supervision of the company, violations of environmental laws and existing complaints from citizens regarding the impact of facilities on the environment. The Responsibility and Openness block focused on identifying the manager's awareness of the existence of non-financial reports in the field of sustainable development and the global reporting initiative (GRI). Assessing the potential willingness of companies to provide environmental information at the request of interested parties, posting quantitative data on the volume of pollution and the real costs of protecting the environment in the public domain was of particular importance in the checklist. Based on the processing of checklists, less than 1/3 of the companies provided answers to official inquiries. Over 70% ignored the request to participate in the study, 21.5% of respondents refused to provide information citing trade secrets or the accessibility of the requested information on the official websites of environmental authorities and territorial statistics agencies.

Only 6 companies submitted completed checklists: LLK-International LLC oil company, Volzhsky Pipe Plant JSC pipe plant, KF LLC research and production enterprise, Municipal Unitary Enterprise Plumbing and Sewer Facilities (sewage treatment facilities for biological reporting of wastewater from the urban district - Volzhsky city), chemical Volzhsky Orgsintez JSC and synthetic rubber company EKTOS-Volga.

Only Volzhsky Pipe Plant JSC and LLK-International LLC are aware of non-financial reporting and are ready to support the business openness initiative by generating reports on sustainable development.

3. Conclusion

In general, studying the official sites of regional objects of the 1st category of negative impact allows us to draw the following conclusions:

- 107 facilities of Negative Environmental Impact belong to 70 enterprises, of which only 72% have official sites;
- less than 2% of enterprises partially publish environmentally relevant information in the public domain;
- the information presented on the websites of 87% of organizations does not allow concluding the impact on environmental components and the effectiveness of environmental policies of enterprises since it does not contain quantitative data.
- the information available in annual corporate reports does not contain quantitative data directly on the Negative Environmental Impact facility which is a branch or structural unit of a large company;
- less than 10% of companies showed interest and participated in the study, 70% ignored official requests for information;
- none of the regional enterprises prepares reports on sustainable development following the requirements of international standards of public non-financial reporting.

The vast majority of regional enterprises do not consider the importance of the environmental component in the formation of modern information content, while this could increase their competitiveness in the international market.

Acknowledgment

The research has been funded by the Russian Foundation for Fundamental Research and the Administration of the Volgograd region, project No. 19-410-340015 r_a "Information openness as a tool of environmental responsibility of business in the structure of environmental safety of the region".

References

[1] *International Survey of Corporate Responsibility Reporting 2017, KPMG International.* URL: https://assets.kpmg/content/dam/kpmg/xx/pdf/2017/10/kpmg-survey-of-corporate-responsibility-reporting-2017.pdf (Accessed 7 November 2019)

[2] Barter N 2015 *Social and Environmental Accountability Journal* **35(3)** 197-198

[3] Bogdanov V D, Ilysheva N N, Baldesku E V and Zakirov U S 2016 *J. Economy of Region* **1** 93-104

[4] Buallay A 2019 *Journal of Applied Accounting Research* **20(4)** 481-496

[5] *Latest reports added to the Database, Global Reporting Initiative.*

IOP Conf. Series: Materials Science and Engineering **828** (2020) 012034 doi:10.1088/1757-899X/828/1/012034

URL: https://database.globalreporting.org (Accessed 1 November 2019)

[6] *National Register and Library of corporate non-financial reports, Russian Union of Industrial Manufacturers and Entrepreneurs.* URL: http://рспп.рф/simplepage/natsionalnyy-registr-i-biblioteka-korporativnykh-nefinansovykh-otchetov/ (Accessed 6 November 2019)

[7] *Resolution Of the government of the Russian Federation N 1029 of 28.09.2015 "About the approval of criteria of reference of the objects having negative impact on environment, to objects of I, II, III and IV categories".* URL: http://www.consultant.ru/cons/cgi/online.cgi?req=doc&base=LAW&n=186693&fld=134&dst=1000000001,0&rnd=0.06941106421398646#022502355793117834 (Accessed 27 October 2019)

[8] *Resolution Of the government of the Russian Federation N 903 of 28.08.2015 "About the approval of criteria of definition of the objects which are subject to Federal state ecological supervision".* URL: http://www.consultant.ru/cons/cgi/online.cgi?req=doc&ts=2755160305821760756472083&cacheid=EF6956C7B49D3B0B5A37E6D595E95F59&mode=splus&base=LAW&n=185330&rnd=1F8882CCC1FF3718B9AE0C483F0C36FA#s7x7gxwemy (Accessed 27 October 2019)

[9] *Resolution Of the government of the Russian Federation N 426 of 08.05.2014 (as amended 21.03.2019) "On Federal state environmental supervision".* URL: http://www.consultant.ru/cons/cgi/online.cgi?req=doc&ts=17936526920570726205522407 8&cacheid=5001290340B2AF3F65147B7F71A887AE&mode=splus&base=LAW&n=321417&rnd=1F8882CCC1FF3718B9AE0C483F0C36FA#bjquwlq64rk (Accessed 27 October 2019)

[10] *Statistics of objects of negative impact on the environment included in the registers, Software and hardware for accounting of objects of negative impact on the environment.* URL: https://onv.fsrpn.ru/#/public/stats/registry/category (Accessed 27 October 2019)

[11] *Report "On the state of the environment of the Volgograd region in 2017"* 2019. URL: https://oblkompriroda.volgograd.ru/current-activity/analytics/reports/ (Accessed 21 October 2019)

[12] *Report "On the state of the environment of the Volgograd region in 2015"* 2019. URL: https://oblkompriroda.volgograd.ru/current-activity/analytics/reports/ (Accessed 21 October 2019)

[13] *The order of Ministry of Russia N 154 from 18.04.2018 "About approval of the list of objects that have a negative impact on the environment, relating to category I, the contribution of which to the total emissions, discharges of polluting substances in the Russian Federation is not less than 60 percent."* URL: http://www.consultant.ru/cons/cgi/online.cgi?req=doc&ts=179365269205707262055224078&cacheid=CB7FFC50E2924BBDB5B154B1573F92EA&mode=splus&base=LAW&n=301627&rnd=1F8882CCC1FF3718B9AE0C483F0C36FA#3w8f8ljs3c (Accessed 15 October 2019)

[14] *The profile of the company JSC "Volzhsky abrasive works", Interfax-ERA.* URL: https://interfax-era.ru/predpriyatiya/oao-volzhskiy-abrazivnyy-zavod. (Accessed 3 November 2019)

[15] Lizunkov V, Politsinskaya E, Malushko E, Kindaev A, Minin M 2018 *International journal of energy economics and policy* **8 (3)** pp 250-257

[16] *Enterprises of Volgograd region, Interfax-ERA.* URL: https://interfax-era.ru/predpriyatiya/?region=12. (Accessed 3 November 2019)

[17] Shamne N L, Milovanova M V, Malushko E Yu 2019 *IOP Conference Series: Materials Science and Engineering* **483(1)** 012081 DOI: 10.1088/1757-899X/483/1/012081

IOP Conf. Series: Materials Science and Engineering **828** (2020) 012035 doi:10.1088/1757-899X/828/1/012035

Accessibility of urban and rural environment of the Kola North of Russia (Murmansk region) for disabled people

G V Zhigunova and Yu A Afonkina

Murmansk Arctic State University, 15, Kapitana Egorova St., Murmansk, 183038, Russia

E-mail: Galina-zhigunova@yandex.ru

Abstract. In today's Russian society, people with disabilities cannot successfully join society because of the existing social barriers in various spheres of activity. As a result, they have limited access to social resources and a higher social status. In this regard, ensuring the independent life of disabled people is an urgent task for Russian society. The present research aims at identifying the level of accessibility of the social environment in the North of Russia for people with disabilities. Research methods are represented with general scientific methods of cognition, analysis of scientific literature, regulatory documents, data of secondary sociological research, and results of the method of observation conducted by the authors in 2018-2019. The analysis showed that today there are social barriers in all spheres of activity, and the level of infrastructure's accessibility in rural settlements is much worse than in urban environment. The obtained results can be used in urban planning and social practice with the aim to transform the social environment and ensure its accessibility for people with disabilities.

1. Introduction

Ensuring the independent life of people with disabilities is important due to a number of factors with the key one being the availability of the social environment.

Accessible social environment for people with disabilities is a set of social, material, psychological, information and other conditions that meet the needs of these people and allow having an independent lifestyle.

The level of accessibility of social environments for people with disabilities is an important criterion for the development of public institutions and civil society. The extent to which people with disabilities are given the opportunity to participate in public processes is an indicator for judging the observance of legally enshrined human rights, as well as the dominant social values.

Studies conducted in Russia and abroad indicate the relevance of the problem of ensuring the independent life of persons with disabilities.

According to researchers, disabled people in Croatia face such obstacles as inaccessible transport, inadequate adaptation of buildings, restrooms, elevators for disabled people, the poor financial situation [1]; in Greece, Spain and Lithuania, there are cultural and institutional barriers limiting access to inclusive education for young people [2]. According to the results of the study by J. Park and S. Chowdhury, in New Zealand the main obstacles for citizens with physical disabilities in the urban environment are associated with terminals, stops, services and walkways. The main obstacles for visually impaired citizens include poor presentation of information and obstacles on walkways [3, p. 361].

American scientists point to adverse socio-psychological environment as a serious obstacle to the independent life of disabled people that contributes to their labeling and discrimination[4].

According to the study by S.G. Terskova, E.S. Gavrilova and E.E. Yurkova, in Russia, most services and facilities for people with disabilities are not accessible. In Russian cities people with disabilities face the following difficulties: high street curbs (61.9%); lack of ramps (40.2%); unequipped walkways (28.9%), etc.[5].

In addition to problems with the accessibility of buildings, the researchers also note barriers in various spheres of activity. In particular, today, barriers of the educational environment appear not only in the form of insufficiency of material and technical means and special equipment, but also in the form of absence of necessary tutor support for disabled people, lack of adaptive educational programs and individual approaches to education of disabled people [6, p. 40].

Access to social infrastructure should necessarily allow individuals with persistent health problems for maximum participation in the life of society in the places of residence. In the absence of this access, it is appropriate to talk about discrimination, widespread in the world [7].

Creation of a barrier-free environment for people with disabilities in the Russian Federation became a conscious public and legislative activity only in the late 20th –early 21st centuries. During this period, Russia begins to develop a legal framework for the elimination of social barriers and the realization of the rights and freedoms of people with disabilities.

In 2012, Russia ratified the Convention on the rights of persons with disabilities (adopted by the UN General Assembly on December 13, 2006) and set a course for compliance with international standards with the regard to the population layer under study. Health care, culture, transport and pedestrian infrastructure, information and communication, education, social protection, employment, sports and physical culture have been identified as priority spheres of disabled people's life. These spheres are currently being transformed to ensure accessibility of facilities and services for persons with disabilities.

However, social barriers still remain a significant problem for modern Russian society, impeding the independent life of persons with disabilities in many spheres of life.

Lack of access to many social resources has serious socio-economic consequences not only for people with disabilities, but also for society. First of all, it is high social dependence and low quality of life of disabled people, as well as intolerance, disunity and disintegration of social groups.

There is no doubt that the inaccessible social environment structures and stratifies urban space. As a result of architectural and social barriers, it is possible to easily isolate people with disabilities from certain urban facilities or completely isolate them from society.

According to E.K. Naberushkina, it is possible to allocate the following markers of city space barriers for people with disabilities: a) enclosure: disabled persons are assigned a certain place, including social pharmacies, social shops; b) functional accommodation: the state system places people with disabilities in boarding schools, specialized enterprises, "locks" them in their own apartments as having scarce social capital; c) stratification of urban institutions of consumption, establishing an unspoken hierarchy and social distances by introducing systems of "face control", electronic access, pricing and corporate policies [8, p. 17].

Social environment can be understood as a set of social conditions of human life and objective factors that affect his consciousness and behavior. Having numerous spheres, the social environment acts as a mediating link between the individual and the personality.

On the basis of the presented concepts, we define the social life environment of disabled people as a normal environment, equipped with their special needs, allowing them having an independent lifestyle. Each environment has resources for inclusion, but there may be barriers as well. Accordingly, the inclusive barriers faced by persons with disabilities stem from the resource scarcity of social environments.

2. Results

Examination of social infrastructure with the aim to identify the level of its accessibility for disabled people was carried out under guidance of G.V. Zhigunova in 2018-2019 in the cities of the Murmansk region: Apatity, Kandalaksha, Kola, Monchegorsk, Severomorsk, as well as in rural settlements: Alakurtti, Varzuga, Korzunovo, Lovozero, Pushnoy, Teriberka. The sample is targeted and provides representation of all municipalities in the region.

Speaking in general about the studied region, it can be noted that the Murmansk region is a highly urbanized territory of the Arctic zone of the Russian Federation. As of January 1, 2019, 690 thousand people lived in cities and urban settlements of the region, and only 58.1 thousand people lived in rural areas [9]. As of April 1, 2019, 28,496 disabled people and 2,690 disabled children lived in the Murmansk region [10]. The disability rate is 4.76% – the lowest in the North-Western and Central Federal districts of Russia. The places of residence of these persons are unevenly distributed both in urban and rural settlements [11].

The subject of observation is the possibility of unhindered access to people with disabilities to urban and rural environment. Special attention is paid to facilities for people with muscle-skeleton disorders, wheelchair users and visually impaired people since these categories of people have the greatest difficulties with access to infrastructure. Attention was paid to availability and quality of ramps, railings, call buttons, doors and doorways, space in front of the door; tactile stripes, Braille plates, etc.

Infrastructure objects under study refer to priority (in accordance with the state program "Accessible environment") and non-priority spheres of activity, including institutions of social protection and social services, education and further education, health care, culture, physical culture and sports, transport and communications; shops, supermarkets, shopping malls, banks, service centres, post offices, other institutions (administration buildings, departments, Registrar's office, police, notary offices, industrial buildings, etc.); playgrounds; parks and gardens.

According to the research results, the objects can be divided into 3 groups: accessible, conditionally accessible and inaccessible.

As of April 30, 2019, 571 objects were examined and analyzed, 439 of them are located in urban settlements, and 132 – in rural settlements. In rural settlements, all objects were examined, while in cities – selective examination was carried out.

According to the results, 30.3% of accessible infrastructure objects were identified for persons with disabilities in urban settlements, 14.6% of conditionally accessible, and 55.1% of inaccessible. The major part of the accessible facilities is designed for people with muscle-skeleton disorders and for wheelchair users. Only 6% of the objects are accessible to the visually impaired.

Only 6% of the objects are accessible in rural settlements, 7,6% – are conditionally accessible, and 86,4% – are inaccessible. Only 1% of the objects are accessible to the visually impaired.

List of social infrastructure objects grouped by the spheres of activity depending on the organized accessible environment is presented in Figure 1 (constructed on the basis of index positions).

As can be seen from the constructed indices, attempts to create an accessible social environment have been made in cities in all spheres of activity, while in rural settlements no attempts have been made to create facilities for disabled people in shops, post offices, catering establishments, playgrounds, physical culture and sports organizations (index -1). The level of accessibility of educational, cultural, administrative, credit institutions and various departments in rural settlements is much worse than in urban settlements. Health care is the only sphere in rural settlements characterized by a certain level of buildings' accessibility.

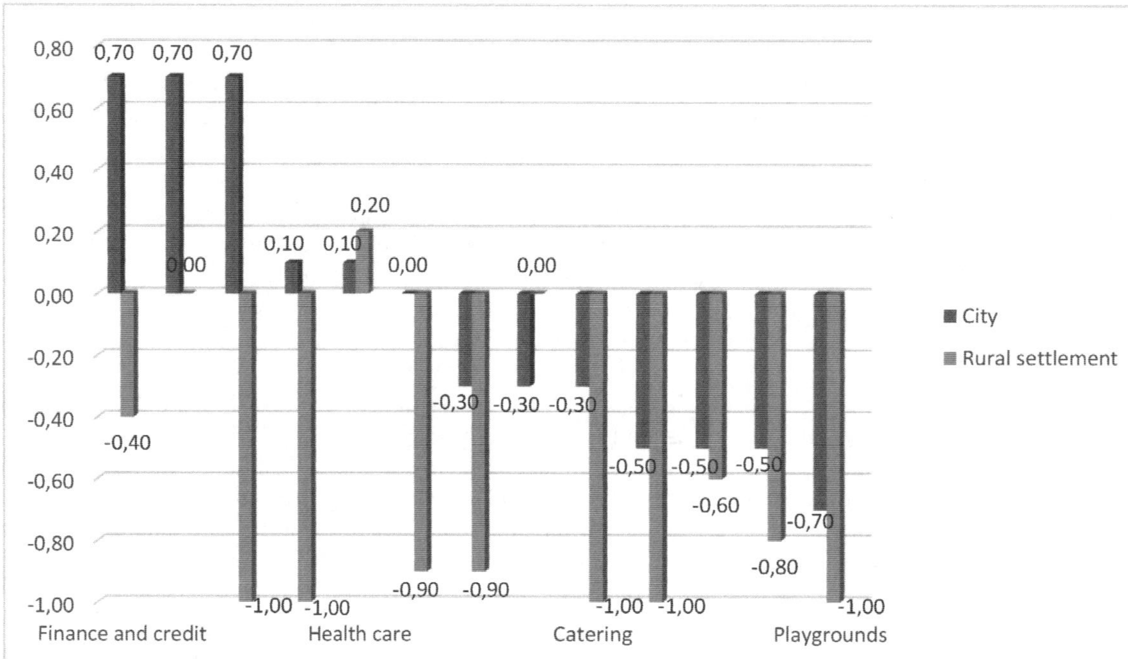

Figure 1. The level of accessibility of social infrastructure objects by spheres of activity

As we can see in Figure 1, no objects in any of the spheres under consideration are fully accessible. Such urban objects as finance and credit institutions (non-priority sphere), social protection institutions (priority sphere) and post offices (non-priority sphere) are more accessible as compared to others. The hypothesis that institutions of the priority spheres are more accessible has not been confirmed.

General index of accessibility (or rather inaccessibility) makes -0,248 for urban settlements, -0,804 – for rural settlements, which indicates unsatisfactory conditions of the environment in terms of its accessibility for people with disabilities. However, it should be noted that these conditions are three times worse in rural settlements as compared to urban ones.

The main problems of the region's urban infrastructure include the lack of ramps or non-compliance with the requirements for their quality and installation, insufficient width of doorways, inconvenient doors and door handles, lack of handrails, call buttons, tactile strips and tactile plates, the presence of obstacles on the way from the ramps to the doors or insufficient space for wheelchairs [12].

In rural settlements, the main problems include the lack of ramps and other special equipment for creating an accessible environment for disabled people, the presence of thresholds and inconvenient doors and door handles, the lack of asphalt or paved paths and sidewalks.

A common problem for all settlements is the poor condition of sidewalks and pedestrian crossings, the lack of tactile lanes, exits from curbs, the poor condition of yard areas.

3. Conclusion

Summarizing the results of the observation, we can conclude that in the Murmansk region, social and environmental barriers for people with disabilities exist in all spheres of activity. The social infrastructure of the cities meets the requirements of accessibility for people with disabilities by only a third, very unevenly in different areas. In rural settlements, the situation is ten times worse [13]. All of these confirm the authors' hypothesis.

The lag of villages behind inclusive environmental transformations can be explained by the peripheral location of rural settlements, the priority of urban development in the Murmansk region as a highly urbanized region, the lack of regional programs on creating a comfortable social environment in rural areas, insufficient funding, insufficient understanding of the importance of social transformations to ensure an independent life for people with disabilities, etc. All this points to the need of implementing targeted measures on transformation of social infrastructure objects into an accessible environment for disabled people, both in cities and in rural settlements. These transformations will not only enable people with disabilities to exercise their civil rights on an equal basis with healthy citizens, but also improve the level of comfort of the territories for all their residents.

Acknowledgment
The research has been funded by The Russian Foundation for Fundamental Research (grant No. 18-011-00712).

References
[1] Babic M M and Dowling M 2015 *Disability & Society* **30 (4)** 614-629
[2] Genova A 2015 *Disability & Society* **30 (7)** 1042-54
[3] Park J and Chowdhury S 2018 *Journal of transport & health* **10** 361-68
[4] Shier M, Graham J R and Jones M E 2009 *Disability & Society* **24 (1)** 63-75
[5] Terskova S G, Gavrilova E S and Yurkova E E 2016 *Humanities research: electronic scientific and practical journal*. URL: http://human.snauka.ru/2016/02/14031#_ftn1
[6] Zbirovskaya E P and Tarasenko E A 2016 *Vestnik RGGU. Series: Psychology. Pedagogy: Education* **4** 33-41
[7] Rafferty A 2019 *Work, employment and society* Article 0950017019865692. DOI: https://doi.org/10.1177/0950017019865692
[8] Naberushkina E K 2013 *Mobile citizenship of disabled people in the social space of the city: abstract (*Saratov) 40 p.
[9] *Population size of the municipalities of the Murmansk region as of January 1, 2019, and the average indicator for 2018*. Murmanskstat: website. URL: http://murmanskstat.gks.ru/wps/wcm/connect/rosstat_ts/murmanskstat/ru/statistics/population/ (accessed: 14.06.2019).
[10] *Number of children with disabilities*. Murmanskstat: website. URL: https://sfri.ru/analitika/chislennost/chislennost-detei?territory=47000000
[11] Lizunkov V, Politsinskaya E, Malushko E, Kindaev A, and Minin M 2018 *International journal of energy economics and policy* **8 (3)** 250-257
[12] Ergunova O, Lizunkov V, Blaginin, V, Politsinskaya E, and Shaykina O 2018 *International Journal of Energy Economics and Policy* **8(1)** 122-127
[13] Murugova E V 2017 *Bulletin of Volgograd State University-series 2-linguistics* **16 (2)** 197–200

AUTHOR INDEX

Institute of Physics
Dirac House, Temple Back
Bristol BS1 6BE UK

ISSN: 1757-8981
ISBN 978-1-7138-3163-1

9 781713 831631

CURRAN ASSOCIATES INC.
proceedings.com